计算机系列教材

刘兆毓 郑家农 闫金平 刘华群 武华 编著

计算机英语
（第五版）

清华大学出版社
北京

内 容 简 介

本书是计算机及相关专业的专业英语教材，是按计算机的知识结构层次编写的。内容涵盖以下三个部分：计算机及计算机网络硬件结构（第1章和第3章），计算机软件（第2章、第6章为系统软件，第7、8、9、10、11章为应用软件）和因特网应用（第4、5章）。在具体内容的选取上，除了基础知识以外，尽量吸取最先进的技术和知识，这其中包括云计算、大数据、3D打印技术、工业4.0和慕课等。鉴于因特网应用发展异常迅猛，本书有19课（约占全书59课的1/3）涉及因特网应用。

为便于读者阅读，书中对一些较难理解和翻译的句子做了注释，对遇到的专业术语也都做了解释。

本书主要用作大学计算机及相关专业本科生、专科生的教材，也可供研究生及参加计算机水平考试的考生使用。

本书封面贴有清华大学出版社激光防伪标签，无标签者不得销售。
版权所有，侵权必究。举报：010-62782989，beiqinquan@tup.tsinghua.edu.cn。

图书在版编目（CIP）数据

计算机英语 / 刘兆毓等编著. —5版. —北京：清华大学出版社，2018（2024.9重印）
（计算机系列教材）
ISBN 978-7-302-47561-3

Ⅰ. ①计⋯　Ⅱ. ①刘⋯　Ⅲ. ①电子计算机－英语－教材　Ⅳ. ①TP3

中国版本图书馆CIP数据核字（2017）第277440号

责任编辑：袁勤勇
封面设计：常雪影
责任校对：李建庄
责任印制：曹婉颖

出版发行：清华大学出版社
网　　址：https://www.tup.com.cn，https://www.wqxuetang.com
地　　址：北京清华大学学研大厦A座　　邮　编：100084
社 总 机：010-83470000　　邮　购：010-62786544
投稿与读者服务：010-62776969，c-service@tup.tsinghua.edu.cn
质 量 反 馈：010-62772015，zhiliang@tup.tsinghua.edu.cn
课 件 下 载：https://www.tup.com.cn，010-83470236
印 装 者：三河市天利华印刷装订有限公司
经　　销：全国新华书店
开　　本：185mm×260mm　　印　张：23.5　　字　数：572千字
版　　次：1992年12月第1版　　2018年1月第5版　　印　次：2024年9月第10次印刷
定　　价：59.00元

产品编号：064594-03

前言

《计算机英语（第四版）》出版6年以来，计算机和因特网又出现了很多新技术，为了适应这种新的形势，编写了第五版。第五版只保留了第四版中计算机和网络的基础内容，超过80%是新内容。考虑到长期使用本教材的教师和读者的用书习惯，本书编写格式与第四版保持一致，即课文中有注释、关键词和练习题，书后附有习题答案和参考译文。

本书由刘兆毓、郑家农等编著，全书共11章，46节（内含59课），其中郑家农编写了第9~11章，闫金平编写了第1、3章，刘华群编写了第2、6章，武华编写了第4、8章，刘兆毓编写了第5、7章。

本书在编写过程中，引用了刘艺、王春生编写的《计算机英语（第4版）》的部分内容，在此表示感谢。

由于作者水平有限，书中难免有不当之处，敬请批评指正。

编者
2017年10月于北京

目录

PART I COMPUTER ARCHITECTURE AND COMPUTER NETWORK

CHAPTER 1 COMPUTER ORGANIZATION AND UNITS /3
- 1.1 COMPUTER ORGANIZATION /3
- 1.2 MICROPROCESSOR AND SYSTEM BOARD /10
- 1.3 MEMORY /15
- 1.4 SECONDARY STORAGES /19
- 1.5 INPUT AND OUTPUT DEVICES /27

CHAPTER 2 SYSTEM SOFTWARE /40
- 2.1 COMPUTER SOFTWARE /40
- 2.2 SUMMARY OF OPERATING SYSTEM /43
- 2.3 WINDOWS 10 /47
- 2.4 UNIX AND LINUX /52
- 2.5 MAC OS /57
- 2.6 ANDROID /60

CHAPTER 3 COMPUTER NETWORK /65
- 3.1 LOCAL AREA NETWORKS (LANs) /65
  - 3.1.1 ETHERNET /65
  - 3.1.2 WI-FI AND BLUETOOTH /69
  - 3.1.3 HOME NETWORK /74
- 3.2 WIDE AREA NETWORKS (WANs) /76
  - 3.2.1 OVERVIEW OF WIDE AREA NETWORKS (WANs) /76
  - 3.2.2 STORAGE AREA NETWORK (SAN) AND INTERNET AREA NETWORK (IAN) /81
- 3.3 THE INTERNET /83

PART II INTERNET APPLICATIONS

CHAPTER 4 TRADITIONAL INTERNET APPLICATIONS /93

4.1 OVERVIEW OF WORLD WIDE WEB (WWW) /93
    4.1.1 ABOUT WWW /93
    4.1.2 SEARCH ENGINES /97

4.2 E-MAIL /100

4.3 ELECTRONIC COMMERCE AND THE INTERNET OF THINGS /103
    4.3.1 ELECTRONIC COMMERCE AND SOCIAL COMMERCE /103
    4.3.2 INTERNET OF THINGS /111

4.4 COMPUTER AND NETWORK SECURITY /115
    4.4.1 COMPUTER CRIMINALS AND CRIME /115
    4.4.2 PROTECT COMPUTER SECURITY /120

CHAPTER 5 NEW INTERNET APPLICATIONS /123

5.1 INSTANT MESSAGING (IM) /123
    5.1.1 OVERVIEW OF INSTANT MESSAGING (IM) /123
    5.1.2 QQ /126
    5.1.3 FACEBOOK /129
    5.1.4 TWITTER /132
    5.1.5 MICROSOFT SERVICE NETWORK (MSN) /134
    5.1.6 WECHAT /138

5.2 SOCIAL NETWORKING SERVICE (SNS) /140
    5.2.1 OVERVIEW OF SOCIAL NETWORKING SERVICE (SNS) /140
    5.2.2 WIKI /144
    5.2.3 BLOG AND MICROBLOG /146
5.3 CLOUD COMPUTING /151
5.4 BIG DATA /158
5.5 MOOC AND FLIPPED CLASS ROOM /163

PART III PROGRAM DESIGN LANGUAGES AND DATABASES

CHAPTER 6 PROGRAMMING LANGUAGES /173
    6.1 OVERVIEW OF PROGRAMMING LANGUAGES /173
    6.2 C, C++, AND C# /177
    6.3 JAVA /181
    6.4 MARKUP AND SCRIPTING LANGUAGES /185

CHAPTER 7 DATABASE /192
    7.1 DATABASE CONCEPTS /192
    7.2 THE WEB AND DATABASES /197
    7.3 MySQL /201

PART IV APPLICATION SOFTWARE

CHAPTER 8 OFFICE AUTOMATION SOFTWARE /209
    8.1 THE BASICS OF OFFICE AUTOMATION SOFTWARE /209

8.2　MICROSOFT OFFICE 2013　/215
8.3　MICROSOFT OFFICE MOBILE　/220

CHAPTER 9　MULTIMEDIA　/224
9.1　MULTIMEDIA AND ITS MAJOR CHARACTERISTICS　/224
9.2　USAGE/APPLICATION　/227
9.3　MULTIMEDIA NETWORKING　/232

CHAPTER 10　COMPUTER GRAPHICS AND IMAGES　/235
10.1　OVERVIEW　/235
10.2　THE VARIOUS COMPUTER GRAPHICS　/237
10.3　GRAPHICS SOFTWARE (1)　/241
10.4　GRAPHICS SOFTWARE (2)　/246
10.5　DIGITAL IMAGE FILE FORMATS　/249

CHAPTER 11　MODERN INDUSTRIAL AUTOMATION　/253
11.1　OVERVIEW　/253
11.2　USE OF CAD, CAM, AND CAE　/256
11.3　PRODUCT LIFECYCLE MANAGEMENT (PLM)　/260
11.4　INDUSTRY 4.0　/265
11.5　3D PRINTING　/270

ANSWERS TO THE EXERCISES　/277
BIBLIOGRAPHY　/282
参考译文　/283
　　第一部分　计算机体系结构和计算机网络　/283
　　　　第1章　计算机组成和部件　/283
　　　　　　1.1　计算机组成　/283
　　　　　　1.2　微处理器和系统板　/285
　　　　　　1.3　存储器　/286
　　　　　　1.4　二级存储器　/288

目录

  1.5 输入与输出设备 /290

第 2 章 系统软件 /294

  2.1 计算机软件 /294

  2.2 操作系统概述 /296

  2.3 Windows 10 /297

  2.4 UNIX 和 LINUX /299

  2.5 Mac 操作系统 /300

  2.6 安卓操作系统 /301

第 3 章 计算机网络 /303

  3.1 局域网 /303

    3.1.1 以太网 /303

    3.1.2 Wi-Fi 和蓝牙 /304

    3.1.3 家庭网 /305

  3.2 广域网 /305

    3.2.1 广域网概述 /305

    3.2.2 存储区域网和因特网区域网 /307

  3.3 因特网 /308

第二部分 因特网应用 /310

第 4 章 传统因特网应用 /310

  4.1 万维网概述 /310

    4.1.1 关于万维网 /310

    4.1.2 搜索引擎 /311

  4.2 电子邮件 /312

  4.3 电子商务和物联网 /313

    4.3.1 电子商务和社交电子商务 /313

    4.3.2 物联网 /315

  4.4 计算机和网络安全 /316

    4.4.1 计算机罪犯和犯罪 /316

    4.4.2 保护计算机安全 /318

第 5 章 因特网新应用 /319

  5.1 即时消息 /319

　　　　5.1.1　即时消息概述　/319

　　　　5.1.2　QQ　/320

　　　　5.1.3　脸谱　/321

　　　　5.1.4　推特　/322

　　　　5.1.5　微软服务网 MSN　/323

　　　　5.1.6　微信　/324

　　5.2　社交网络服务　/324

　　　　5.2.1　社交网络服务概述　/324

　　　　5.2.2　维基　/325

　　　　5.2.3　博客与微博　/326

　　5.3　云计算　/327

　　5.4　大数据　/329

　　5.5　慕课与翻转课堂　/331

第三部分　程序设计语言和数据库　/333

　第 6 章　程序设计语言　/333

　　6.1　程序设计语言概述　/333

　　6.2　C、C++ 和 C#　/335

　　6.3　Java　/336

　　6.4　标记和脚本语言　/337

　第 7 章　数据库　/339

　　7.1　数据库的概念　/339

　　7.2　万维网与数据库　/341

　　7.3　MySQL　/342

第四部分　应用软件　/343

　第 8 章　办公自动化软件　/343

　　8.1　办公自动化软件基本知识　/343

　　8.2　微软 Office 2013　/345

　　8.3　微软 Office Mobile　/346

　第 9 章　多媒体　/347

　　9.1　多媒体及其主要特点　/347

　　9.2　多媒体应用　/348

　　9.3　多媒体网络技术　/350

目录

第 10 章　计算机图形图像　/351
 10.1　概述　/351
 10.2　各种各样的计算机图形　/351
 10.3　图形软件（1）　/353
 10.4　图形软件（2）　/355
 10.5　数字图像文件格式　/356

第 11 章　现代工业自动化　/357
 11.1　概述　/357
 11.2　CAD、CAM、CAE 的应用　/358
 11.3　产品生命周期管理　/359
 11.4　工业 4.0（第 4 次工业革命）　/361
 11.5　3D 打印　/363

PART I

COMPUTER ARCHITECTURE AND COMPUTER NETWORK

CHAPTER 1
COMPUTER ORGANIZATION AND UNITS

1.1 COMPUTER ORGANIZATION

1. Computer Organization

A computer is a programming, electronic device that accepts input, performs operations or processing on the data, and outputs and stores the results. Because it is programmable, the instructions—called the program—tell the computer what to do. The relationships between these four main computer operations (input, processing, output, and storage) are shown in Figure 1-1.

Figure 1-1　Basic operations within a computer

The corresponding devices to perform these tasks are input devices, processing devices, output devices, and storage devices.

(1) **Input Devices**

An input device is any piece of equipment that supplies materials (input) to the computer. The most common input devices are the keyboard and mouse (See Figure 1-2). Other possibilities include image and bar-code scanners, joysticks, touch screens, digital cameras, electronic pens, fingerprint readers, and microphones. Input devices for a stereo system might be a CD player and antenna.

Figure 1-2　Hardware of a computer system

(2) Processing Unit

The heart of any computer system is the central processing unit (CPU), located inside the computer's main box or system unit.

A processor is composed of two functional units—a control unit and an arithmetic/logic unit—and a set of special workspaces called registers.

Figure 1-3 depicts its structure, in which the Internal CPU Interconnection provides communication among the Control Unit, ALU, and Registers.

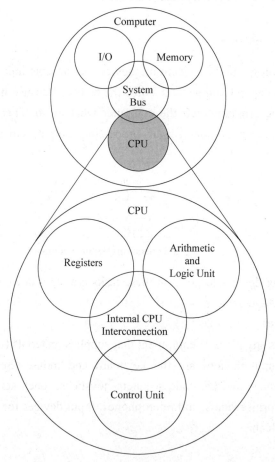

Figure 1-3　Central Processing Unit (CPU)

The control unit is the functional unit that is responsible for supervising the operation of the entire computer system.

The control unit fetches instructions from memory and determines their types or decodes them. It then breaks each instruction into a series of simple small steps or actions. By doing this, it controls the step-by-step operation of the entire computer system.

The arithmetic and logic unit (ALU) is the functional unit that provides the computer with logical and computational capabilities. Data are brought into the ALU by the control unit, and the ALU performs whatever arithmetic or logic operations are required to help carry out the instruction [1].

A register is a storage location inside the processor. Registers in the control unit are used to keep track of the overall status of the program that is running. Control unit registers store information such as the current instruction, the location of the next instruction to be executed, and the operands of the instruction [2]. In the ALU, registers store data items that are added, subtracted, multiplied, divided, and compared. Other registers store the results of arithmetic and logic operations.

(3) **Output Devices**

Like input units, output devices are instruments of interpretation and communication between humans and computer systems of all sizes. These devices take output results from the CPU in machine-coded form and convert them into a form that can be used (a) by people (e.g. a printed and/or displayed report) or (b) as machine input in another processing cycle [3].

In personal computer systems, display screen and desktop printers are popular output devices. Larger and faster printers, many online workstations, and magnetic tape drives are commonly found in larger systems.

(4) **Storage Devices**

Storage is a computer section used primarily for storing information such as instructions, programs and data.

There are two types in storage devices, one is the memory (sometimes called as primary storage), another is the secondary storage. Primary storage is located within the system unit that houses the CPU and other components [4]. Secondary storages include the storage media and drives, We will describe them in section 1.4 of this textbook.

An arbitrary desktop computer (not necessarily a PC) is shown in Figure 1-4. It has a large main memory to hold the operating system, applications and data, and an interface to mass storage devices(disks and DVD/CD-ROMs). It has a variety of I/O devices for user input (keyboard, mouse, and audio), user output (display interface and audio), and connectivity (networking and peripherals). The fast processor requires a system manager to monitor its core temperature and supply voltages ,and to generate a system reset.

2. Types of Computer

There are four types of computers: supercomputers, mainframe computers, midrange computers, and microcomputers.

(1) **Supercomputers** are the most powerful type of computer. These machines are special high-capacity computers used by very large organizations. IBM's Blue Gene is considered by many to be the fastest computer in the world.

(2) **Mainframe computers** occupy specially wired, air-conditioned rooms. Although not nearly as powerful as supercomputers, mainframe computers are capable of great processing

speeds and data storage [5]. For example, insurance companies use mainframes to process information about millions of policyholders.

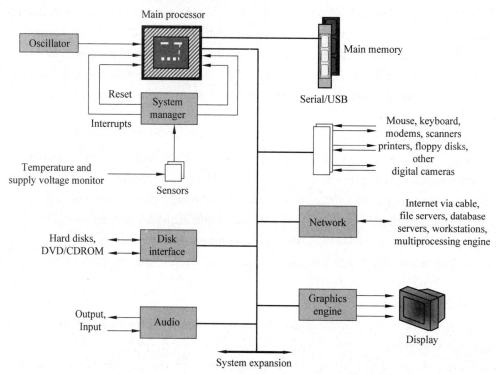

Figure 1-4 Block diagram of a generic computer

(3) **Midrange computers** also referred to as servers, are computers with processing capabilities less powerful than a mainframe computer yet more powerful than a microcomputer[6]. Originally used by medium-sized companies or departments of large companies to support their processing needs, today midrange computers are most widely used to support or serve end users for such specific needs as retrieving data from a database or supplying access to application software.

(4) **Microcomputers** are the least powerful, yet the most widely used and fastest-growing, type of computer. There are four types of microcomputers: desktop, notebook, tablet and handheld computers (See Figure 1-5). Desktop computers are small enough to fit on top of or alongside a desk yet are too big to carry around [7]. Notebook computers, also known as laptop computers, are portable and lightweight and fit into most briefcases. Tablets, also known as tablet computers, are the newest type of computer. They are smaller, lighter, and generally less powerful than notebook. Like a notebook, tablets have a flat screen but typically do not have a standard keyboard. Instead tablets typically use a virtual keyboard that appears on the screen and is touch-sensitive [8]. The best-known tablet is Apple's iPad [9]. Handheld computers are the smallest and are designed to fit into the palm of one hand. These systems

contain an entire computer system, including the electronic components, secondary storage, and input and output devices. Personal Digital Assistants (PDAs)[10] and smartphones are the most widely used handheld computers. Smartphones are cell phones with wireless connections to the Internet. Their growth has been explosive in the past few years.

Desktop   Handheld   Tablet PC   Notebook

Figure 1-5   Microcomputers

NOTES

[1] 这是一个 and 连接的并列句。后一个分句中的 whatever 是关系代词，引导后面的宾语从句。

[2] 长句中 such 引导的同位语中有三个并列的宾语。

[3] 由 and 连接的并列句，后一句中 that 引导的定语从句修饰 form。

[4] that 引导的定语从句，修饰 system unit；house 为动词，原意为"留宿""收容"，此处为"含有""包含"之意。

[5] Although 引导的是让步状语从句。

[6] 连接词 yet 连接的两个句子都是比较句型。

[7] 连接词 yet 连接的两个句子，后一句省略了主语 desktop computers。

[8] that appears…and is 是定语从句，修饰 keyboard。

[9] iPad 是由苹果公司于 2010 年开始发布的平板电脑系列，定位介于苹果的智能手机 iPhone 和笔记本电脑产品之间（屏幕中有 4 个虚拟程序固定栏）。与 iPhone 布局一样，iPad 提供浏览互联网、收发电子邮件、观看电子书、播放音频或视频、玩游戏等功能。

[10] PDA 即掌上电脑，主要用于存储和访问个人信息，如地址、电话以及记事等。

KEYWORDS

architecture	体系结构，结构，层次结构，总体结构，结构格式
programmable	可编程的
input	输入
output	输出
store, storage	存储，存储器
bar-code	条形码
scanner	扫描仪
joystick	操纵杆
touch screen	触摸屏

digital camera	数码相机
electronic pen	电子笔
fingerprint reader	指纹阅读器
stereo system	立体系统
CD(Compact Disk)	高密度磁盘，光盘，激光唱盘
player	播放器
CPU(Central Processing Unit)	中央处理器
main box	主机箱
control unit	控制器，控制部件
ALU(Arithmetic and Logic Unit)	算术／逻辑部件
register	寄存器
memory	存储器
decode	译码
operand	操作数
communication	通信
machine-coded	机器编码
display screen	显示屏
desktop printer	台式打印机
online	在线，联机
workstation	工作站
magnetic tape drive	磁带驱动器
primary storage	主存储器
secondary storage	二级存储器
supercomputer	超级计算机
mainframe computer	大型计算机
midrange computer	中型计算机
server	服务器
microcomputer	微型计算机
desktop computer	台式计算机
portable	便携的，手提的，可移植的
screen	屏幕
keyboard	键盘
notebook computer	笔记本电脑
tablet PC	平板电脑
handheld computer	手持计算机
retrieve	检索，使恢复
database	数据库
access	访问，存取，接入，进入

input device	输入设备
output device	输出设备
PDA(Personal Digital Assistants)	个人数字助理
smartphone	智能电话
cell phone	蜂窝电话，移动电话，手机
wireless	无线（的）

EXERCISES

Match the following terms to the appropriate definitions

1. _____ Computer
2. _____ Instructions
3. _____ cell phones
4. _____ Input device
5. _____ CPU
6. _____ Registers
7. _____ Control unit
8. _____ the smallest computer
9. _____ Output device
10. _____ Storage
11. _____ Supercomputers
12. _____ Mainframe computers
13. _____ servers
14. _____ Microcomputers
15. _____ basic operations within a computer
16. _____ ALU

   a. a computer section stored instructions, programs and data

   b. the most widely used and fastest-growing computers

   c. input, processing, storage, and output

   d. supervises the operation of the entire computer system

   e. have great processing speeds and data storage

   f. a programming electronic device

   g. instrument of interpretation and communication between human and computer systems

   h. tell the computer what to do

   i. the heart of any computer system

   j. has logical and computational capabilities

   k. midrange computers

   l. the most powerful type of computer

   m. storage location inside the processor

n. supplies materials to the computer

o. smartphones

p. handheld

1.2 MICROPROCESSOR AND SYSTEM BOARD

1. Microprocessors

In a microcomputer system, the central processing unit (CPU) or processor is contained on a single chip called the microprocessor. The microprocessor is either mounted onto a carrier package that plugs into the system board or contained within a cartridge that plugs into a special slot on the system board[1] (See Figure1-6). The microprocessor is the "brains" of the computer system. It has two basic components: the control unit and the arithmetic-logic unit. Functions of these two units have been mentioned in section 1.1 of this textbook.

Figure 1-6　Microprocessor carrier package and cartridge

2. Microprocessor chips

Chip capacities are often expressed in word sizes. A word is the number of bits (such as 16, 32, or 64) that can be accessed at one time by the CPU. The more bits in a word, the more powerful—and the faster—the computer is. As mentioned previously, eight bits group together to form a byte. A 32-bit-word computer can access 4 bytes at a time. A 64-bit-word computer can access 8 bytes at a time. Therefore, the computer designed to process 64-bit words is faster.

Older microcomputers typically process data and instructions in millionths of a second or microseconds. Newer microcomputers are much faster and process data and instructions in billionths of a second, or nanoseconds. Supercomputers, by contrast, operate at speeds measured in picoseconds—1000 times as fast as microcomputers(See Figure1-7).

The two most significant recent developments in microprocessors are the 64-bit processor and the dual-core chip. Until recently, 64-bit processors were only used in large mainframe and supercomputers. All of that is changing as 64-bit processors are becoming commonplace in today's more powerful microcomputers.

Unit	Speed
Microsecond	Millionth of a second
Nanosecond	Billionth of a second
Picosecond	Trillionth of a second

Figure 1-7 Processing speeds

A new type of chip, the dual-core chip, can provide two separate and independent CPUs. These chips allow a single computer to run two programs at the same time. For example, access could be searching a large database while the end user is creating a multimedia presentation with PowerPoint [2]. More significantly, however, is the potential for microcomputers to run very large complex programs that previously were run only on mainframe and supercomputers. This requires specifically designed programs that are divided into parts that each CPU could process independently [3]. This approach is called parallel processing.

3. Multi-core processor

A multi-core processor is a single computing component with two or more independent actual central processing units (called "cores"). Figure 1-8 shows a diagram of a generic dual-core processor.

Figure 1-8 A diagram of a generic dual-core processor, with CPU-local
level 1 caches, and a shared, on-die level 2 cache

Manufacturers typically integrate the cores onto a single integrated circuit die (known as a chip multiprocessor or CMP), or onto multiple dies in a single chip package.

Multicore processors may have two cores (dual-core CPUs, for example, AMD Phenom II X2 and Intel Core Duo), four cores (quad-core CPUs, for example, Intel's i7 processors), six cores, eight cores, or more[4].

Multi-core processors are widely used across many application domains including general-purpose, embedded, network, digital signal processing (DSP), and graphics[5].

Commercially, Adapteva Epiphany, a many-core processor architecture which allows up to 4096 processors on-chip, although only a 16 core version has been commercially produced[6].

4. System board

The system board is also known as the motherboard. The system board is the communications medium for the entire computer system. Every component of the system unit connects to the system board. It acts as a data path allowing the various components to communicate with one another. External devices such as the keyboard, mouse, and monitor could not communicate with the system unit without the system board.

On a desktop computer, the system board is located at the bottom of the system unit or along one side. It is a large flat circuit board covered with a variety of different electronic components including sockets, slots, and bus lines (See Figure 1-9).

Figure 1-9 System board

(1) Sockets provide a connection point for small specialized electronic parts called chips. Chips consist of tiny circuit boards etched onto squares of sandlike material called silicon[7]. These circuit boards can be smaller than the tip of your finger. A chip is also called a silicon chip, semiconductor, or integrated circuit. Chips are mounted on carrier packages (See Figure 1-10). These packages either plug directly into sockets on the system board or onto cards that are then plugged into slots on the system board. Sockets are used to connect the system board to a variety of different types of chips, including microprocessor and memory chips.

(2) Slots provide a connection point for specialized cards or circuit boards. These cards provide expansion capability for a computer system. For example, a modem card plugs into a slot on the system board to provide a connection to the Internet.

CHAPTER 1 COMPUTER ORGANIZATION AND UNITS

Figure 1-10   Chips mounted onto a carrier package

(3) Connecting lines called bus lines provide pathways that support communication among the various electronic components that are either located on the system board or attached to the system board[8].

Notebook, tablet PC, and handheld system boards are smaller than desktop system boards. However, they perform the same functions as desktop system boards.

NOTES

[1] 由 either…or…构成的并列句，每个句子里都有由 that 引导的定语从句。

[2] while 连接的是两个并列句子，这一句话与上一句有密切关系，故 access 可译为程序。

[3] 句中有两个 that 引导的定语从句。

[4] AMD Phenom II X2 是 AMD 公司生产的中高端 CPU，其中文译名为 AMD 羿龙 II。Intel Core Duo，即 Core 2 Duo。Intel（英特尔公司）沿用奔腾系列的命名规则，将新系列"酷睿"（Core）芯片命名为 Core 2 Duo。Intel's i7 processors 酷睿 i7 是由 Intel 生产的面向中高端用户的 CPU 系列标识。

[5] including…，现在分词短语作定语，修饰 domains。

[6] Adapteva 是一个名不见经传的芯片设计小公司，此次 64 核 CPU 采用的是 Epiphany 架构，是一种用来进行 DSP 数字信号处理的芯片架构。

[7] 句中有两个过去分词短语做定语：etched 和 called。

[8] 长句。句中有两个 that 引导的定语从句。

KEYWORDS

microprocessor	微处理器
chip	芯片
system board	系统板
carrier package	承载插件
cartridge	盒
slot	（插）槽
capacity	容量，能力

word sizes	字长，字尺寸
bit	比特，二进制位
byte	字节
microsecond	微秒
nanosecond	毫微秒，纳秒
picosecond	微微秒，皮秒
core	核，核心，芯子
parallel processing	并行处理
die	电路小片，模片，饼图
DSP(Digital Signal Processing)	数字信号处理
embedded	嵌入式的
motherboard	主板，母板
integrated circuit	集成电路
data path	数据通路（路径）
socket	插座，套接字，网络应用程序接口

EXERCISES

Multiple Choices

1. Microprocessor _____.
    a. is the "brains" of the computer system
    b. is mounted onto a carries package
    c. is contained within a cartridge
    d. has two basic components: the control unit and arithmetic-logic unit
2. We often express chip capacities in word size, word size can be _____.
    a. one byte   b. two bytes   c. three bytes   d. four bytes
3. We usually measure the processing speed in today's computer according to the _____.
    a. thousandth of a second   b. millionth of a second
    c. billionth of a second   d. trillionth of a second
4. The 64-bit processor can be used in _____.
    a. large mainframe computers   b. older microcomputers
    c. supercomputers   d. more powerful microcomputers
5. With dual-core chip a microcomputer can _____.
    a. run two programs at the same time
    b. run very large complex programs
    c. take parallel processing
    d. be used to access a large database while creating a multimedia presentation with PowerPoint
6. Multicore processor may have _____ cores.

    a. one   b. two   c. four   d. sixteen
7. Dual-core processor has _____.
    a. one CPU core b. two CPU core c. one-level cache d. two-level cache
8. Multicore processors can be used for _____ applications.
    a. network   b. DSP   c. graphics   d. embedded
9. System board is _____.
    a. used for desktop computer only
    b. the communications medium for the entire computer system
    c. known as the motherboard
    d. a data path allowing the various components to communicate with one another
10. A connection point on a system board can be provided by _____.
    a. chips   b. slots
    c. electronic components d. sockets
11. We can call a chip as _____.
    a. silicon chip   b. carrier package
    c. integrated circuit   d. semiconductor
12. Slots can be used for _____.
    a. specialized cards   b. specialized circuit boards
    c. expansion capability of a computer d. modem card

1.3 MEMORY

1. Memory System Desiderata

The memory system has three desiderata.
(1) Size: infinitely large, no constraints on program or data set size.
(2) Speed: infinitely fast, latency equal to the fastest memory technology available [1].
(3) Cost: per bit cost should approach the lowest-cost technology available.

Clearly, these specifications cannot all be achieved as they are mutually exclusive. However, with the semiconductor and magnetic memory technology of today, these specifications are closely approximated.

2. Memory

Memory is a holding area for data, instructions, and information. Like microprocessors, memory is contained on chips connected to the system board. There are three well-known types of memory chips: random-access memory (RAM), read-only memory (ROM), and complementary metal-oxide semiconductor (CMOS).

(1) RAM

Random-access memory (RAM) chips hold the program (sequence of instructions) and data that the CPU is presently processing. Before data can be processed or a program can be run, it must be in RAM. For this reason, RAM is sometimes referred to as primary storage. RAM is called temporary or volatile storage because everything in most types of RAM is lost as soon as the microcomputer is turned off [2]. It is also lost if there is a power failure or other disruption of the electric current going to the microcomputer.

There are two types of RAM: Dynamic RAM (DRAM) and Static RAM (SRAM).

① DRAM

Recent advances in DRAM chips have produced three types of DRAM chips:

- Synchronous DRAM (SDRAM) is faster than conventional RAM chips and more expensive. SDRAM chips streamline operations by coordinating or synchronizing the movement of data and instructions between the chip and other components in the system unit[3].
- Double data rate SDRAM (DDR SDRAM), also known as SDRAM II, is faster, more reliable, and more expensive than SDRAM. DDR SDRAM chips are able to transfer twice (double) as much data in the same amount of times as SDRAM.
- Direct DRAM is the fastest and most expensive.

Almost all of today's microcomputers use a combination of DRAM chips.

② SRAM

Static RAM (SRAM), like DRAM, requires a constant supply of power. Compared to DRAM, SRAM does not require as much power, is faster, and is more reliable. SRAM is also more expensive and typically used for specialized applications. One of these applications is for cache memory or RAM cache[4].

③ CACHE

Cache (pronounced "cash") memory improves processing by acting as a temporary high-speed holding area between the memory and the CPU. In a computer with a cache (not all machines have one), the computer detects which information in RAM is most frequently used. It then copies that information into the cache. When needed, the CPU can quickly access the information from the cache.

There are three different types or levels of cache:

- Level 1 (L1), also known as primary cache and internal cache, is built into the microprocessor chip.
- Level 2 (L2), also known as external cache, is slower than L1 but has a greater capacity. In older computers with older microprocessors, the Level 2 cache is located on a chip that is plugged into the system board. New computers with newer microprocessors have the L2 caches built into the microprocessor. This arrangement, sometimes referred to as advanced transfer cache, provides a faster response than

cache located on the system board[5].
- Level 3 (L3), the newest type of cache, works with special microprocessor L2 caches. L3 caches are not built into the microprocessor. Rather, L3 caches use SDRAM chips located on the system board.

Most of today's microcomputers have two or three types of cache. The most powerful have all three types.

④ FLASH RAM

Flash RAM or flash memory chips can retain data even if power is disrupted. This type of RAM is the most expensive and used primarily for special applications such as digital cell telephones, digital video cameras, and portable computers.

(2) ROM

Read-only memory (ROM) chips have programs built into them at the factory. Unlike RAM chips, ROM chips are not volatile and cannot be changed by the user. "Read only" means that the CPU can read or retrieve data and programs written on the ROM chip. However, the computer cannot write—encode or change—the information or instructions in ROM.

ROM chips typically contain special instructions for detailed computer operations. For example, ROM instructions are needed to start a computer, give keyboard keys their special control capabilities, and put characters on the screen.

(3) CMOS

A complementary metal—oxide semiconductor (CMOS) chip provides flexibility and expandability for a computer system. It contains essential information that is required every time the computer system is turned on. The chip supplies such information as the current date and time, amount of RAM, type of keyboard, mouse, monitor, and disk drives. Unlike RAM, it is powered by a battery and does not lose its contents when the power is turned off. Unlike ROM, its contents can be changed to reflect changes in the computer system such as increased RAM and new hardware devices.

NOTES

[1] latency 即等待时间，潜伏时间。此处指存储器访问时间。

[2] 在 because 引导的原因状语从句中，as soon as 引导的是时间状语从句。

[3] streamline 为动词，意为使（工作、生产等）简化而更具效率。

[4] cache memory 就是 cache；RAM cache 一般作为 L2 cache 使用，在苹果计算机中指 disk cache。

[5] transfer cache, 传输缓存，其中关键是 transfer，它能控制 cache 和其中的内容。

KEYWORDS

RAM（random-access memory） 随机存储器

ROM（read-only memory） 只读存储器
CMOS（Complementary Metal-Oxide Semiconductor）
互补金属氧化物半导体
volatile 易失性
DRAM（Dynamic RAM） 动态随机存储器
SRAM（Static RAM） 静态随机存储器
cache 高速缓冲存储器
flash memory 快闪存储器，闪存

EXERCISES

Fill in the blanks with appropriate words or phrases found behind this exercise.

1. There are three specifications in the memory system, they are _____.
2. An area that holds data, instructions, and information is called as _____.
3. There are three types of memory chips, they are _____.
4. Which chips are sometimes referred to primary storage? _____.
5. RAM is called _____ storage as everything in most types of RAM is lost as soon as the microcomputer is turned off.
6. There are two types of RAM: _____.
7. There are three types of DRAM: _____.
8. _____ acts as a temporary high-speed holding area between the memory and the CPU.
9. _____ is also known as external cache.
10. _____ is also known as primary cache.
11. _____ is faster than _____.
12. The newest type of cache is _____.
13. The most powerful microcomputers have _____ of cache.
14. _____ can retain data even if power is disrupted.
15. _____ chips are not volatile.
16. _____ are needed to start a computer.
17. CMOS chip is powered by _____.
18. _____ supplies such information as the current date, time and so forth.

    a. RAM chips
    b. Cache
    c. CMOS chip
    d. Flash RAM
    e. L3 cache
    f. size, speed, and cost
    g. L2 cache
    h. temporary or volatile

i. RAM, ROM, and CMOS
j. ROM instructions
k. all three types
l. DRAM and SRAM
m. memory
n. a battery
o. L1 cache
p. L1 cache, L2 cache
q. ROM
r. SDRAM, DDR SDRAM and Direct DRAM

1.4　SECONDARY STORAGES

As most RAM provides only temporary or volatile storage, therefore we need more permanent or nonvolatile storage for data and programs. We also need external storage because users need much more storage capacity than is typically available in a computer's primary or RAM memory [1].

Secondary storage provides permanent or nonvolatile storage. Using secondary storage devices such as a hard disk drive, data and programs can be retained after the computer has been shut off. This is accomplished by writing files to and reading files from secondary storage devices.

1. FLOPPY DISKS

Floppy disks are removable storage media. They are typically used to store and transport relatively small word processing, spreadsheet, and other types of files.

2. HARD DISKS

Compared to floppy disks, hard disks are able to store and retrieve information much faster and have a greater capacity.

There are three types of hard disks: internal hard disk, hard-disk cartridge, and hard-disk pack.

(1) Internal hard disk

An internal hard disk is located inside the system unit. For most microcomputer systems, one of the internal hard disk drives is designated as the C drive. It is used to store programs and large data files. For example, nearly every microcomputer uses its internal hard disk to store its operating system and major applications such as Word and Excel[2].

(2) Hard-disk cartridges

While internal hard disks provide fast access, they have a fixed amount of storage and cannot be easily removed from the system cabinet[3]. Hard-disk cartridges are as easy to remove as a cassette from a videocassette recorder.

Hard-disk cartridges are used primarily to complement an internal hard disk. Because the cartridges are easily removed, they are particularly useful to protect or secure sensitive information.

Hard-disk cartridges for desktop computers have typical capacities of 20 to 100 GB (gigabytes). Credit card-size hard-disk cartridges called PC Card hard disks are available for notebook computers with typical capacities up to 5 gigabytes.

(3) Hard-disk packs

Hard-disk packs are removable storage devices used to store massive amounts of information. Their capacity far exceeds the other types of hard disks. Although you may never have seen one, it is almost certain that you have used them[4]. Microcomputers that have access to the Internet, minicomputers, or mainframes often have access to external hard-disk packs through communication lines. Banks and credit card companies use them to record financial information.

Three ways to improve the performance of hard disks are disk caching, redundant arrays of independent disks (RAID), and file compression/decompression.

3. OPTICAL DISCS

Today's optical discs can hold over 50 gigabytes of data. That is the equivalent of millions of typewritten pages or a medium-sized library all on a single disc[5].

(1) Compact disc

Compact disc, or as it is better known, CD, is one of the most widely used optical formats. CD drives are standard on many microcomputer systems. Typically, CD drives can store from 650 MB (megabytes) to 1 GB (gigabyte) of data on one side of a CD.

There are three basic types of CDs: read only, write once, and rewritable.

① **Read only-CD-ROM,** which stands for compact disc-read-only memory, is similar to a commercial music CD. Read only means it cannot be written on or erased by the user. Thus, you as a user have access only to the data imprinted by the publisher. CD-ROMs are used to distribute large databases and references[6]. They also are used to distribute large software application packages.

② **Write once-CD-R**, which stands for CD-recordable, can be written to once. After that they can be read many times without deterioration but cannot be written on or erased. CD-R drives often are used to archive data and to record music downloaded from the Internet.

③ **Rewriteable-CD-RW,** which stands for compact disc rewritable. Also known as erasable optical discs, these discs are very similar to CD-Rs except that the disc surface is not permanently altered when data is recorded[7]. Because they can be changed, CD-RWs are often used to create and edit multimedia presentations.

(2) Digital versatile disc (DVD)

DVD stands for digital versatile disc or digital video disc. This is a newer format that has replaced CDs as the standard optical disc. DVDs are very similar to CDs except that more data can be packed into the same amount of space. DVD discs can store 4.7GB to 17GB on a single DVD disc-17 times the capacity of CDs. There are three basic types of DVDs, similar to CDs: read only, write once, and rewriteable.

(3) High-definition disc

While CDs and DVDs are the most widely used optical discs today, the future belongs to discs of even greater capacity. While DVD discs have sufficient capacity to record standard-definition movies and music, they are insufficient for recording high-definition video, which requires about four times as much storage[8]. This next generation of optical disc is called as hi def (high definition), with a far greater capacity than DVDs. Like CDs and DVDs, hi def has three basic types: read only, write once, and rewriteable.

4. Other types of secondary storage

For the typical microcomputer user, the three basic storage options—floppy disk, hard disk, and optical disc—are complementary, not competing. Almost all microcomputers today have at least one floppy-disk drive, one hard-disk drive, and one optical drive. For many users, these secondary storage devices are further complemented with more specialized storage such as solid-state storage, Internet hard drives, and magnetic tape.

(1) Solid-state storage

Each of the secondary storage devices discussed thus far has moving parts. For example, hard disks rotate and read/write heads move in and out. Unlike these devices, solid-state storage devices have no moving parts. Data and information are stored and retrieved electronically directly from these devices much as they would be from conventional computer memory[9]. While this type of storage is more expensive than the others, it is more reliable and requires less power. For these reasons, this technology is becoming widely used for specialized secondary storage.

① **Flash memory cards** are credit card-sized solid-state storage devices widely used in notebook computers. Flash memory also is used in a variety of specialized input devices to capture and transfer data to desktop computers. For example, flash memory is used to store images captured from digital cameras and then to transfer the images to desktop and other computers. Flash memory is used in digital media players like the iPod to store and play music and video files.

② **USB drives,** also known as USB flash drives, are so compact that they can be transported on a key ring or a necklace. These drives conveniently connect directly to a computer's USB port to transfer files and have typical capacities of 2 GB. As of January 2013, drives of up to 512 gigabytes (GB) were available. A one-terabyte (TB) drive was unveiled at the 2013 Consumer Electronics Show and became available later that year[10]. Storage capacities as large as 2 TB are planned, with steady improvements in size and price per capacity expected[11]. Some allow up to 100,000 write/erase cycles, depending on the exact type of memory chip used, and have a 10-year shelf storage time[12].

(2) Internet hard drives

Special service sites on the Web provide users with storage. This storage is called an Internet hard drive (See Figure1-11).

Figure 1-11　An Internet hard drive site

Advantages of Internet hard drives compared to other types of secondary storage include low cost and the flexibility to access information from any location using the Internet. Because all information must travel across the Internet, however, access speed is slower. Another consideration is that users are dependent on the availability and security procedures of the service site. Because of these limitations, Internet hard drives are typically used as a specialized secondary storage device and not for storing highly personalized or sensitive information.

(3) Magnetic tape

To find a particular song on an audiotape, you may have to play several inches of tape. Finding a song on an audio compact disc, in contrast, can be much faster. You select the track, and the disc player moves directly to it. That, in brief, represents the two different approaches to external storage. The two approaches are called sequential access and direct access.

Tapes provide slower sequential access. Although slower to access specific information, magnetic tape is an effective and commonly used tool for backing up data.

5. Mass storage devices

Mass storage refers to the tremendous amount of secondary storage required by large organizations. Mass storage devices are specialized high-capacity secondary storage devices designed to meet organizational demands for data.

Most large organizations have established a strategy called an enterprise storage system to promote efficient and safe use of data across the networks within their organizations (See Figure 1-12). Some of the mass storage devices that support this strategy are

（1） File servers

（2） RAID systems

（3） Tape library

（4） DVD-ROM and CD-ROM jukeboxes

（5） Organizational Internet storage

Figure 1-12　Enterprise storage system

NOTES

[1] more…than…构成的比较句，在 than 后面省略了 that。

[2] Word 和 Excel 分别是微软公司的字处理软件和电子表格软件。

[3] While 引导的是让步状语从句。

[4] Although 引导的是让步状语从句，主句中 that 引导的是主语从句。

[5] all on a single disc，介词短语做状语。

[6] distribute 原意为分配、供应，此处是指将 CD-ROM 用作大型数据库备份。

[7] Also known… 为过去分词短语做状语。 except that 引导的是状语从句；when 引导的是时间状语从句，此句话是指记录在 CD-RW 上面的数据是可以更改的。

[8] 此句与上一句都是由 while 引导的让步状语从句开始，后一句中 which 引导的是非限定性定语从句。

[9] much as 后面是由虚拟语气构成的方式状语从句；they 代表 data and information。

[10] Consumer Electronics Show(CES)，国际著名的电子学和技术商业展，用于吸引世界上的主要公司和业界专业人士，每年一月份举办。

[11] size and price per capacity，按单位存储容量计算的存储芯片面积和价格。

[12] write/erase cycles，写/擦除周期。由于受写-擦除次数的限制，这表示相对有限的闪存寿命。

KEYWORDS

shelf storage	备用存储器
external storage	外部存储器
storage capacity	存储容量
floppy disk drive	软（磁）盘驱动器
storage medium	存储媒体
hard disk	硬（磁）盘
hard-disk cartridge	盒式硬盘
hard-disk pack	硬盘组
cassette	盒，磁带盒
videocassette recorder	盒式录像机
disk cache	磁盘高速缓存
RAID (Redundant Arrays of Independent Disks)	独立磁盘冗余阵列
file compression	文件压缩
file decompression	文件解压缩
optical disc	光盘
GB (gigabytes)	吉字节
compact disc	高密度磁盘，光盘
CD-ROM（Compact Disc-Read-Only-Memory）	光盘只读存储器
CD-R（CD-Recordable）	可录制光盘
CD-RW（Compact Disc Rewritable）	可重写光盘
DVD（Digital Versatile Disc，Digital Video Disc）	数字通用光盘，数字视频光盘
high-definition discs	高清晰光盘
definition	清晰度，分辨率，定义，规定
hi def（high definition）	高清晰度，高分辨率
solid-state storage	固态存储器，半导体存储器
magnetic tape	磁带
read/write head	读/写磁头
flash memory card	闪存卡

USB drive	U 盘驱动器
TB(TeraByte)	太字节，1TB=1024GB
audio compact disc	高密度音盘
sequential access	顺序访问（存取）
mass storage	海量存储器
file server	文件服务器
tape library	磁带库
jukebox	光盘机，光盘自动换盘机，点播机

EXERCISES

1. Multiple Choices

(1) Secondary storage is _____.

    a. permanent

    b. temporary

    c. volatile

    d. nonvolatile

(2) The following storages are belong to the secondary storages: _____.

    a. hard disks

    b. optical disks

    c. solid-state storages

    d. magnetic tapes

(3) Hard disks have the following types, they are _____.

    a. floppy disks

    b. internal hard disks

    c. hard-disk cartridge

    d. hard-disk packs

(4) Point out that which disk is removable one: _____.

    a. hard-disk pack

    b. internal hard disk

    c. floppy disk

    d. hard-disk cartridge

(5) Hard-disk cartridges are used to _____.

    a. replace the floppy disk

    b. protect personal privacy

    c. complement an internet hard disk

    d. secure sensitive information

(6) The following ways can be used for improving the performance of hard disk: _____.

    a. RAID

b. compression / decompression

  c. disk caching

  d. CD-ROM

(7) Optical disk has the following types: _____.

  a. DVD

  b. compact disk

  c. solid-state storage

  d. high-definition disk

(8) Read only-CD-ROM _____.

  a. stands for CD-recordable

  b. cannot be written on by user

  c. is similar to a commercial music CD

  d. is used for distributing large software application packages

(9) Rewriteable-CD-RW _____.

  a. is known as erasable optical disk

  b. has a permanently altered surface

  c. is used to create and edit multimedia presentations

  d. stands for compact disk rewritable

(10) DVD _____.

  a. stands for digital versatile disk

  b. stands for digital video disk

  c. has higher capacity of storage than CD

  d. is a newer format

(11) High-definition disk _____.

  a. can not be used to record high-definition video

  b. can be used to record high-definition video

  c. is called hi def

  d. can be used to record standard-definition movies and music

(12) Solid-state storage devices _____.

  a. have not moving parts

  b. have moving parts

  c. are more expensive than other secondary storages

  d. require less power

(13) Flash memory is widely used in _____.

  a. notebook computers

  b. digital media players

  c. digital cameras

  d. a variety of specialized input devices

(14) USB drives _____.
   a. have very small size and higher capacities of storage
   b. have already replaced the floppy disk
   c. can be connected directly to a computer's USB port
   d. have higher price
(15) Accessing Internet drives, we _____.
   a. can get a higher access speed than accessing a internal drives
   b. should access a special Web service site
   c. should consider information security
   d. can get low cost
(16) Magnetic tape is _____.
   a. a sequential access device
   b. a direct access device
   c. an effective tool for backing up data
   d. a slower speed device
(17) Enterprise storage system should include _____.
   a. File server
   b. RAID system
   c. Internet backup
   d. Tape library

2. True / False
   (1) _____ Most RAM is a permanent or non volatile storage.
   (2) _____ We need external storage for much more storage capacity.
   (3) _____ Hard disks are permanent or nonvolatile storage.
   (4) _____ Using RAM, data and programs can be retained after the computer has been shut off.
   (5) _____ Floppy disks are typically used to store and transport relatively small files.
   (6) _____ There are three types of hard disks.
   (7) _____ Internal hard disk is a removable disk.
   (8) _____ For most PCs, one of the internal hard disk drives is designated as the C drive.

1.5 INPUT AND OUTPUT DEVICES

1. Overview

Input and output equipment allows people to communicate with computers. An input device converts data that humans can understand into a form that the computer can process[1].

Such a device translates the letters, numbers, and other natural-language symbols that humans conventionally use in reading and writing into the binary 0s and 1s that the computer can process. Input devices can also be used to input other types of data, such as photographs, speech, and video.

Output devices, on the other hand, convert the processed 0s and 1s back into a form understandable to humans. These devices typically present output on the screen or paper. Output devices produce results in either hard-copy or soft-copy form. The term hard copy generally refers to output permanently recorded onto an easily portable medium such as paper. The term soft copy generally refers to output that appears temporarily in a form with limited portability, such as on a computer screen[2].

Some devices, such as a modem, can be used for both input and output.

2. Input Devices

(1) Keyboard Entry

One of the most common ways to input data is by keyboard. Keyboards convert numbers, letters, and special characters that people understand into electrical signals. These signals are sent to, and processed by, the system unit. Most keyboards use an arrangement of keys given the name QWERTY. This name reflects the keyboard layout by taking the letters of the first six alphabetic characters found on the fourth row of keys.

There are a wide variety of different keyboard designs. They range from the full-sized to miniature and from rigid to flexible[3]. Most common types are :

- Traditional Keyboards (Figure 1-13)   • Flexible Keyboards (Figure 1-14)
- Ergonomic Keyboards (Figure 1-15)   • Wireless keyboards
- PDA keyboards

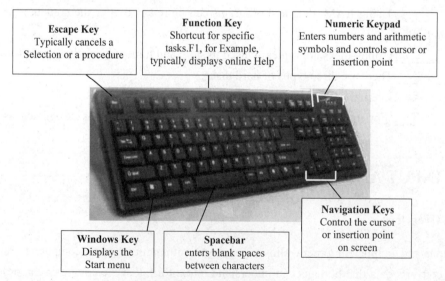

Figure 1-13 Traditional keyboard

CHAPTER 1　COMPUTER ORGANIZATION AND UNITS

Figure 1-14　Flexible keyboard

Figure 1-15　Ergonomic keyboard

(2) Pointing Devices

Pointing devices provide a comfortable interface with the system unit by accepting pointing gestures and converting them into machine-readable input. There are a wide variety of different pointing devices, including the mouse, joystick, touch screen, light pen, and stylus.

① Mice [4]

A mouse controls a pointer that is displayed on the monitor. The mouse pointer usually appears in the shape of an arrow. It frequently changes shape, however, depending on the application. A mouse can have one, two, or more buttons, which are used to select command options and to control the mouse pointer on the monitor. Some mice have a wheel button that can be rotated to scroll through information that is displayed on the monitor[5]. Although there are several different mouse types, there are three basic designs: mechanical mouse, optical mouse, cordless or wireless mouse.

② Joysticks

A joystick is the most popular input device for computer games. You control game actions by varying the pressure, speed, and direction of the joystick. Additional controls, such as buttons and triggers, are used to specify commands or initiate specific actions (See Figure 1-16).

Figure 1-16　Video game joystick elements: 1. stick, 2. base, 3. trigger, 4. extra buttons,
5. autofire switch, 6. throttle, 7. hat switch (POV hat), 8. suction cup

③ Touch Screen

A touch screen is a particular kind of monitor with a clear plastic outer layer. Behind this layer are crisscrossed invisible beams of infrared light. This arrangement enables someone to select actions or commands by touching the screen with a finger[6]. Touch screens are easy to use, especially when people need information quickly. They are commonly used in restaurants, automated teller machines (ATMs), and information centers (See Figure 1-17).

Figure 1-17   A touch screen: a consumer application

④ Light Pen

A light pen is a light-sensitive penlike device. The light pen is placed against the monitor. This closes a photoelectric circuit and identifies the spot for entering or modifying data. For example, light pens are used to edit digital images and drawings.

⑤ Stylus

A stylus is a penlike device commonly used with tablet PCs and PDAs (See Figure 1-18). A stylus uses pressure to draw images on a screen. A stylus interacts with the computer through handwriting recognition software. Handwriting recognition software translates handwritten notes into a form that the system unit can process.

Figure 1-18   Stylus

(3) Scanning Devices

Scanners move across text and images. Scanning devices convert scanned text and images into a form that the system unit can process. There are four types of scanning devices:

optical scanners, card readers, bar code readers and character and mark recognition devices.

① Optical Scanners

An optical scanner, also known simply as a scanner, accepts documents consisting of text and/or images and converts them to machine-readable form. These devices do not recognize individual letters or images. Rather, they recognize light, dark, and colored areas that make up individual letters or images. Typically, scanned documents are saved in files that can be further processed, displayed, printed, or stored for later use. There are two basic types of optical scanners: flatbed and portable.

② Card Readers

Nearly everyone uses a credit card, debit card, access (parking or building) card, and/or some type of identification card. These cards typically have the user's name, some type of identification number, and signature embossed on the card. Additionally, encoded information is often stored on the card as well. Card readers interpret this encoded information. There are two basic types:

- Magnetic Card Reader

By far the most common is the magnetic card reader. The encoded information is stored on a thin magnetic strip located on the back of the card. When the card is swiped through the magnetic card reader, the information is read.

- Radio Frequency Card Readers

Radio frequency card readers are not as common but more convenient because they do not require the card to actually make contact with the reader. The card has a small RFID (radio frequency identification) microchip that contains the user's encoded information. Whenever, the card is passed within a few inches of the card reader, the user's information is read.

③ Bar Code Readers and QR Code

You are probably familiar with bar code readers or scanners from grocery stores (See Figure 1-19). These devices are either handheld wand readers or platform scanners[7]. They contain photoelectric cells that scan or read bar codes, or the vertical zebra-striped marks printed on product containers[8].

Figure 1-19 A bar code reader is used to record product codes

QR code (abbreviated from Quick Response Code) is the trademark for a type of matrix barcode (or two-dimensional barcode) first designed for the automotive industry in Japan. A barcode is a machine-readable optical label that contains information about the item to which it is attached[9]. A QR code uses four standardized encoding modes (numeric, alphanumeric, byte / binary, and kanji) to efficiently store data; extensions may also be used.

The QR Code system became popular outside the automotive industry due to its fast readability and greater storage capacity compared to standard UPC barcodes[10]. Applications include product tracking, item identification, time tracking, document management, and general marketing.

A QR code consists of black modules (square dots) arranged in a square grid on a white background, which can be read by an imaging device (such as a camera) and processed using Reed–Solomon error correction until the image can be appropriately interpreted[11]. The required data are then extracted from patterns present in both horizontal and vertical components of the image (See Figure 1-20).

Figure 1-20　A sample of QR code

④ Character and Mark Recognition Devices

Character and mark recognition devices are scanners that are able to recognize special characters and marks. They are specialty devices that are essential tools for certain applications. Three types are

- Magnetic-ink character recognition (MICR).
- Optical-character recognition (OCR).
- Optical-mark recognition (OMR).

(4) Image Capturing Devices

Like traditional copy machines, image capturing devices are optical scanners that can make a copy from an original. For example, an optical scanner can make a digital copy of a photograph. Image capturing devices, on the other hand, create or capture original images. These devices include digital cameras and digital video cameras.

(5) Audio-Input devices

Audio-input devices convert sounds into a form that can be processed by the system unit. By far the most widely used audio-input device is the microphone. Audio input can take many forms, including the human voice and music.

Voice recognition systems use a microphone, a sound card, and special software. These systems allow users to operate computers and to create documents using voice commands. Portable voice recognition systems are widely used by doctors, lawyers, and others to record dictation (See Figure 1-21). These devices are able to record for several hours before connecting to a computer system to edit, store, and print the dictated information. Some systems are even able to translate dictation from one language to another, such as from English to Japanese.

Figure 1-21　A portable voice recognition system

3. Output Devices

(1) Monitors

The most frequently used output device is the monitor. Also known as display screens, monitors present visual images of text and graphics. The output is often referred to as soft copy. Monitors vary in size, shape, and cost. Almost all, however, have some basic distinguishing features.

The most important characteristic of a monitor is its clarity. Clarity refers to the quality and sharpness of the displayed images. It is a function of several monitor features, including resolution, dot pitch, refresh rate, and size.

- Resolution is one of the most important features. Images are formed on a monitor by a series of dots or pixels (picture elements). Resolution is expressed as a matrix of these dots or pixels. For example, many monitors today have a resolution of 1600 pixel columns by 1200 pixel rows for a total of 1,920,000 pixels. The higher a monitor's resolution (the more pixels), the clearer the image produced. See Figure 1-22 for the most common monitor resolutions[12].
- Dot (pixel) pitch is the distance between each pixel. Most newer monitors have a dot pitch of .31 mm(31/100th of a millimeter) or less. The lower the dot pitch (the shorter the distance between pixels), the clearer the images produced.

Standard	Pixels
SVGA	800×600
XGA	1,024×768
SXGA	1,280×1,024
UXGA	1,600×1,200
QXGA	2,048×1,536

Figure 1-22 Resolution standards

- Refresh rate indicates how often a displayed image is updated or redrawn on the monitor. Most monitors operate at a rate of 75 hertz, which means that the monitor is redrawn 75 times each second. Images displayed on monitors with refresh rates lower than 75 hertz appear to flicker and can cause eye strain. The faster the refresh rate (the more frequently images are redrawn), the better the quality of images displayed.
- Size is measured by the diagonal length of a monitor's viewing area. Common sizes are 15, 17, 19 and 21 inches. The smaller the monitor size, the better the quality of images displayed.

There are several types of monitor, among which the most common monitors are Cathode-Ray Tubes (CRTs) and Flat Panel Monitors. Almost all of today's flat-panel monitors are LCD (Liquid Crystal Display).

There are also several other types of monitors.
- E-books are handheld, book-sized devices that display text and graphics.
- Data projectors are specialized devices similar to slide projectors.
- High-definition television (HDTV)

(2) Printers

You probably use a printer with some frequency to print homework assignments, photographs, and Web pages. Printers translate information that has been processed by the system unit and present the information on paper. Printer output is often called hard copy.

Printers have some basic features, they are:
- Resolution for a printer is similar to monitor resolution. It is a measure of the clarity of images produced. Printer resolution, however, is measured in dpi (dots per inch) (See Figure 1-23). Most printers designed for personal use average 1,200 dpi. The higher the dpi, the better the quality of images produced.

Figure 1-23 Dpi comparison

- Color capability is provided by most printers today. Users typically have the option to print either with just black ink or with color.

- Speed is measured in the number of pages printed per minute. Typically, printers for personal use average 15 to 19 pages per minute for single-color (black) output and 13 to 15 pages per minute for color output.
- Memory within a printer is used to store printing instructions and documents waiting to be printed. The more memory in a printer, the faster it will be able to create large documents.

There are many types of printer, they are:
- Ink-Jet Printers
- Laser Printers
- Thermal Printers
- Other printers: dot-matrix printers, plotters, photo printers, and potable printers.

(3) Audio-Output Devices

Audio-output devices translate audio information from the computer into sounds that people can understand. The most widely used audio-output devices are speakers and headphones. These devices are connected to a sound card in the system unit. The sound card is used to capture as well as play back recorded sounds. Audio-output devices are used to play music, vocalize translations from one language to another, and communicate information from the computer system to users.

Creating voice output is not anywhere near as difficult as recognizing and interpreting voice input[13]. In fact, voice output is quite common. It is used with many soft-drink machines, telephones, and cars. It is used as a reinforcement tool for learning, such as to help students study a foreign language.

4. Combination Input and Output Devices

Many devices combine input and output capabilities. Sometimes this is done to save space. Other times it is done for very specialized applications. Common combination devices include fax machines, multifunctional devices, Internet telephones, and terminals.

NOTES

[1] 句中有两个 that 引导的定语从句，下一句中也有两个 that 引导的定语从句。

[2] refers to output 句后面是由 that 引导的定语从句；上一句 refers to output 后面的过去分词 recorded…构成的短语做定语。

[3] full-sized, rigid keyboard 为原尺寸和刚性键盘，就是常用的矩形键盘，即传统键盘。flexible keyboard 是指可折叠、可卷起来的软键盘，是一种小型键盘，适合于装入袋中保存。

[4] mice 是 mouse 的复数形式。

[5] 句中有两个 that 引导的定语从句。

[6] select actions，选择动作，实为选择菜单项。

[7] platform scanner 与 flatbed scanner 二者是通用的，都是平板式或平台式扫描仪。

[8] that 引导的定语从句中，宾语为 bar codes, or the vertical…。

[9] 句中有 that 和 to which 引导的定语从句，分别修饰 label 和 item，it 代表 information。

[10] UPC barcodes，UPC 码，是美国统一代码委员会制定的一种商品用条码，主要用于美国和加拿大地区，在从美国进口的商品上可以看到。

[11] which 引导的是非限定性定语从句。Reed-Solomon，里德-所罗门码（Reed-Solomon 码），简称 RS 码，是 1980 年用数论方法实现的。它实际上是多进制的 BCH 码（是一类重要的纠错码）。这种纠错编码技术能使编码器集成电路的元件数减少一个数量级。

[12] 五种分辨率标准：SVGA（Super Video Graphics Array），超级视频图形阵列；XGA（eXtended Graphics Array），扩展图形阵列；XGA 2 的高分辨率标准是 1024×768 像素。 SXGA（Super XGA），超级扩展图形阵列；广泛用于 17 英寸和 19 英寸 LCD 监视器中。UXGA（Ultra XGA）的分辨率标准是 1600×1200。QXGA（Quarter XGA），分辨率比 XGA 高 4 倍。

[13] not anywhere near 与 nowhere near 用法相同，意为一点也不，离……很远。此处 not anywhere near as difficult as 可译为不像……那样困难。

KEYWORDS

hard copy	硬拷贝
soft copy	软拷贝
alphabetic	字母的
pointing	定位，定点，指向
pointer	指针
gesture	手势，动作，光电指令
joystick	操作杆
touch screen	触摸屏
light pen	光笔
stylus	记录笔
button	按钮
mechanical mouse	机械鼠标
optical mouse	光学鼠标
cordless (wireless) mouse	无线鼠标
light-sensitive	光敏的
scanner	扫描仪
optical scanner	光学扫描仪
bar code reader	条形码阅读器
recognition device	识别设备

CHAPTER 1　COMPUTER ORGANIZATION AND UNITS

QR(Quick Response) code	快速响应码
matrix	矩阵，行列
label	标记，标号，标签，标识符
mode	方法，模式，方式
alphanumeric	字母数字的
Kanji	汉字（日）
marketing	购物，在市场做生意
pattern	模型，型，样式，图案，样本，模范
flatbed	平板
radio frequency card reader	射频读卡机
RFID (Radio Frequency Identification)	射频识别，电子标签
microchip	微芯片
wand reader	棒式阅读器
MICR(Magnetic-Ink Character Recognition)	磁墨水字符识别
OCR(Optical-Character Recognition)	光学字符识别
OMR(Optical-Mark Recognition)	光学标记识别
image capturing	图像捕获
digital camera	数码相机
digital video camera	数字视频相机
audio-input	音频输入
voice recognition	语音识别
sound card	声卡
clarity	清晰度，明晰，透明
sharpness	清晰度，锐度
resolution	分辨率
dot pitch	点距
refresh rate	刷新速率
pixel	像素
redraw	重新刷新
CRT(Cathode-Ray Tube)	阴极射线管
Flat Panel Monitor	平面监视器
LCD (Liquid Crystal Display)	液晶显示器
data projector	数据投影仪
HDTV(High-Definition Television)	高清晰电视
dpi (dots per inch)	每英寸点数
ink-jet printer	喷墨打印机
laser printer	激光打印机
thermal printer	热敏打印机

dot-matrix printer　　　　　　点阵式打印机
plotter　　　　　　　　　　　　绘图仪
photo printer　　　　　　　　　照片打印机

EXERCISES

Fill in the blanks with appropriate words or phrases found behind this exercise

1. _____ device translates the letters, numbers and other natural-language symbols into 0s and 1s.
2. _____ devices convert the processed 0s and 1s back into a form understandable to humans.
3. _____ allow people to communicate with computers.
4. Output devices produce results in either _____ form.
5. Output on a screen is called as _____.
6. Output on a paper is called as _____.
7. _____ is a pattern to be used to arrange the keys of a keyboard.
8. Mouse, joysticks, touch screen, light pen and stylus are belong to _____.
9. According to applications we can change _____ of the mice pointer.
10. Joystick is the most popular input device for _____.
11. _____ is a matrix barcode.
12. When we use a light pen, it must be placed against the _____.
13. A stylus interacts with the computer through _____.
14. _____ convert scanned text and images into a form that the system unit can process.
15. There are two basic types of optical scanners: _____.
16. A card reader which does not require the card to actually make contact with the reader is a _____.
17. In most grocery stores we usually find _____ to scan the bar code.
18. A device that is able to recognize special characters and marks is _____.
19. Digital cameras are belong to _____.
20. QR code was designed for _____ originally.
21. We can describe the clarity of a monitor by _____.
22. _____ is a picture element.
23. _____ is expressed as a matrix of series of pixels.
24. _____ is the distance between each pixel.
25. If refresh rate is lower than 75 hertz _____ will appear.
26. The smaller the monitor size, the better the _____ of images displayed.
27. Almost all of today's flat-panel monitors are _____.
28. Printers have some basic features, they are _____.
29. The _____ is used to capture as well as play back recorded sounds.

CHAPTER 1 COMPUTER ORGANIZATION AND UNITS

30. _____ is one of the combination input and output devices.
31. In information centers we usually find _____.
32. Doctors and lawyers often use portable _____ to record dictation.

   a. LCDs
   b. QWERTY
   c. Resolution
   d. Input
   e. touch screen
   f. bar code reader
   g. Scanners
   h. Pixel
   i. flicker
   j. Input and output devices
   k. sound card
   l. soft-copy
   m. Output
   n. voice recognition systems
   o. monitor
   p. flatbed and portable
   q. shape
   r. Radio Frequency Card Reader
   s. pointing devices
   t. image capturing devices
   u. quality
   v. computer games
   w. a character and mark recognition devices
   x. hard-copy
   y. resolution, dot pitch, refresh rate, and size
   z. hard-copy or soft-copy
   aa. Dot pitch
   bb. Internet telephone
   cc. hand-writing recognition program
   dd. resolution, color capability, speed, and memory
   ee. QR code
   ff. Automotive industry

CHAPTER 2 SYSTEM SOFTWARE

2.1 COMPUTER SOFTWARE

As already mentioned, the term computer software refers to the programs or instructions used to tell the computer hardware what to do. Software is generally purchased in some sort of a physical package. Such a software package may consist of the program, as well as instructional and help materials, on CD or DVD discs, printed operating instructions and user manuals, and a printed license to use the software—all or some of which are inside a shrink-wrapped box or plastic case[1]. You can buy software in a store, through mail order, or over the Internet. If you download the software from the Internet, you won't receive a physical package. Instead, the components (program, license, user's manual, and so forth) are downloaded directly to your computer in an electronic format.

Computers use two basic types of software: Application software and system software. Application software helps you perform a specific task. System software refers to the operating systems and all utility programs that manage computer resources at a low level[2]. Figuratively speaking, application software sits on top of system software. Without the operating system and system utilities, the computer cannot run any application program.

1. Application Software

Application Software is widely referred to as productivity software. Application software is comprised of programs designed for an end user. Three types of application software are general-purpose, specialized, and mobile apps.

General-purpose applications are widely used in nearly all career areas. They are the kinds of programs you have to know to be considered computer competent[3]. One of these general-purpose applications is a browser to navigate, explore, and find information on the Internet. The three most widely used browsers are Mozilla's Firefox, Microsoft's Internet Explorer, and Google's Chrome[4].

Specialized applications include thousands of other programs that are more narrowly focused on specific disciplines and occupations. Two of the best known are graphics and web authoring programs.

Mobile apps, or mobile applications, are small programs designed for mobile devices such as smartphones, tablet computers, and other mobile devices. There are over half a million apps. The most popular mobile apps are for text messaging, Internet browsing, and connecting to social networks.

2. System Software

System software is a group of programs that coordinate and control the resources and operations of a computer system. System software enables the many components of the computer system to communicate. There are three categories of system software: operating systems, utilities, and language translators.

(1) Operating System

Operating systems provide an interface between the user or application program and the computer hardware. There are many brands and versions of operating system software. Each of these is designed to work with one or more particular processors. For example, an operating system like Windows is designed to work with a processor made by Intel. Many IBM PC compatible computers contain this brand of processor. Most Macintosh computers contain a processor manufactured by Motorola. The Windows operating system does not work with this Motorola processor.

(2) Utilities

Utility programs help you perform housekeeping chores. You use these programs to complete specialized tasks related to managing the computer's resources, file management, and so forth. Some utility programs are part of the operating system, and others are self-contained programs[5]. Some examples of utility program functions are as follows:

You format a disk—a disk formatting utility provides the instructions to the computer on how to do this.

You copy a file from the hard drive to a floppy disk—the file management utility provides the instructions to the computer.

You do a backup of the hard drive—you use the backup utility.

(3) Language Translators

Language Translators convert English-like software programs into machine language that the computer can understand. Vista Multimedia hires a programmer to write a software program to inventory all of the items in the store[6]. The programmer writes the program statements using a programming language called Visual Basic. A program statement directs the computer to perform a specified action.

The computer, however, cannot read the Visual Basic programming statements because they are written in a language that we understand. This is where the language translator takes over[7]. The translator changes each of the Visual Basic programming statements into machine language. A single statement in a high-level language can represent several machine-language instructions. Now the statements can be executed and the Vista Multimedia inventory can be processed.

NOTES

[1] 长句。句中介词 of 的宾语为多个并列，最后 all or some of which 引导的是非限定性定语从句。

[2] low level 软件是指软件层次结构中的最底层的软件，通常是对硬件进行操作的。

[3] you have to…引导的是定语从句，修饰 programs。

[4] Mozilla's Firefox，中文名通常称为"火狐"或"火狐浏览器"（正式缩写为 Fx，非正式缩写为 FF），是一个开源网页浏览器，使用 Gecko 引擎，支持多种操作系统，如 Windows、Mac 和 Linux。Internet Explorer，是美国微软公司推出的一款网页浏览器，10、11 版本），简称 IE。在 IE7 以前，中文直译为"网络探路者"，但在 IE7 以后官方便直接俗称"IE 浏览器"。2015 年 3 月微软确认将放弃 IE。Google Chrome，又称 Google 浏览器，详见 4.1.2 节。

[5] self-contained program，自包含程序。

[6] Vista Multimedia 为一公司名。

[7] where 引导的是表语从句，take over 接收，占领，此处表明语言翻译器起作用。

KEYWORDS

software package	软件包
mail order	邮购
download	下载
application software	应用软件
system software	系统软件
operating system	操作系统
utility program	实用程序，公用程序
productivity software	生产力软件
end user	最终（终端，直接）用户
browser	浏览器
smartphone	智能电话
social network	社交网
authoring	创作，编辑
language translator	语言翻译器（程序）
backup	备份
machine language	机器语言
Visual Basic	可视 Basic 语言，简称 VB

EXERCISES

True/False

1. _____ Generally, a software we bought in a store is on CDs or DVDs discs.
2. _____ You will receive a physical package if you download a software from the

Internet.
3. _____ There are two basic types of software in the computers.
4. _____ System software helps you perform a specific task.
5. _____ All utility programs belong to the application software.
6. _____ With the operating system and system utilities, the computer can run any application program.
7. _____ Browser is one of the most used general-purpose application software.
8. _____ A computer system is controlled by its operating system.
9. _____ There are four categories of system software.
10. _____ Operating system is designed to work with one or more particular processors.
11. _____ Windows operating system is designed to work with a processor made by Motorola.
12. _____ Utility programs help you to manage the computer's resources.
13. _____ Using a application program, you can format a disk.
14. _____ A computer can understand a English-like program without language translator.
15. _____ A single statement in a high-level language can represent several machine language instructions.
16. _____ Computer software refers to the programs or instructions used to tell the computer hardware what to do.

2.2 SUMMARY OF OPERATING SYSTEM

Operating systems have developed over the past thirty years for two main purposes. First, they provide a convenient environment for the development and execution of programs. Second, operating systems attempt to schedule computational activities to ensure good performance of the computing system.

As we have stated, operating systems are normally unique to their manufacturers and the hardware in which they are run. Generally, when a new computer system is installed, operational software suitable to that hardware is purchased. Users want reliable operational software that can effectively support their processing activities.

Though operational software varies with manufacturers, it has similar characteristics[1]. Modern hardware, because of its sophistication, requires that operating systems meet certain specific standards. For example, considering the present state of the field, an operating system must support some form of online processing. Functions normally associated with operational software are:

(1) Job management;
(2) Resource management;
(3) Control of I/O operations;

(4) Error recovery;

(5) Memory management.

1. Job Management

A very important responsibility of any operational software is the scheduling of jobs to be handled by a computer system. This is one of the main tasks of the job management function. The operating system sets up the order in which programs are processed, and defines the sequence in which particular jobs are executed. The term "job queue" is often used to describe the series of jobs awaiting execution. The operating system weighs[2] a variety of factors in creating the job queue. These include which jobs are currently being processed, the system's resources being used, which resources will be needed to handle upcoming programs, the priority of the job compared to other tasks, and any special processing requirements to which the system must respond.

The operational software must be able to assess these factors and control the order in which jobs are processed.

Batch systems allowed automatic job sequencing by a resident monitor and improved the overall utilization of the computer greatly. The computer no longer had to wait for human operation. CPU utilization was still low, however, because of the slow speed of the I/O devices relative to the CPU. Offline operation of slow devices was tried. Buffering was another approach to improving system performance by overlapping the input, output, and computation of a single job. Finally, spooling allowed the CPU to overlap the input of one job with the computation and output of other jobs[3].

Spooling also provides a pool of jobs which have been read and are waiting to be run. This job pool supports the concept of multiprogramming. With multiprogramming, several jobs are kept in memory at one time; the CPU is switched back and forth between them in order to increase CPU utilization and to decrease the total real time needed to execute a job.

Multiprogramming, which was developed to improve performance, also allows time sharing. Time-shared operating systems allow many users (from one to several hundred) to use a computer system inter-leavingly at the same time. Other operating systems types include real-time systems and multiprocessor systems.

2. Resource Management

The management of resources in a computer system is another major concern of the operating system. Obviously, a program cannot use a device if that hardware is unavailable. As we have seen, the operational software oversees the execution of all programs. It also monitors the devices being used. To accomplish this, it establishes a table in which programs are matched against the devices they are using or will use[4]. The operating system checks this table to approve or deny use of a specific device.

3. Control of I/O Operations

Allocation of a system's resources is closely tied to the operational software's control of I/O operations. As access is often necessary to a particular device before I/O operations may begin, the operating system must coordinate I/O operations and the devices on which they are performed. In effect, it sets up a directory of programs undergoing execution and the devices they must use in completing I/O operations. Using control statements, jobs may call for specific devices[5]. This lets users read data from specific sites or print information at selected offices. Taking advantage of this facility, data read from one location may be distributed throughout computerized system.

To facilitate execution of I/O operations, most operating systems have a standard set of control instructions to handle the processing of all input and output instructions. These standard instructions, referred to as the input/output control system (IOCS), are an integral part of most operating systems. They simplify the means by which all programs being processed may undertake I/O operations.

In effect, the program undergoing execution signals the operating system that an I/O operation is desired, using a specific I/O device. The controlling software calls on the IOCS software to actually complete the I/O operation. Considering the level of I/O activity in most programs, the IOCS instructions are extremely vital.

NOTES

[1] …varies with… 随……不同而不同；Though 引导的是让步状语从句。

[2] weighs 作动词，意为"权衡"。

[3] spooling 来源于 spool（Simultaneous Peripheral Operations On Line，外部设备同时联机操作），称假脱机（操作）。

[4] 这是一个复合句，"to accomplish this"是动词不定式作目的状语；"in which"到句末是带前置词的限制性定语从句，修饰宾语 table，其中 they are using or will use 是省略了引导词 that 的限制性定语从句，修饰"devices"。

[5] "using control statements"是现在分词短语，作伴随状语。

KEYWORDS

job management	作业管理
resource management	资源管理
job queue	作业队列
input/output control system(IOCS)	输入输出控制系统
batch system	批处理系统
time-shared operating system	分时操作系统
offline operation	脱机操作

real-time system	实时系统
buffering	缓冲技术
spooling system	假脱机系统
multiprogramming	多道程序设计
program order	程序顺序
job sequence	任务排序，加工序列

EXERCISES

1. True/ False.

(1) _____ Operating systems are not unique to their manufactures.

(2) _____ When you purchase a new computer, an operating software suitable to the hardware must be bought.

(3) _____ The series of jobs awaiting execution is called a queue.

(4) _____ A program can use a device if that hardware is unavailable.

(5) _____ Operating system sets up a directory of programs undergoing execution and the devices.

(6) _____ IOCS simplifies the ways of I/O operations.

(7) _____ In order to create the job queue the operating system must control I/O operation.

(8) _____ Error recovery does not belong to operating system.

(9) _____ One of the five functions of an operating system is memory management.

(10) _____ Many operating systems have similar characteristics.

2. Fill in the blanks with appropriate words and phrases found behind this exercise.

(1) Buffering can overlap _____.

(2) Operating systems should meet certain specific standards as modern computer hardware's _____.

(3) The main task of the job management is_____.

(4) In order to create the job queue the operating system should_____.

(5) If more than one job require to be processed the operating system should_____.

(6) Batch systems automate job sequencing by a _____.

(7) _____ allowed the CPU to overlap the I/O operation with the computation.

(8) Time sharing can be got by _____.

(9) _____ are other operating system types.

(10) To monitor the devices being used the operating system establishes a table in which programs are matched against _____.

    a. the scheduling of jobs

    b. real time systems and multiprocessor systems

    c. compare the priority of these jobs

d. the input, output, and computation of a single job.

e. resident monitor

f. the device being used

g. sophistication

h. spooling

i. weighing a variety of factors

j. multiprogramming

2.3 WINDOWS 10

Windows 10 is an upcoming operating system developed by Microsoft as part of the Windows NT family of operating systems. First presented in April 2014 at the Build Conference, it is scheduled to be released in mid-2015, and is currently in public beta testing through the Windows Insider program[1]. During its first year of availability, upgrades to Windows 10 will legally be offered at no charge for licensed consumer users of Windows 7 and Windows 8.1.

The goal of Windows 10 is to unify the Windows PC, Windows Phone, Windows Embedded and Xbox One product families around a common internal core[2]. These products will share a common, "universal" application architecture and Windows Store ecosystem that expands upon the Windows Runtime platform introduced by Windows 8[3]. Windows 10 will also introduce a new bundled web browser, Microsoft Edge, to replace Internet Explorer[4].

Figure 2-1 shows a Screenshot of Windows 10.

Figure 2-1 Screenshot of Windows 10

1. User interface and desktop

Windows 10's user interface is an evolution of Windows 8's, it changes its behavior depending on the type of device being used and available input methods. When a keyboard is attached, users are asked if they want to switch to a user interface mode that is optimized for mouse and keyboard, or stay within the touch-optimized mode. A new iteration of the Start menu is used, with an application list and the "All apps" button on the left side, and live tiles on the right. The menu can be resized, and expanded into a full-screen display, which is the default option in touch environments.

A new virtual desktop system known as Task View was added. Clicking the Task View button on the taskbar or swiping from the left side of the screen displays all open windows and allows users to switch between them, or switch between multiple workspaces[5]. Windows Store apps, which previously could only be used full-screen, can now be used in desktop windows or full-screen mode[6]. Program windows can now be snapped to quadrants of the screen by dragging them to the corner [7]. When a window is snapped to one side of the screen, the user is prompted to choose a second window to fill the unused side of the screen (called "Snap Assist")[8].

2. Features

A major aspect of Windows 10 is a focus on harmonizing user experiences between different classes of devices, along with addressing shortcomings in the Windows user interface that was first introduced in Windows 8[9].

The Windows Store app ecosystem has been revised into "Windows apps". They are made to run across multiple platforms and device classes, including smartphone, tablet, Xbox One, and other compatible Windows 10 devices. Windows apps share code across platforms, have responsive designs that adapt to the needs of the device and available inputs, can synchronize data between Windows 10 devices (including notifications, credentials, and allowing cross-platform multiplayer for games), and will be distributed through a unified Windows Store[10]. Developers can allow "cross-buys", where purchased licenses for an app apply to all of the user's compatible devices, rather than only the one they purchased on (i.e. a user purchasing an app on PC is also entitled to use the smartphone version at no extra cost) [11].

Windows 10 will also allow Web apps and desktop software (using either Win32 or .NET Framework) to be packaged for distribution on Windows Store[12]. Desktop software distributed through Windows Store will be packaged using the App-V system to allow sandboxing[13].

3. Windows 10 Editions

As in the past, we will offer different Windows editions that are tailored for various device families and uses. These different editions address specific needs of our various customers, from consumers to small businesses to the largest enterprises.

- Windows 10 Home is the consumer-focused desktop edition. It offers a familiar and personal experience for PCs, tablets and 2-in-1s.
- Windows 10 Mobile is designed to deliver the best user experience on smaller, mobile, touch-centric devices like smartphones and small tablets.
- Windows 10 Pro is a desktop edition for PCs, tablets and 2-in-1s. Building upon both the familiar and innovative features of Windows 10 Home, it has many extra features to meet the diverse needs of small businesses.
- Windows 10 Enterprise builds on Windows 10 Pro, adding advanced features designed to meet the demands of medium and large sized organizations.
- Windows 10 Education builds on Windows 10 Enterprise, and is designed to meet the needs of schools – staff, administrators, teachers and students.
- Windows 10 Mobile Enterprise is designed to deliver the best customer experience to business customers on smartphones and small tablets.

NOTES

[1] 长句，其中 First presented…为过去分词构成的时间状语从句，主句为 it is…，是并列句，主语 it 代表 Windows 10。Build 是微软主持的年会，旨在推进 Windows、Windows Phone、微软 Azure 和微软其他技术的使用。beta 是希腊 alphabet 的第二个字母，beta test 是软件开发中紧跟 alpha 第一阶段的第二阶段。在此阶段，软件功能基本实现，但还会有很多已知或未知的瑕疵(bugs)要发现和处理。Windows Insider program 是微软公司的程序，可使用户去访问微软软件的开发者。2014 年 9 月 30 日，该程序与 Windows 10 一起公布。2015 年 2 月 12 日用该程序为用户提供针对 Windows 10 的手机版预览。

[2] Windows Phone (WP)是微软为其智能手机开发的移动操作系统系列，取代了以前的 Windows Mobile 和 Zune 操作系统。由于微软新的商标策略和 Windows 10 的移动版在 2015 年面世，因此 Windows Phone 商标被淘汰。Windows Embedded 是微软用于嵌入式系统的操作系统系列。Xbox One 是微软在 2013 年 5 月 21 日公布并投放市场的家庭游戏控制台，它的前身是 Xbox 360。Xbox One 是 Xbox 系列中的第三个控制台。

[3] Windows Store 是微软发布和销售软件的平台，形象地说是一个商店。Windows Runtime (WinRT)是 2012 年 9 月在 Windows Server 2012 上首次提出的平台性质的应用体系结构。

[4] Microsoft Edge（代号 Spartan）是微软开发的万维网浏览器。官方首次公布是在 2015 年 1 月 21 日。作为预览的第一个公共版本是 2015 年 3 月 30 日公布的，它作为 Windows 10 PCs 智能手机和平板电脑的默认浏览器，取代了 IE。

[5] 长句。主语为现在分词短语 Clicking…or swiping…，谓语为 displays…and allows…。

[6] Windows Store app 是在 Windows 8 下的一种新型应用软件，是为触摸环境设计并优化的，比桌面应用软件更专业一些。

[7] Program windows 应为 Program for windows。

[8] Snap Assist 是 Windows 10 桌面侧边新加入的按钮，称为抓取助手，它可将多个不同桌面的应用展示在此，并和其他应用程序自由组合成多任务模式。

[9] 长句。主句是 A major aspect of Windows 10 is…，along with…介词短语作状语，其中 that 引导的定语从句修饰 interface。

[10] 长句。句子结构为 Windows apps share…，have…，can…，and will be…。其中 have…句中有 that 引导的定语从句。responsive design 响应式设计，实为 responsive Web design(RWD)，响应式页面设计，又称自适应页面设计，回应式页面设计，该设计可使网站在多种浏览设备（从桌面电脑显示器到移动电话或其他移动设备）上阅读和导航，同时减少缩放、平移和滚动。

[11] where 引导的是非限定性定语从句。cross-buys 是由 Sony 公司在 2012 年推出的，它可以让一个游戏在多个游戏网站上使用。

[12] Win32 是 WinAPI（正式应称为 Windows API）中的一个层次。WinAPI 是微软 Windows 操作系统中的应用程序界面的一套核心程序。.NET Framework 是微软开放的软件架构，主要运行在微软 Windows 上，它包括一个架构类库（Framework Class Library (FCL)）的大型类库。

[13] App-V（Application Virtualization）微软的应用程序虚拟化，它是将应用程序转换为集中管理的服务。Sandbox，即沙盒、沙箱，是一种安全机制，为运行中的程序提供隔离环境，通常是为来源不可信、具有破坏力或无法判定程序意图的那些程序提供实验用。

KEY WORDS

codename	代号
availability	可用性，有效性，利用率，利用度
upgrade	升级，提高
licensed	特许的，许可的
architecture	体系结构，结构，层次结构，总体结构，结构格式
family	族，类，系列，种类
phased out	逐渐淘汰
console	控制台
embedded	嵌入式的
platform	平台
browser	浏览器
touch	触摸
button	按钮
taskbar	任务栏（条）

tile	平铺（并列）显示，铺切
resize	调整（重设）大小
framework	框架，构架，架构，体制组织
trackpad	跟踪垫（板），轨迹板
snap	抓取，快照，取图，快速移动
icon	图标，图符
default	默认，缺省
addressing	编址，寻址，定址
screenshot	屏幕快照
window(s)	窗口，视窗
Snap Assist	抓取助手

EXERCISES

Multiple Choices

1. September 30, 2014 Microsoft released _____.
   a. Windows 7   b. Windows 10
   c. Build Conference   d. Windows Insider
2. Windows 10 has unified _____.
   a. Xbox One   b. Windows Phone
   c. Windows Embedded   d. Windows PC
3. Windows 10 introduced _____.
   a. IE   b. Microsoft Edge
   c. a new web browser   d. a bundled web browser
4. According to _____. Windows 10's user interface changes behavior of using computer.
   a. available input methods   b. available output methods
   c. the type of device being used   d. the type of software being used
5. The Start menu in Windows 10 has _____.
   a. Live tiles   b. an application list
   c. "an app" button only   d. the "all apps" button
6. Task View button of Windows 10 _____.
   a. is known as virtual desktop system
   b. used for displaying all open windows
   c. used for switching between windows
   d. used for switching between multispaces
7. The Windows Store app ecosystem of Windows 10 can run _____.
   a. across multiple device classes   b. across multiple platforms
   c. smartphone   d. a platform only
8. Windows apps _____.
   a. can share code across platform

b. have responsive design

   c. can synchronize data between Windows 10 devices

   d. will be distributed through a unified Windows Store

9. Data in Windows apps can be _____.

   a. notifications   b. credentials   c. input devices   d. games

10. App-V system _____.

   a. stands for Application Visualization   b. stands for Application Virtualization

   c. stands for Application Videocam   d. is used to package a Desktop software

11. Windows 10 Enterprise _____.

   a. builds on Windows 10 Mobile   b. used for small sized organizations

   c. builds on Windows 10 Pro   d. used for large sized organizations

12. Windows 10 Pro is _____.

   a. a desktop edition   b. used for PCs

   c. used for tablets   d. used for small businesses

2.4 UNIX AND LINUX

1. UNIX

UNIX is an operating system originally developed by Dennis Ritchie and Ken Thompson at AT&T Bell Laboratories that allows a computer to handle multiple users and programs simultaneously. Since its development in the early 1970s, UNIX has been enhanced by many individuals and particularly by computer scientists at the University of California, Berkeley (known as Berkeley Software Distribution UNIX, or BSD UNIX) [1]. This operating system is available on a wide variety of computer systems, ranging from personal computers to mainframes, and is available in other related forms[2]. AIX is an implementation that runs on IBM workstations, A/UX is a graphical version that runs on Macintosh computers; Solaris runs on Intel microprocessors[3].

Features:

(1) The UNIX system can support multi-users and multi-tasks

(2) The UNIX System Kernel

The kernel is the heart of the UNIX operating system, responsible for controlling the computer's resources and scheduling user jobs so that each one gets its fair share of the resources. Programs interact with the kernel through special functions with well-known names, called system calls.

(3) The shell

The shell is a command interpreter that acts as an interface between users and the operating system. When you enter a command at a terminal, the shell interprets the command

and calls the program you want.

(4) Device-Independent Input and Output

Devices (such as a printer or terminal) and disk files all appear as files to UNIX programs. When you give the UNIX operating system a command, you can instruct it to send the output to any one of several devices or files. This diversion is called output redirection.

2. Linux

Linux is a UNIX-like and mostly POSIX-compliant computer operating system assembled under the model of free and open-source software development and distribution[4]. The defining component of Linux is the Linux kernel, an operating system kernel first released on 5 October 1991 by Linus Torvalds. The Free Software Foundation uses the name GNU/Linux to describe the operating system, which has led to some controversy[5].

Linux was originally developed as a free operating system for Intel x86–based personal computers, but has since been ported to more computer hardware platforms than any other operating system[6]. It is the leading operating system on servers and other big iron systems such as mainframe computers and supercomputers, but is used on only around 1.5% of desktop computers[7]. Linux also runs on embedded systems, which are devices whose operating system is typically built into the firmware and is highly tailored to the system. Android, the most widely used operating system for tablets and smartphones, is built on top of the Linux kernel[8].

(1) Hardware support

Linux kernel is a widely ported operating system kernel; it runs on a highly diverse range of computer architectures, including the hand-held ARM-based iPAQ and the IBM mainframes System z9 or System z10—covering devices ranging from mobile phones to supercom-puters, as showing as Figure 2-2[9].

Figure 2-2   Linux is ubiquitously found on various types of hardware

(2) Uses

Beside the Linux distributions designed for general-purpose use on desktops and servers, distributions may be specialized for different purposes including: computer architecture support, embedded systems, stability, security, localization to a specific region or language, targeting of specific user groups, support for real-time applications, or commitment to a given desktop environment[10]. As of 2015, over four hundred Linux distributions are actively developed, with about a dozen distributions being most popular for general-purpose use.

(3) Desktop

The popularity of Linux on standard desktop computers and laptops has been increasing over the years. Currently, most distributions include a graphical user environment, with the two most popular environments being GNOME (which can utilize additional shells such as the default GNOME Shell and Ubuntu Unity) and the KDE Plasma Desktop[11].

No single official Linux desktop exists: rather desktop environments and Linux distributions select components from a pool of free and open-source software with which they construct a GUI implementing some more or less strict design guide[12]. GNOME, for example, has its human interface guidelines as a design guide, which gives the human–machine interface an important role, not just when doing the graphical design, but also when considering people with disabilities, and even when focusing on security[13].

Figure 2-3 shows the Visible software components of the Linux desktop stack.

Figure 2-3　Visible software components of the Linux desktop stack

Visible software components of the Linux desktop stack include the display server, widget engines, and some of the more widespread widget toolkits. There are also components not directly visible to end users, including D-Bus and PulseAudio[14].

NOTES

[1] BSD 即 Berkeley Software Distribution，原意为伯克利软件发行中心，可直译为伯克利版软件或伯克利 UNIX。

[2] 由 and 连接的并列句，前一句中 ranging…为分词短语做状语。

[3] 此处介绍了 UNIX 的几个变种，其中 Macintosh 机器上的 A/UX 是苹果公司开发的 UNIX， Solaris 是运行在 SUN 公司工作站上的操作系统，其 CPU 使用的是 Intel 微处理器。

[4] POSIX，Portable Operating System UNIX，可移植的 UNIX 系统。

[5] Free Software Foundation (FSF)即自由软件基金会，是 1985 年成立的非营利组织，致力于自由软件的推广。GNU(Gnu's Not UNIX)是一个自由软件组织。Which 引导的是非限定性定语从句。

[6] 并列句，后一句省略了主语 Linux。Intel x86 是 Intel 公司生产的微处理器芯片，x 代表 802，803，804…。

[7] big iron 在计算机领域中是指大型、昂贵、非常快速的计算机，常用于描述像格雷（Grays）那样的超级计算机，也用来指大型商业 IBM 计算机等。

[8] Android 是用于智能电话和平板电脑等移动设备的操作系统。

[9] ARM-based 是 Arm 公司的微处理器和微控制器集成电路，是基于各种 32 位的 ARM 处理器芯片。iPAQ 是指 2000 年 4 月由康柏（Compaq）公司首次公布的袖珍电脑和个人数字助理（PDA），此名称是借用康柏公司早期 iPAQ 台式个人计算机名字。

[10] 长句。Beside 此处为介词,由它引出的介词短语做状语,主句为 distributions may be…。Linux distributions 是 Linux 发行版，是为一般用户预先集成好的 Linux 操作系统及各种应用软件，通常包括桌面环境办公套件，媒体播放器，数据库等应用软件。

[11] GNOME 是一个桌面环境，它全部由自由和开源软件组成，其目标操作系统是 Linux，但也支持其他大多数 BSD 衍生软件。GNOME Shell 是 GNOME 桌面环境的正式用户界面。Ubuntu Unity 中，Unity 是由 Canonical 有限公司为其 Ubuntu 操作系统开发的 GNOME 桌面环境中的图形外壳。KDE 是一个国际自由软件团体，它是制作在 Linux、FreeBSD、Solaris、Microsoft Windows 和 OS X 系统上的，集成化、跨平台的应用软件系列。这种称为 Plasma Desktop 的，是一个桌面环境。

[12] with which 引导的定语从句，修饰 components。

[13] 长句，which 引导的是非限定性定语从句。

[14] D-Bus 实质上是一个适用于桌面应用进程间的通信机制，即所谓的 IPC（Inter-Process Communication）机制。 PulseAudio（以前叫 Polypaudio）是一个跨平台的、可通过网络工作的声音服务，其一般使用于 Linux 和 FreeBSD 操作系统。

KEYWORDS

UNIX	一种计算机操作系统
workstation	工作站
multi-user	多用户
multi-task	多任务
job	作业
kernel	内核

interface	接口，界面
system call	系统调用
interpreter	解释程序
disk file	磁盘文件
redirection	重定向
device-independent	设备无关的
shell	外壳（程序）、命令解释程序
Linux	以 Linux 命名的操作系统
Gnu's Not UNIX (GNU)	一个自由软件组织
embedded system	嵌入式系统
firmware	固件
tailor	设计，裁剪，加工，处理，编（特）制
port	移植，端口
localization	局部化，地方化，本地化
security	安全，安心，防护，保障
widget	小装置（窗口），窗口软件设计工具，界面构造（组件），图形设备，数据类型
GUI(Graphic User Interface)	图形用户界面
human-machine interface	人机界面

EXERCISE

Fill in the blanks with appropriate words or phrases found behind this exercise.

1. UNIX can be used on _____.
2. UNIX System can support _____ users and _____ tasks.
3. The UNIX allows each user to run _____ job at a time.
4. _____ are the special functions that interact programs with the kernel of UNIX.
5. The shell is an interface between _____.
6. You can instruct _____ to send the output to any one of several devices or files.
7. Linux is a _____ operating system.
8. FSF uses the name _____ to describe Linux.
9. _____ is the leading operating system on servers.
10. _____ has a built-in operating system.
11. _____ is built on top of the Linux kernel.
12. _____ is a graphical user environment.

    a. UNIX-like

    b. UNIX

    c. embedded system

    d. PCs, minicomputers, mainframes

e. users, and the operating system

f. GNOME

g. more than one

h. Linux

i. Android

j. GNU/Linux

k. system calls

l. many,many

2.5 MAC OS

Mac OS is a series of graphical user interface-based operating systems developed by Apple Inc. (formerly Apple Computer, Inc.) for their Macintosh line of computer systems. The Macintosh user experience is credited with popularizing the graphical user interface.

Early versions of the Mac OS were compatible only with Motorola 68000-based Macintoshes. As Apple introduced computers with PowerPC hardware, the OS was ported to support this architecture as well[1]. Mac OS 8.1 was the last version that could run on a 68000-class processor (the 68040). Mac OS X, also known as "Tiger", which has superseded the "Classic" Mac OS, is compatible with both PowerPC and Intel processors through to version 10.5 ("Leopard")[2]. Version 10.6 ("Snow Leopard") supports only Intel processors.

Mac OS can be divided into two families:
- The Mac OS Classic family, which was based on Apple's own code
- The Mac OS X operating system, developed from Mac OS Classic family, and NeXTSTEP, which was UNIX-based[3].

1. "Classic" Mac OS

The "classic" Mac OS is characterized by its total lack of a command line; it is a completely graphical operating system. Versions of Mac OS up through System 4 only ran one application at a time. Even so, it was noted for its ease of use. Mac OS gained cooperative multitasking with System 5, which ran on the Mac SE and Macintosh II[4].

2. Mac OS X

Mac OS X is the newest of Apple Inc.'s Mac OS line of operating systems. Although it is officially designated as simply "version 10" of the Mac OS, it has a history largely independent of the earlier Mac OS releases.

The operating system is the successor to Mac OS 9 and the "classic" Mac OS. It is a UNIX operating system, based on the NeXTSTEP operating system and the Mach kernel which Apple acquired after purchasing NeXT Computer[5]. Mac OS X also makes use of the

BSD code base [6]. There have been six significant releases of the client version, the most recent being Mac OS X 10.6, referred to as Snow Leopard. On Apple's October 20th 2010 "Back to the Mac" event, Mac OS X 10.7 Lion was previewed, showing improvements and additions including a Mac App Store [7].

As well as the client versions, Mac OS X has also had six significant releases as a server version, called Mac OS X Server. The server versions are architecturally identical to the client versions, with the differentiation found in their inclusion of tools for server management, including tools for managing Mac OS X-based workgroups, mail servers, and web servers, amongst other tools [8]. It is currently the default operating system for the Xserve server hardware, and as an optional feature on the Mac Mini, as well as being installable on most other Macs [9]. Unlike the client version, Mac OS X Server can be run in a virtual machine using emulation software such as Parallels Desktop.

Mac OS X is also the basis for iOS, (previously iPhone OS) used on Apple's iPhone, iPod Touch, and iPad, as well as being the basis for the operating system used in the Apple TV [10].

3. OS X Yosemite

OS X Yosemite is the eleventh major release of OS X, Apple Inc.'s desktop and server operating system for Macintosh computers.

OS X Yosemite was announced and released to developers on June 2, 2014, at the WWDC 2014 and it was released to public beta testers on July 24, 2014 [11]. Yosemite was released to consumers on October 16, 2014.

Many of Yosemite's new features focus on the theme of continuity, increasing its integration with other Apple services and platforms such as iOS and iCloud [12]. The Handoff functionality allows the operating system to integrate with iOS 8 devices over Bluetooth LE and Wi-Fi; users can place and answer phone calls using their iPhone as a conduit, send and receive text messages, activate personal hotspots, or load items being worked on in a mobile app (such as Mail drafts or Numbers spreadsheets) directly into their desktop equivalent [13].

NOTES

[1] As 引导的是时间状语从句；PowerPC 是用 IBM 公司第二代 RISC（精简指令系统计算机）技术制作的 PC，又称威力电脑。 Power 是 Performance optimized with enhanced risc 的缩略语。

[2] 长句。句子结构为 Mac OS X is…，中间插入同位语 also known…，以及由 which 引导的非限定性短语从句。

[3] NeXTSTEP，是由 Next Computer 公司开发的面向对象的多任务操作系统，运行在专用工作站，如 NexTcube 上。

[4] Mac SE 是 20 世纪 80 年代末由苹果公司生产的个人电脑，后来发展为 Macintosh。

[5] which 引导的定语从句，修饰 kernel。Mach kernel 是一个操作系统内核，由卡内

基·梅隆大学开发，主要是为了支持分布式并行计算方面研究。

[6] BSD 的解释见 2.4 节 NOTES[1]。

[7] Mac App Store，由苹果公司开发，用于 Mac OS X 应用的分布式数字平台。

[8] 长句。句中有两个状语，分别由 with…介词短语和 including…现在分词短语构成。

[9] Xserve，是苹果公司设计的一种机架式计算机系列。 Mac Mini 是苹果公司采用 OS X Lion 操作系统的计算机，它配有高速处理器，雷电式（Thunderbolt）传输，轻松快捷双倍速图形功能，为 3D 游戏提供更高像素，还可以连接高清电视。

[10] iPhone、iPod Touch 和 iPad 是苹果公司的智能电话、便携式媒体播放器和平板电脑。

[11] WWDC 原名为 World Wide Developers Conference，是苹果公司在加利福尼亚州召开的年会，主要是为软件开发者展示新的软件和技术。有关 beta 测试（beta tester）的说明，请见 2.3 节中的 NOTES[1]。

[12] iCloud，苹果公司提供的云存储和云计算服务，初始空间有 5GB，可以扩展。

[13] 长句，两个句子之间用分号隔开。Bluetooth LE,蓝牙低功耗技术，其耗电量比其他蓝牙产品低很多，也可以和智能手机相连接。有关蓝牙技术和 Wi-Fi，参见 3.1.2 节。

KEYWORDS

compatible	兼容的
port	移植，端口
architecture	体系结构
client	客户
default	默认，缺省
virtual machine	虚拟机
emulation	仿真
conduit	导管，线管

EXERCISES

True/False

1. _____ Mac OS series were developed by Microsoft Inc.
2. _____ Early versions of the Mac OS were compatible only with Motorola 68000-based Macintosh.
3. _____ PowerPC was introduced by Apple Computer Inc.
4. _____ Mac OS X 10.5 is also called as "Snow Leopard".
5. _____ Mac OS 8.1 was the last version running on a 68040 processor.
6. _____ The Mac OS Classic family was based on Apple's own code.
7. _____ The Mac OS X was UNIX-based.
8. _____ The "classic" Mac OS uses command line interface.

9. _____ Mac OS System 4 can run multitasking.
10. _____ Mac OS System 5 can run on the Mac SE and Macintosh II.
11. _____ Mac OS X is the newest of Apple Inc.'s Mac OS line of operating system.
12. _____ Mac OS X has a history largely dependent of the earlier Mac OS releases.
13. _____ The Mach kernel comes from the NeXT computer Inc.
14. _____ BSD code base is used in Mac OS X.
15. _____ Mac OS X has six significant releases both of the client and server versions.
16. _____ Mac OS X is a server version only.
17. _____ The different between client version and server version of Mac OS is the latter has the tools for server management.
18. _____ Mac OS X can be used to manage Mac OS X-based Web servers.
19. _____ Mac OS X Server is currently the default OS for the Xserve server hardware.
20. _____ Mac OS X 10.7 can be run in a virtual machine.
21. _____ Mac OS X is the basic for iPhone OS.
22. _____ OS X Yosemite is the Apple Inc.'s desktop and server operating system for Macintosh computers.
23. _____ Yosemite increases its integration with other Apple services and platforms.
24. _____ Handheld functionality allows the operating system to integrate with iOS 8 devices over Bluetooth LE and Wi-Fi.

2.6 ANDROID

Android is an operating system for touchscreen mobile devices such as smartphones and tablet computers. It is developed by the Open Handset Alliance—a consortium of hardware, software, and telecommunication companies led by Google[1].

Android consists of a kernel based on the Linux kernel, with middleware, libraries and APIs written in C and application software running on an application framework which includes Java-compatible libraries based on Apache Harmony[2].

As of July 2013 the Google Play store has had over one million Android applications ("apps") published, and over 50 billion applications downloaded. A developer survey conducted in April—May 2013 found that 71% of mobile developers develop for Android[3]. At Google I/O 2014, the company revealed that there were over one billion active monthly Android users, up from 538 million in June 2013. As of 2015, Android has the largest installed base of all general-purpose operating systems[4].

Android's source code is released by Google under open source licenses, although most Android devices ultimately ship with a combination of open source and proprietary software, including proprietary software developed and licensed by Google.

Android is popular with technology companies which require a ready-made, low-cost and customizable operating system for high-tech devices. Android's open nature has encouraged a large community of developers and enthusiasts to use the open-source code as a foundation for community-driven projects, which add new features for advanced users or bring Android to devices which were officially released running other operating systems[5]. The operating system's success has made it a target for patent litigation as part of the so-called "smartphone wars" between technology companies.

Current features and specifications in handset layouts include

(1) Platform

The platform is adaptable to larger, VGA, 2D graphics library, 3D graphics library based on OpenGL ES 2.0 specifications, and traditional smartphone layouts[6].

(2) Storage

SQLite, a lightweight relational database, is used for data storage purposes.

(3) Connectivity

Android supports connectivity technologies including GSM/EDGE, IDEN, CDMA, EV-DO, UMTS, Bluetooth, Wi-Fi, and etc[7].

(4) Messaging

SMS and MMS are available forms of messaging, including threaded text messaging and now C2DM is also a part of Android Push Messaging service[8].

(5) Web browser

The web browser available in Android is based on the open-source WebKit layout engine, coupled with Chrome's V8 JavaScript engine[9].

(6) Java support

While most Android applications are written in Java, there is no Java Virtual Machine in the platform and Java byte code is not executed[10].

(7) Media support

Android supports the following audio/video/still media formats: MP3, MIDI, WAV, JPEG, PNG, GIF, BMP and so forth[11].

(8) Streaming media support

RTP/RTSP streaming, HTML progressive download, Adobe Flash Streaming (RTMP) and HTTP Dynamic Streaming are supported[12].

(9) Additional hardware support

Android can use televisions (Android TV), car (Android Auto), wrist watches (Android Wear), still/digital cameras, game consoles, touchscreens, GPS, etc.

(10) Video calling

Android does not support native video calling, but some handsets have a customized version of the operating system that supports it.

There are another supports in handset layouts, they are multiple language, multitasking, multi-touch, bluetooth, voice based features, and screen capture.

NOTES

[1] Open Handset Alliance，开放手机联盟，是美国 Google 公司于 2007 年 11 月 5 日宣布组建的一个全球性的联盟组织。

[2] 本段就是一句话，句中大部分是 with…引导的状语，其中还有 running…现在分词短语和 which 引导的定语从句。Linux 内容见 2.4 节。 Apache Harmony 是 Apache 软件基金会主导的开放源代码专题，是自由 JAVA 实现计划（Free Java implementations）的一部分。有关 Java 请见 6.3 节。

[3] conducted…过去分词短语做定语，修饰 survey。Found 为过去时，做谓语。 that 引导的是宾语从句。

[4] Google I/O 2014，是谷歌在美国加州旧金山召开的软件开发者年会，其主旨是在使用谷歌和开放万维网技术，如 Android、Chrome、Chrome OS、Google APIs、Google Web Kit、App Engine 等方面，采用高科技构建万维网、移动和企业应用等问题。

[5] which add…or bring…是非限定性定语从句，后一个 which 引导的是定语从句，修饰 devices。

[6] VGA（Video Graphics Array）即视频图形阵列，是 IBM 在 1987 年随 PS/2 机推出的使用模拟信号的一种视频传输标准。这个标准对于现今的个人计算机市场已经十分过时。即便如此，VGA 仍然是最多制造商支持的一个标准。

OpenGL ES (OpenGL for Embedded Systems) 是 OpenGL 三维图形 API 的子集，针对手机、PDA 和游戏主机等嵌入式设备设计。

[7] GSM 是全球移动通信系统（Global System for Mobile Communications）的缩写，俗称"全球通"。EDGE 是英文 Enhanced Data Rate for GSM Evolution 的缩写，即增强型数据速率 GSM 演进技术。EDGE 是一种从 GSM 到 3G 的过渡技术。

IDEN（集成数字增强型网络）系统是摩托罗拉公司研制和生产的一种数字集群移动通信系统。CDMA 是码分多址的英文缩写（Code Division Multiple Access），它是在数字技术的分支——扩频通信技术上发展起来的一种崭新而成熟的无线通信技术。

EV-DO 是英文 Evolution-Data Optimized 或者 Evolution-Data Only 的缩写，有时也写作 EVDO 或者 EV。其中 CDMA 2000 1x EV-DO 是一种可以满足移动高速数据业务的技术。UMTS（Universal Mobile Telecommunications System）即通用移动通信系统。

Bluetooth、Wi-Fi 见 3.1.2 节。

[8] SMS 是一种存储和转发服务。也就是说，短消息并不是直接从发送人发送到接收人，而是通过 SMS 中心进行转发。MMS 为 Multimedia Messaging Service 的缩写，中文译为多媒体短信服务，即彩信。C2DM 是谷歌的一种服务，它允许开发者从服务器向 Android 应用或 Chrome apps 发送数据。

[9] WebKit 是在 Web 浏览器中用来绘制网页的排版引擎软件。WebKit 目前是 Apple Safari 及 Google Chrome 等浏览器的主要引擎。Chrome's V8 JavaScript engine 是谷歌公司为万维网浏览器 Google Chrome 开发的一个开源 JavaScript 引擎。有关 Google Chrome，请见 4.1.2 节。

[10] Java Virtual Machine（JVM）是一个抽象计算机，它由三个概念构成：规范、实现和范例。

[11] MIDI 是 Musical Instrument Digital Interface 的缩写，意思是音乐设备数字接口。这种接口技术的作用就是使电子乐器与电子乐器，电子乐器与计算机之间通过一种通用的通信协议进行通信，这种协议自然就是 MIDI 协议了。WAV 为微软公司开发的一种声音文件格式，它符合 RIFF（Resource Interchange File Format）文件规范，用于保存 Windows 平台的音频信息资源，被 Windows 平台及其应用程序所广泛支持。JPEG 是联合图像专家组（Joint Photographic Experts Group）的首字母缩写，是一种广泛适用的压缩图像标准格式。PNG 即图像文件存储格式，其目的是试图替代 GIF 和 TIFF 文件格式，同时增加一些 GIF 文件格式所不具备的特性。GIF（Graphics Interchange Format）是图像互换格式。BMP 是英文 Bitmap（位图）的简写，它是 Windows 操作系统中的标准图像文件格式，能够被多种 Windows 应用程序所支持。

[12] RTP（Real-time Transport Protocol），实时传送协议是一个网络传输协议。RTSP（Real Time Streaming Protocol），实时流协议是应用级协议，控制实时数据的发送。HTML（HyperText Markup Language，超文本标记语言）是一种专门用于创建 Web 超文本文档的编程语言。HTTP Dynamic Streaming（动态流）已逐渐成为因特网上主流的媒体形式，占据着不可替代的位置。

KEYWORDS

middleware	中间件
API (Application Program Interface)	应用程序接口
framework	框架，构架，架构，体制组织
smartphone	智能手机，智能电话
lightweight relational database	小型关系型数据库
streaming media	流媒体
threaded text messaging	线程文本消息
source code	源代码
proprietary software	专有软件，专利软件
alliance	联合，联盟
consortium	会，社，联盟，合作，财团，联营企业，国际性企业
telecommunication	电信，远程通信
layout	布局，布置，格式方案，规划，设计，草图，轮廓

EXERCISES

Fill in the blanks with using appropriate words or terms found behind this exercise.

1. Android is developed by _____.

2. Android has been deployed with _____.

3. SQLite is a _____.

4. Android's connectivity includes _____ and so forth.

5. Android's Cloud To Device Messaging can be abbreviated to _____.

6. The Web browser available in Android is based on _____.

7. Android has a customized version of the operating system that supports _____.

8. Function of wrist watches can be performed with _____ software.

9. Android has the largest installed base of all _____.

10. Most Android devices ship with a combination of _____ software.

11. The Open Handset Alliance is a _____ of hardware, software, and telecommunication companies.

12. The open-source code used for Android is as a _____ for community-driven projects.

13. Android can be brought to devices running _____.

14. "Smartphone wars" are caused by Android's _____.

    a. video calling

    b. the open-source WebKit layout engine

    c. other operating system

    d. open source and proprietary

    e. Android Wear

    f. success

    g. foundation

    h. general-purpose operating system

    i. C2DM

    j. GSM, CDMA, Wi-Fi, Bluetooth

    k. lightweight relational database

    l. consortium

    m. the Open Handset Alliance

    n. middleware, libraries and APSs

CHAPTER 3   COMPUTER NETWORK

3.1   LOCAL AREA NETWORKS (LANs)

3.1.1   ETHERNET

A local area network (LAN) is a computer network that interconnects computers within a limited area such as a home, school, computer laboratory, or office building, using network media [1]. The defining characteristics of LANs, in contrast to wide area networks (WANs), include their smaller geographic area, and non-inclusion of leased telecommunication lines.

ARCNET, Token Ring and other technology standards have been used in the past, but Ethernet over twisted pair cabling and Wi-Fi are the two most common technologies currently used to build LANs[2].

Network topology describes the layout of interconnections between devices and network segments. At the Data Link Layer and Physical Layer, a wide variety of LAN topologies have been used, including ring, bus, mesh, and star, but the most common LAN topology in use today is switched Ethernet [3]. At the higher layers, the Internet Protocol (TCP/IP) has become the standard, replacing NetBEUI, IPX/SPX, AppleTalk and others[4].

Simple LANs generally consist of one or more switches. A switch can be connected to a router, cable modem, or ADSL modem for Internet access. Complex LANs are characterized by their use of redundant links with switches using the spanning tree protocol to prevent loops, their ability to manage differing traffic types via quality of service (QoS), and to segregate traffic with VLANs [5]. A LAN can include a wide variety of network devices such as switches, firewalls, routers, load balancers, and sensors.

LANs can maintain connections with other LANs via leased lines, leased services, or the Internet using virtual private network (VPN) technologies.

1. Network switch

A network switch (also called switching hub, bridging hub, officially MAC bridge) is a computer networking device that connects devices together on a computer network (Figure 3-1), by using packet switching to receive, process and forward data to the destination device[6]. Unlike less advanced network hubs, a network switch forwards data only to one or multiple devices that need to receive it, rather than broadcasting the same data to each of its ports.

Figure 3-1　Avaya ERS 2550T-PWR, a 50-port Ethernet switch.

2. 10 Gigabit Ethernet

10 Gigabit Ethernet (10GE, 10GbE, or 10 GigE) is a group of computer networking technologies for transmitting Ethernet frames at a rate of 10 gigabits per second (10×10^9 or 10 billion bits per second). It was first defined by the IEEE 802.3ae-2002 standard[7]. Unlike previous Ethernet standards, 10 Gigabit Ethernet defines only full duplex point-to-point links which are generally connected by network switches; shared-medium CSMA/CD operation has not been carried over from the previous generations Ethernet standards. Half duplex operation and hubs do not exist in 10GbE.

3. Terabit Ethernet

Facebook and Google, among other companies, have expressed a need for TbE[8]. Some think that a speed of 400 Gbit/s is a more practical goal than 1Tbit/s (1000Gbit/s). In 2011, researchers predicted Terabit Ethernet (1Tbit/s) in 2015, and 100 Terabit Ethernet (100Tbit/s) by 2020.

NOTES

[1] network media，此处指传输媒体，或传输介质，如双绞线、光纤和无线传播等。

[2] 长句。ARCNET 原文是 Attached Resource Computer NETwork，是一种局域网通信协议，在 20 世纪 80 年代微型计算机网络中较为常用。随着以太网的出现和快速普及，ARCNET 已退出历史舞台。Token Ring 即令牌环，是环形局域网采取的工作方式，现已不多用。Wi-Fi，见 3.1.2 节。

[3] Data Link Layer and Physical Layer 分别是国际标准化组织（ISO）制定的开放系统互连协议（OSI）七层中的第二层和第一层，即数据链路层和物理层。

[4] TCP/IP（Transmission Control Protocol/Internet Protocol）即传输控制协议/互联网协议，是因特网最重要的协议。NetBEUI 和 IPX/SPX 是早期局域网 Novell 使用的协议。AppleTalk 是苹果公司为其 Macintosh 计算机开发的网络协议簇，有很强的网络功能。

[5] 长句。主句为 Complex LANs are characterized by…，by 后面为两个并列的方式状语。spanning tree protocol，生成树协议，即生成树算法。QoS（Quality of Service）即服务质量，在计算机网络中，是流量工程中的术语，是一种控制机制。VLAN（Virtual LAN）即虚拟局域网。

[6] 长句。主句是 A network switch is…，其中 that 引导的定语从句，修饰 device，by using…分词短语做状语。

[7] IEEE 802.3 是电气电子工程师协会（Institute of Electrical and Electronic Engineers）IEEE 802 课题为局域网制定的标准。这一标准采用 CSMA/CD 访问方法，是以太网的正式标准。

[8] Facebook 即脸谱网站，详见 5.1.3 节。

KEYWORDS

LAN (Local Area Network)	局域网
WAN(Wide Area Network)	广域网
telecommunication	电信，远程通信
twisted pair	双绞线（对）
Wi-Fi (Wireless Fidelity)	无线保真
network topology	网络拓扑(结构)
ring	环
bus	总线
mesh	网格
star	星形
Switched Ethernet	交换式以太网
switch	交换机
router	路由器
cable modem	电缆调制解调器
ADSL(Asymmetric Digital Subscriber Line)	非对称数字用户线
link	链接，链路
redundant links	冗余链路
spanning tree	生成树
loop	环，环路
traffic	通信数据量，业务量
QoS(Quality of Service)	质量保证
firewall	防火墙
load balance	负载均衡
sensor	传感器
leased	租用
VPN(Virtual Private Network)	虚拟专用网
hub	集线器
bridge	桥，网桥
MAC(Medium Access Control)	媒体访问（接入，存取）控制
packet switching	分组交换
forward	转发
broadcasting	广播

port	端口，移植
Gigabit	吉比特，吉位
Tb(Terabit)	太位
frame	帧
full-duplex	全双工，双向同时通信
half-duplex	半双工通信，双向交替通信
point-to-point	点对点
shared-medium	共享媒体

EXERCISES

Fill in the blanks with appropriate words or phrases found behind this exercise.

1. A LAN is a computer network that connects_____.
2. Network media we used here include_____.
3. The two most common technologies currently used to build LANs are_____.
4. _____describes the layout of interconnections between devices and network segments.
5. _____is the most common LAN topology in use today.
6. Spanning tree topology is used for_____.
7. VLAN stands for_____.
8. Network devices consisting a LAN can be_____.
9. Switching hub, bridging hub, and MAC bridge are the same as_____.
10. Network switch uses_____to transmit data.
11. Official Ethernet standard uses_____access method.
12. _____is the term used for traffic engineering in computer network.

    a. switches, firewalls, routers, load balancers, and sensors

    b. QoS

    c. twisted pair, optical fiber, and wireless

    d. Network topology

    e. packet switching

    f. complex LANs

    g. computers within a limited area

    h. network switch

    i. CSMA/CD

    j. Virtual LANs

    k. Switched Ethernet

    l. twisted pair cabling and Wi-Fi

3.1.2 WI-FI AND BLUETOOTH

1. Wi-Fi

Wi-Fi (or, incorrectly but commonly, WiFi) is a local area wireless technology that allows an electronic device to participate in computer networking using 2.4GHz UHF and 5GHz SHF ISM radio bands[1].

The Wi-Fi Alliance defines Wi-Fi as any "wireless local area network" (WLAN) product based on the Institute of Electrical and Electronics Engineers' (IEEE) 802.11 standards. However, the term "Wi-Fi" is used in general English as a synonym for "WLAN" since most modern WLANs are based on these standards. "Wi-Fi" is a trademark of the Wi-Fi Alliance. The "Wi-Fi CERTIFIED" trademark can only be used by Wi-Fi products that successfully complete Wi-Fi Alliance interoperability certification testing[2].

Many devices can use Wi-Fi, e.g. personal computers, video-game consoles, smartphones, digital cameras, tablet computers and digital audio players. These can connect to a network resource such as the Internet via a wireless network access point. Such an access point (or hotspot) has a range of about 20 meters (66 feet) indoors and a greater range outdoors. Hotspot coverage can comprise an area as small as a single room with walls that block radio waves, or as large as many square kilometers achieved by using multiple overlapping access points[3]. Figure 3-2 depicts the communication course between a notebook and a printer through an access point.

Figure 3-2 Depiction of a device sending information wirelessly to another device, both connected to the local network, in order to print a document.

Wi-Fi can be less secure than wired connections, such as Ethernet, because an intruder does not need a physical connection. Web pages that use SSL are secure but unencrypted internet access can easily be detected by intruders[4]. Because of this, Wi-Fi has adopted various encryption technologies. The early encryption WEP proved easy to break[5]. Higher quality protocols (WPA, WPA2) were added later[6]. An optional feature added in 2007, called Wi-Fi Protected Setup (WPS), had a serious flaw that allowed an attacker to recover the router's password[7]. The Wi-Fi Alliance has since updated its test plan and certification program to ensure all newly certified devices resist attacks.

2. Bluetooth

Bluetooth is a wireless technology standard for exchanging data over short distances (using short-wavelength UHF radio waves in the ISM band from 2.4 to 2.485GHz) from fixed and mobile devices, and building personal area networks (PANs) [8]. Invented by telecom vendor Ericsson in 1994, it was originally conceived as a wireless alternative to RS-232 data cables. It can connect several devices, overcoming problems of synchronization.

Bluetooth is managed by the Bluetooth Special Interest Group (SIG), which has more than 25,000 member companies in the areas of telecommunication, computing, networking, and consumer electronics[9]. The IEEE standardized Bluetooth as IEEE 802.15.1, but no longer maintains the standard. The Bluetooth SIG oversees development of the specification, manages the qualification program, and protects the trademarks. A manufacturer must make a device to meet Bluetooth SIG standards, and market it as a Bluetooth device.

Bluetooth is a standard wire-replacement communications protocol primarily designed for low-power consumption, with a short range based on low-cost transceiver microchips in each device. Because the devices use a radio (broadcast) communications system, they do not have to be in visual line of sight of each other, however a quasi optical wireless path must be viable[10]. Range is power-class-dependent, but effective ranges vary in practice; See Figure 3-3.

Class	Max. permitted power (mW)	(dBm)	Typ. range (m)
1	100	20	~100
2	2.5	4	~10
3	1	0	~1

Figure 3-3 Class of Bluetooth

3. Bluetooth VS Wi-Fi

Bluetooth and Wi-Fi have some similar applications: setting up networks, printing, or transferring files. Wi-Fi is intended as a replacement for high speed cabling for general local area network access in work areas. This category of applications is sometimes called wireless local area networks (WLAN). Bluetooth was intended for portable equipment and its applications. The category of applications is outlined as the wireless personal area network (WPAN). Bluetooth is a replacement for cabling in a variety of personally carried applications in any setting and also works for fixed location applications such as smart energy functionality in the home (thermostats, etc.) [11].

Wi-Fi and Bluetooth are to some extent complementary in their applications and usage. Wi-Fi is usually access point-centered, with an asymmetrical client-server connection with all traffic routed through the access point, while Bluetooth is usually symmetrical, between two Bluetooth devices[12]. Bluetooth serves well in simple applications where two devices need to connect with minimal configuration like a button press, as in headsets and remote controls, while Wi-Fi suits better in applications where some degree of client configuration is possible and high speeds are required, especially for network access through an access node[13].

NOTES

[1] that 引导的定语从句，修饰 technology，using…分词短语作状语。UHF（Ultra High Frequency）、SHF（Super-High Frequency）都是超高频，但频带范围不同，UHF 为 300MHz～3GHz，SHF 为 3GHz～30GHz。ISM（Industrial Science Medicine）频段是由 ITU-R（ITU Radio communication Sector，国际电信联盟无线电通信组）定义的。

[2] that 引导的定语从句，修饰 products。

[3] or 连接的并列长句，两句中分别有 that 引导的定语从句和过去分词 achieved 短语构成的定语从句。

[4] SSL（Secure Sockets Layer），安全套接层，计算机网上提供通信安全的加密协议。

[5] WEP（Wired Equivalent Privacy），与有线网等效的保密协议，是早期（1999 年 9 月批准的）为 IEEE 802.11 设计的无线网络安全算法。

[6] WPA（Wi-Fi Protected Access）和 WPA2 是 Wi-Fi 联盟开发的，用于保护无线计算机网络安全的安全协议和认证程序。

[7] WPS（Wi-Fi Protected Setup）是一个网络安全标准，目的是使用户较容易地保护无线家庭网络。

[8] 句中 for exchanging…，and building…是目的状语从句。

[9] Bluetooth Special Interest Group（SIG）即蓝牙专门兴趣组，是非营利组织，主要任务是监督蓝牙标准的发展，蓝牙技术和商标的授权等情况。

[10] line of sight 即视距（传播），是电磁辐射或声波传播的一个特性，即直线传播特性。

[11] 由 and 连接的并列句。

[12] 并列长句。第一个 with 引导的是方式状语从句；第二个 with 引导的是定语，修饰 connection。

[13] 并列长句，两个句子中都有 where 引导的定语从句。

KEYWORDS

Wi-Fi (Wireless Fidelity)　　　　　　　　无线保真
UHF(Ultra High Frequency)　　　　　　　特高频
SHF(Super-High Frequency)　　　　　　　超高频
ISM(Industrial Science Medicine)　　　　　工业、科学和医用频段
WLAN(Wireless Local Area Network)　　　虚拟局域网

certified	已证明无误的，合格的
video-game console	电视游戏控制台
smartphone	智能手机
digital camera	数码相机
tablet computer	平板电脑
digital audio player	数字音频播放器
access point	接入点，存取点，访问点
hotspot	热点
radio wave	无线电波
overlap	重叠，部分重叠
intruder	侵入者，干扰者
SSL(Secure Sockets Layer)	安全套接层
detect	检测，探测，检波
encryption	加密
unencryption	解密
flaw	缺陷，瑕疵，裂缝，故障
router	路由器
password	密码，口令
certification	证明（书），认可
Bluetooth	蓝牙
PAN(Personal Area Network)	个人区域网
telecom	电信，远程通信
vendor	自动售货机，卖主，厂商，供应商
qualification	资格，限定，合格性
protect	保护，防御
consumption	消费，消耗
transceiver	收发器，收发信机
microchip	微[芯]片
line of sight	视线，视距
thermostat	自动调温器
complementary	互补的，附加的
optical wireless	光无线电
symmetrical	对称的
asymmetrical	非对称的
client-server	客户-服务器
configuration	配置，排列，结构，外形，形状
suit	一套，一组，适合于
access node	访问节点，接入节点

EXERCISES

Multiple choices

1. Wi-Fi_____.
   a. is a wireless technology of LAN
   b. uses 2.4GHz UHF bands
   c. can be written to WiFi
   d. complies IEEE 802.11 standards

2. A wireless network access point_____.
   a. can cover about 20 meters indoors
   b. is known as hotspot
   c. can cover about several square kilometers
   d. can cover about 100 meters indoors

3. Wi-Fi network _____.
   a. is secure enough
   b. is less secure
   c. is secure with using SSL technology
   d. is less secure with unencrypted internet access

4. Devices that can use Wi-Fi include _____.
   a. tablet computer
   b. video-game consoles
   c. digital cameras
   d. smartphones

5. Higher quality protocol of security is_____.
   a. WEP
   b. WPA
   c. WPA2
   d. WPS

6. Bluetooth is used for exchanging data_____.
   a. over short distances
   b. over longer distances
   c. between mobile devices only
   d. between fixed and mobile devices

7. Members of SIG are in the area of_____.
   a. computing
   b. telecommunication
   c. networking
   d. consumer electronics

8. Tasks of SIG are_____.
   a. managing the qualification program
   b. protecting the trademarks
   c. keeping watch on the development of the Bluetooth standard
   d. maintaining the IEEE standard

9. Devices used in Bluetooth are_____.
   a. higher power consumption
   b. low power consumption
   c. low cost devices
   d. with a broadcast communication system

10. The effective range of Bluetooth communication is_____ meters.
    a. one
    b. ten
    c. one hundred
    d. one thousand

11. Wi-Fi and Bluetooth have some similar applications_____.
    a. transferring file
    b. asymmetrical client/server connection
    c. setting up network
    d. printing

12. Bluetooth and Wi-Fi are to some extent complementary in_____.

a. Wi-Fi is access point-centered
b. Bluetooth is two devices accessed
c. Wi-Fi suits for network access via an access node
d. Bluetooth suits for simple applications

3.1.3 HOME NETWORK

A home network or home area network (HAN) is a type of local area network with the purpose to facilitate communication among digital devices present inside or within the close vicinity of a home[1]. Devices capable of participating in this network, for example, smart devices such as network printers and handheld mobile computers, often gain enhanced emergent capabilities through their ability to interact [2]. These additional capabilities can be used to increase the quality of life inside the home in a variety of ways, such as automation of repetitive tasks, increased personal productivity, enhanced home security, and easier access to entertainment (See Figure 3-4).

Figure 3-4 An example of a simple home network

A home network usually relies on one of the following equipment to establish physical layer, data link layer, and network layer connectivity both internally amongst devices and externally with outside networks:

- A modem is usually provided by an ISP to expose an Ethernet interface to the WAN via their telecommunications infrastructure[3]. In homes these usually come in the form of a DSL modem or cable modem.
- A router manages network layer connectivity between a WAN and the HAN. Most home networks feature a particular class of small, passively-cooled, table-top device with an integrated wireless access point and 4 port Ethernet switch[4]. These devices

aim to make the installation, configuration, and management of a home network as automated, user friendly, and "plug-and-play" as possible.
- A network switch is used to allow devices on the home network to talk to one another via Ethernet. While the needs of most home networks are satisfied with Wi-Fi or the built-in switching capacity of their router, certain situations require the introduction of a distinct switch[5]. For example:
  —When the router's switching capacity is exceeded. Most home routers expose only 4 to 6 Ethernet ports.
  —When Power over Ethernet is required by devices such as IP cameras and IP phones[6].
  —When distant rooms have a large amount of wired devices in close proximity.
- A wireless access point is required for connecting wireless devices to a network. Most home networks rely on one "Wireless Router" combination device to fill this role.
- A network bridge connecting two network interfaces to each other, often in order to grant a wired-only device, e.g. Xbox, access to a wireless network medium [7].

NOTES

[1] present…形容词短语做后置定语，修饰 devices。

[2] 主句由于插入 for example，……而被分开。主句结构为 Devices…often gain…

[3] expose 原意为暴露，此处可译为提供、安排等，以下同。

[4] 长句。feature 此处为动词，意为"以……为特征"。

[5] while 引出的是让步状语从句，主句是 certain situations require…

[6] IP Phone，又称 VoIP，是为了将电话业务在因特网上运行而采用的 Voice over IP 技术。Power Over Ethernet（POE，有源以太网），POE 技术能在确保现有结构化布线安全的同时保证现有网络的正常运作，最大限度地降低成本。

[7] Xbox，是微软公司制作和管理的视频游戏商标，它代表一个视频游戏控制台系列。

KEYWORDS

facilitate	使容易，促进，助长
vicinity	附近，邻近，接近
HAN(Home Area Network)	家庭网
participate	参加，参与，分享，分担
emergent	出现的，露出的，紧急的
productivity	生产力
entertainment	招待，娱乐，乐趣
physical layer	物理层

data link layer	数据链路层
network layer	网络层
modem	调制解调器
ISP(Internet Service Provider)	因特网服务提供商
expose	暴露
infrastructure	基础设施，基础结构
DSL(Digital Subscriber Line)	数字用户线
installation	安装
plug-and-play	即插即用
IP camera	IP 相机
IP Phone	IP 电话
proximity	邻近，接近
wireless access point	无线接入点
wireless router	无线路由器
network bridge	网桥

EXERCISES

True/False

1. _____ The purpose of a home network is to facilitate communication among digital devices present inside of a home.
2. _____ Network printers and handheld mobile computer in a HAN can get capabilities through their ability to interact.
3. _____ We can find the Bluetooth technology in Figure 3-4.
4. _____ If we want to connect an Ethernet to the WAN, we must use DSL modem.
5. _____ Router is the network layer device.
6. _____ We need a router in HAN to talk to one another via Ethernet.
7. _____ When the router's switching capability is exceeded, we should require a distinct switch.
8. _____ To connect wireless devices to a network, we need a wireless access point.
9. _____ Network bridge often used for connecting either a wired device or a wireless device.
10. _____ Xbox is a trademark of video game produced by Microsoft.

3.2 WIDE AREA NETWORKS (WANs)

3.2.1 OVERVIEW OF WIDE AREA NETWORKS (WANs)

A wide area network (WAN) is a network that covers a wide geographical area. Many

WANs link together two or more geographically dispersed LANs. The Internet, by this definition, is a very large WAN. WANs may be publicly accessible, like the Internet, or be privately owned and operated.

One of many methods that can be used to categorize wide area networks is with respect to the flow of information on a transmission facility[1]. If we use this method to categorize wide area networks, we can group them into three types: circuit switched, dedicated and packet switched.

1. Packet Switched Networks

Most modern WANs send data in pieces or packet. The packets are sent individually over the WAN and then reassembled after they reach their destination.

The maximum length of the packet is established by the network. Longer transmissions are broken-up into multiple packets. Each packet contains not only data but also a header with control information (such as priority codes and source and destination addresses). The packets are sent over the network node-to-node. At each node, the packet is stored briefly then routed according to the information in its header[2].

There are two popular approaches to packet switching: datagram and virtual circuit (See Figure 3-5).

Figure 3-5   Packet switching approaches

(1) Datagram Approach

In the datagram approach to packet switching, each packet is treated independently from all others. Even when one packet represents just a piece of a multi-packet transmission, the network (and network layer functions) treats it as though it exited alone[3]. Packets in this technology are referred to as datagrams.

This approach can cause the datagrams of a transmission to arrive at their destination out of order. It is the responsibility of the transport layer in most protocols to reorder the datagram before passing them on to the destination port.

(2) Virtual Circuit Approach

In the virtual circuit approach to packet switching, the relationship between all packets belonging to a message or session is preserved. A single route is chosen between sender and receiver at the beginning of the session. When the data are sent, all packets of the transmission travel one after another along that route.

So what is the difference between circuit switching and virtual circuits? Although circuit switching can use multiplexing at the end-user level, no multiplexing is done at the switches. However, in the virtual circuit approach we can have multiplexing at the switches.

Today, virtual circuit transmission is implemented in two formats: Switched Virtual Circuit (SVC) and Permanent Virtual Circuit (PVC).

2. Dedicated Networks

This is a dedicated network connected by leased lines. Leased line is a communications line reserved for the exclusive use of a leasing customer without inter-exchange switching arrangements. Leased or private lines are dedicated to the user[4]. Their advantage is that the terminal or computer is always physically connected to the line. Very short response times are met with this service. Others, built by Internet service providers, provide connections from an organization's LAN to the Internet[5]. Network protocols including TCP/IP deliver transport and addressing functions. Protocols including Packet over SONET/SDH, MPLS, ATM and Frame Relay are often used by service providers to deliver the links that are used in WANs[6]. X.25 was an important early WAN protocol, and is often considered to be the "grandfather" of Frame Relay as many of the underlying protocols and functions of X.25 are still in use today by Frame Relay[7].

3. Router

Routers have been known by several names. Back in ancient times when what is now the Internet was called the ARPANET, routers were called IMPs, for internet message processors[8]. More recently, routers were called gateways; remnants of this nomenclature can still be found in terms such as Border Gateway Protocol (BGP) and Interior Gateway Routing Protocol (IGRP) . In the Open System Interconnection (OSI) world, routers are known as Intermediate System (IS) [9].

Router, as a name, is probably the most descriptive of what the modern version of these devices do. A router sends information along a route—a path—between two networks. This path may traverse a single router or many routers. Furthermore, in internetworks that have multiple paths to the same destination, modern routers use a set of procedures to determine and use the best route. Should that route become less than optimal or entirely unusable, the router selects the next-best path[10]. The procedures used by the router to determine and select the best route and to share information about network reachability and status with other routers are referred to collectively as a routing protocol [11] .

Just as a data link may directly connect two devices, a router also creates a connection between two devices. The difference is that, as Figure 3-6 shows, whereas the communication path between two devices sharing a common data link is a physical path, the communication path provided by routers between two devices on different network is a higher-level, logical path[12].

This concept is vitally important for understanding a router's function. Notice that the logic path, or route, between the devices in Figure 3-6 traverses several types of data links: a 1Tb Ethernet, two serial links, and a 10Gb Ethernet. As noted earlier, to be delivered on the physical path of a data link, data must be encapsulated within a frame, a sort of digital envelope. Likewise, to be delivered across the logical path of a routed inter-network, data must also be encapsulated and the digital envelope used by routers is a packet.

Figure 3-6　A router creates a logical path between networks

NOTES

　　[1] that 引导的定语从句，修饰 methods； with respect to…，关于。

　　[2] route，路由，在广域网中广泛使用的路由技术。

　　[3] 主句中两个 it 均代表 packet，as though…为方式状语从句。network layer 是指计算机网络体系结构中的层次结构。分组（packet）是在网络层上。

　　[4] private 和 dedicated 都指专用。

　　[5] others，此处指专用网的另外一些功能。句中 built by…是过去分词短语作状语。

　　[6] SONET/SDH（Synchronous Optical NETwork/Synchronous Digital Hierarchy），同步光纤网/同步数字系列，是在使用激光器的光纤上传送多个数字比特流的标准化多路复用协议。MPLS（Multi-Protocol Label Switching），多协议标记交换。它是高性能电信网中的一种控制机制。ATM（Asynchronous Transfer Mode），异步传输模式，是电信网中的交换技术。Frame Relay，帧中继是一种用于连接计算机系统的面向分组的通信方法。

　　[7] X.25 是国际电信联盟电信组（ITU-T）为分组交换网通讯建立的标准协议族。

　　[8] when 引导的是定语从句，修饰 times。ARPANET 为 Advanced Research Projects Agency Network 之缩写，即高级研究计划局网，简称 ARPA 网，后发展成因特网。IMP 应为接口报文处理机（Interface Message Processor）。

　　[9] OSI 开放系统互连，是国际标准化组织 ISO 制定的网络互连参考标准。

　　[10] 由 Should 构成的假设句，且主谓颠倒。

　　[11] 长句。主句结构为 The procedures…are referred to…，其中 used by…为过去分词短语作定语，修饰 procedures。

　　[12] 长句。that 引导的是两个并列的表语从句，且两个从句中的主语都是 communication path。

KEYWORDS

WAN (Wide Area Network)	广域网
packet switching network	分组交换网
priority code	优先级代码
source address	源地址
destination address	目的地址
datagram	数据报
virtual circuit	虚电路
(SVC) Switched Virtual Circuit	交换式虚电路
(PVC) Permanent Virtual Circuit	永久式虚电路
multiplex	多路复用
leased line network	专(用)线(路)网
TCP/IP(Transmission Control Protocol/ Internet Protocol)	传输控制协议/互连网协议
addressing	编址，寻址，定址
frame relay	帧中继
access line	接入线
route	路由，路径
router	路由器
backbone	主干网、骨干网
trunk	干线，[局内] 中继线，信息通路
encapsulate	封装，（信息）打包

EXERCISES

Multiple Choices

1. A Wide Area Network can_____.
   - a. cover a wide geographical area
   - b. link two or more LANs
   - c. link two LANs only
   - d. be created by individual users
2. According to the flow of information on a transmission facility we can categorize WAN into _____network.
   - a. circuit switched
   - b. packet switched
   - c. local area
   - d. leased line
3. In packet switched networks data send in_____.
   - a. byte
   - b. pieces
   - c. frame
   - d. packet
4. The most important items contained in a packet should be_____.
   - a. data to be transmitted
   - b. header of the packet
   - c. tail of the packet
   - d. date
5. The popular approaches we used in the packet switched network are_____.

    a. circuit switching   b. datagram   c. virtual circuit   d. real circuit

6. In the datagram mode when datagrams arrive at their destination, the datagrams_____.
   a. may be out of order
   b. must be in order
   c. should be reordered
   d. may lose
7. In the virtual circuit mode all packets sent by a user at a time belong to_____.
   a. a message
   b. a session
   c. different messages
   d. different sessions
8. Main characteristics of the leased line networks are_____.
   a. the line is private for the user
   b. the user has himself inter-exchange
   c. very short response time
   d. the user computer is always physically connected to the line
9. Network protocols often used by ISP are_____.
   a. TCP/IP   b. SONET/SDH   c. MPLS   d. ATM
10. Router has the following name_____.
    a. ARPANET
    b. IMP (Interface Message Processor)
    c. gateway
    d. IS (Intermediate System)
11. In figure 3-6 a router creates a logical path between networks, these networks include_____.
    a. 1Tb Ethernet   b. serial link   c. datagram   d. 10Gb Ethernet
12. Encapsulation is used here for data delivering on_____.
    a. the circuit switching
    b. the physical path
    c. the analog circuit
    d. the logical path

3.2.2 STORAGE AREA NETWORK (SAN) AND INTERNET AREA NETWORK (IAN)

1. Storage Area Network(SAN)

A Storage Area Network (SAN) is a dedicated network that provides access to consolidated, block level data storage[1]. SANs are primarily used to enhance storage devices, such as disk arrays, tape libraries, and optical jukeboxes, accessible to servers so that the devices appear like locally attached devices to the operating system[2]. A SAN typically has its own network of storage devices that are generally not accessible through the local area network (LAN) by other devices[3]. The cost and complexity of SANs dropped in the early 2000s to levels allowing wider adoption across both enterprise and small to medium sized business environments[4].

A SAN does not provide file abstraction, only block-level operations[5]. However, file systems built on top of SANs do provide file-level access, and are known as SAN file systems or shared disk file systems.

2. Internet Area Network(IAN)

An Internet Area Network (IAN) is a concept for a communications network that connects voice and data endpoints within a cloud environment over IP, replacing an existing local area network (LAN), wide area network (WAN) or the public switched telephone network (PSTN).

Seen by proponents as the networking model of the future, an IAN securely connects endpoints through the public Web, so that they can communicate and exchange information and data without being tied to a physical location [6].

Unlike a LAN or a WAN, transports, the IAN eliminates a geographic profile for the network entirely because the applications and communications services have become virtualized. Endpoints need only be connected over a broadband connection across the Internet.

Hosted in the cloud by a managed services provider, an IAN platform offers users secure access to information from anywhere, at anytime, via an Internet connection. Users also have access to telephony, voicemail, e-mail, and fax services from any connected endpoint. For businesses, the hosted model reduces IT and communications expenses, protects against loss of data and disaster downtime, while realizing a greater return on their invested resources through increased employee productivity and reduction in telecom costs[7].

NOTES

[1] block level data，块级数据。块有时也称为物理记录，通常用于数据传输和数据存储中。块是一串字节或比特，一般包含最大长度的整个记录，即块的规模。

[2] so that 引导的是结果状语从句。

[3] that 引导的是同位语从句。

[4] allowing…现在分词短语作定语，修饰 levels。

[5] file abstraction，文件抽象，文件抽取。

[6] 长句。 Seen by…过去分词短语作状语，主句是 an IAN…，so that…目的状语从句。

[7] 长句，其中 while realizing…为现在分词短语作伴随状语。

KEYWORDS

SAN (Storage Area Network)	存储区域网
IAN (Internet Area Network)	因特网区域网
disk array	磁盘阵列
tape library	磁带库

jokebox	光盘机，光盘自动换盘机，点播机
abstraction	抽象，提取
shared disk file system	共享磁盘文件系统
endpoint	端点，终点
cloud environment	云环境
PSTN (Public Switched Telephone Network)	公共电话交换网
profile	配置文件，简表，概貌
virtualize	虚拟化
broadband	宽频带，宽带
voicemail	语音邮件
fax	传真[机]
hosted model	托管模型
IT (Information Technology)	信息技术
downtime	故障时间，停机时间

EXERCISES

True/False

1. _____ A storage area network is a general-purpose network.
2. _____ Devices used in SAN should be disk arrays, tape libraries, and optical jukeboxes.
3. _____ A SAN typically has its own network of storage devices.
4. _____ A SAN can be accessed easily by the outside devices.
5. _____ Right now SANs are more suitable for small and medium sized business environments.
6. _____ With SAN we can make file abstraction.
7. _____ SAN file systems can provide file-level access.
8. _____ IAN is a communication network built on a cloud environment over IP.
9. _____ In order to exchange information and data through an IAN, we should connect to a physical location.
10. _____ Applications and communication services in IAN have become virtualized.
11. _____ An IAN platform is hosted in the cloud.
12. _____ For businesses, the hosted model increases IT and communications expenses.
13. _____ The hosed model protects against loss of data and disaster downtime.
14. _____ IAN will replace LAN, WAN or PSTN.

3.3 THE INTERNET

The Internet is the largest and most well-known computer network in the world. It is technically a network of networks, since an individual user connects to a network set up by their access provider or Internet service provider (ISP), which in turn is connected to a larger network, which may be connected to an even larger network [1]. Altogether, this network of networks is referred as the Internet. Since all the networks on the Internet are interconnected, any computer with Internet access can communicate with any other computer on the Internet, regardless of the ISP used.

1. The Domain Name System (DNS)

The Domain Name System (DNS) was created to centralize the task of making changes to the network name to address assignments and to automate the task of performing the translation functions[2]. In the early days of the Internet, a central location (SRI NIC at Stanford Research Institute in Melo Park, California) was responsible for maintaining a HOSTS file that contained the name of every host on the Internet along with its address [3]. Administrators had to communicate changes to SRI NIC, and these changes were incorporated into the file periodically. Of course, this meant that the file then had to be distributed to every single host so that it could have the updated version [4].

DNS uses a hierarchical distributed architecture that is spread across many computers throughout the Internet. A root server holds information about top level domains (such as .COM, .EDU, and .GOV), and each domain throughout the Internet has a domain name server that is responsible for the computer names and addresses used in that domain [5]. Client computers query DNS servers when they need to get the address for a hostname. If the local DNS knows the address, it returns it to the client computer. If it does not, it sends the query up the chain of DNS servers until a DNS server is found that can resolve the name, provided that it is indeed a valid name [6].

The topmost entry in the DNS hierarchy is called the root domain and its represented by the period character (.). Underneath this root domain are the top-level directories that fall into two groups: geographical and organizational. Geographical domains are used to specify specific countries. For example, .au for Australia and .cn for China. Under each of the geographical domains, you can find organizational domains.

Organizational domains you might be familiar with include the following[7].

(1) com   Used for commercial organizations.

(2) edu   Used for educational institutions.

(3) gov   Used for U.S. Government entities.

(4) mil   Used for U.S. military organizations.

(5) Int   International organizations.

(6) net   Used for Network organizations such as Internet service providers.

(7) org   Used for nonprofit organizations.

(8) arpa   Used for inverse address lookups.

The structure of the Domain Name System is similar to an inverted tree. In Figure 3-7 you can see that at the top is the root domain with the com through cn domains underneath [8]. Under the com domain are individual business organizations that each their own domain. Under any particular domain, there can be sub-domains.

Figure 3-7 Domain Name System is a distributed, hierarchical structure

2. Connecting To The Internet

(1) Dial-Up Connections

Dial-Up Connections usually work over regular telephone lines. To connect to the Internet your modem (or other appropriate interface devices) dials up and connects to a modem attached to your ISP's computer[9]. While you are connected, your PC is assigned a temporary IP address for the current session[10]. At the end of each Internet session, you disconnect from your ISP's computer to allow another user to connect in your place[11]. Standard dial-up Internet service uses a conventional dial-up modem rated at a maximum data transfer rate of 56Kb/s.

(2) ADSL Connection

Another type of dial-up connection is ADSL which is the most common way to access the Internet today. Sometimes ADSL is called as broadband accessing, because it provides wider bandwidth than that the regular modem provides.

ADSL was first developed in the 1980s the telecommunications industry's answer to the cable industry' request to support video on demand. In the middle 1990s, however, it was quickly recognized as a viable technology to enable access to high-speed services such as the Internet[12]. ADSL delivers asymmetric transmission rates typically up to 9Mb/s downstream (from the CO to the premises) and 16Kb/s to 640Kb/s upstream (from the premises to the CO) as shown in Figure 3-8. Like all copper transmission systems, the higher the bit rate, the shorter the range. A limitation of ADSL transmission is distance: It can only be used within three miles of a telephone switching station, and the speed degrades as the distance gets closer and closer to the three-mile limit.

(3) Dedicated Connection

Unlike dial-up connections that only connect to your ISP's computer when you need to access the Internet, dedicated connections keep you continually connected to the Internet. With a dedicated connection, your PC is typically issues a static (non-changing) IP address to be used to transfer data back and forth via the Internet.

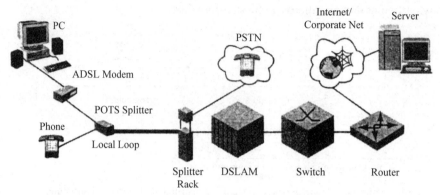

Figure 3-8 ADSL architecture

Types of dedicated Internet connections include connecting through a school or office LAN, as well as ADSL, cable, satellite, and fixed wireless connections.

(4) Wireless Connection

Wireless connections do not use a solid substance to connect sending and receiving devices. Rather, they use the air itself. Primary technologies used for wireless connections are infrared, broadcast radio, microwave, satellite and mobile wireless connections.

① Infrared

Infrared uses infrared light waves to communicate over short distances. It is sometimes referred to as line-of-sight communication because the light waves can only travel in a straight line. This requires that sending and receiving devices must be in clear view of one another without any obstructions blocking that view. One of the most common applications is to transfer data and information from a portable device such as a notebook computer or PDA to a desktop computer.

② Broadcast radio

Broadcast radio uses radio signals to communicate with wireless devices. For example, cellular telephones and many Web-enabled devices use broadcast radio to place telephone calls and/or to connect to the Internet [13]. Some end users connect their notebook or handheld computers to a cellular telephone to access the Web from remote locations. Most of these Web-enabled devices follow a standard known as Wi-Fi (wireless fidelity). This wireless standard is widely used to connect computers to each other and to the Internet.

③ Microwave

Microwave communication uses high-frequency radio waves. Like infrared, microwave communication provides line-of-sight communication because microwaves travel in a straight line. Because the waves can be transmitted only over relatively short distances, thus, microwave is a good medium for sending data between buildings in a city or on a large college campus.

Bluetooth is a short-range wireless communication standard that uses microwaves to transmit data over short distances of up to approximately 33 feet. Unlike traditional microwaves, Bluetooth does not require line-of-sight communication. Rather, it uses radio waves that can pass through nearby walls and other nonmetal barriers.

④ Satellite

Satellite can be used to send and receive large volumes of data. Uplink is a term relating to sending data to a satellite. Downlink refers to receiving data from a satellite. The major drawback to satellite communication is that bad weather can sometimes interrupt the flow of data.

One of the most interesting applications of satellite communications is for global positioning. A network of 24 satellites owned and managed by the Defense Department continuously sends location information to earth. Global Positioning System (GPS) devices use that information to uniquely determine the geographical location of the device.

⑤ Mobile Wireless Connection

Unlike satellite and fixed wireless connections, which use a cable to connect the modem to some type of fixed transceiver, mobile wireless connections allow the device to be moved from place to place. Consequently, most handheld PCs and other mobile devices (like Web-enable cell phones) use a mobile wireless connection and access the Internet through the same wireless network as cell phones and messaging devices.

NOTES

[1] 长句。句首 It 代表 Internet，since 引导的是原因状语从句，而从句中又包含两个由 which 引导的非限定性定语从句。

[2] to centralize…and to automate… 为状语，后面都有介词短语 of…构成的定语，分别修饰两个 the task。此句说明 DNS 的功能是将网络中的主机名翻译成它的 IP 地址。

[3] HOSTS 文件，也是用于将主机名或者正式域名解析为相关的 IP 地址的，功能与 DNS 一样，只不过 HOSTS 比 DNS 出现得早，且早已被 DNS 取代了。SRI NIC 即斯坦福研究院网络信息中心。

[4] 第一个 that 引导的是宾语从句；后面的 so that 引导的是目的状语从句。

[5] 最高层的域名是顶级域名，以行业为顶级域名的如.com（商业），.edu（教育）等是指美国本土的，其他则以国家和地区为顶级域名，如 cn（中国）、hk（中国香港）等。

[6] 句中 it 代表 local DNS，until 引导的时间状语从句中，that can resolve… 为定语从句，修饰 DNS server。provided… 为状语，其中从属连接词 that 引导的是宾语从句，从句中 it 为先行代词，真正的主语为 a valid name。

[7] you might be familiar with 为定语从句，修饰 domains。

[8] that 引导的是宾语从句，句中 at the top 为主语；with the com through cn domains 为状语，underneath 副词，在底下之意。

[9] your modem dials up 为主动语态，实际上是对 modem 进行拨号。后一个 modem 是 ISP 机器上的。attached to…分词短语作定语。

[10] While 引导的是时间状语从句，为被动语态，实际上是完成连接之后，由 ISP 给 PC 指派一个临时 IP 地址。

[11] in your place 指所使用的 IP 地址，此句与上一句的意思紧密相关。

[12] to enable…为实际主语，it 为先行主语。

[13] cellular telephone 即移动电话。

KEYWORDS

access provider	访问提供商，接入提供商
ISP (Internet Service Provider)	因特网服务提供商
DNS (Domain Name System)	域名系统
domain name service	域名服务
host	主机
top level domain	顶级域
domain name server	域名服务器
client	客户
root domain	根域
sub-domain	子域
dial-up	拨号
modem	调制解调器
interface device	接口设备
IP(Internet Protocol)	网际协议，网间协议，互联网协议
transfer rate	传输速率
ADSL(Asymmetric Digital Subscriber Line)	非对称数字用户线
broadband	宽（频）带
VOD (Video On Demand)	视频点播
downstream	下行，顺流，下游
upstream	上行，逆流，上游
CO(Central Office)	中心（交换）局
dedicated connection	专线连接
LAN(Local Area Network)	局域网
satellite	卫星
wireless	无线
infrared	红外线
microwave	微波
line-of-sight communication	视距通信
portable device	便携设备

PDA(Personal Digital Assistant)	个人数字助理
Cellular telephone, cell phone	蜂窝电话
Web	万维网
end user	最终用户
Wi-Fi (Wireless Fidelity)	无线保真度（无线局域网）联盟
uplink	上行链路
downlink	下行链路
global positioning	全球定位

EXERCISES

Multiple Choices

1. The Internet is _____.
   a. a network of networks
   b. a public network
   c. the largest and most well known computer network in the world
   d. a private network

2. DNS is _____.
   a. an abbreviation for Domain Name System
   b. used to translate network name to address assignments
   c. used at SRI NIC only
   d. the successor of HOSTS file

3. DNS has _____.
   a. a flat distributed architecture
   b. a root server
   c. a hierarchical distributed architecture
   d. many root servers

4. Top-level domains in the DNS _____.
   a. have two groups
   b. have three groups
   c. are located on the second layer of a similar inverted tree
   d. don't include domain name of China

5. Following device belongs to the one of dial-up connection _____.
   a. Wi-Fi
   b. modem
   c. dedicated line device
   d. ADSL

6. When we use a dial-up connection to access the Internet, we need_____.
   a. a TV cable
   b. a regular telephone line
   c. a modem
   d. a temporary IP address

7. ADSL works at a rate of_____.
   a. asymmetric transmission
   b. symmetric transmission

    c. asymmetric transmission with 9Mb/s downstream

    d. asymmetric transmission with 16K-64Kb/s upstream

8. With the dedicated connection to connect the Internet we should_____.

    a. use dial-up connection   b. keep a continual connection

    c. use a static IP address   d. keep a temporary connection

9. Following technologies are used for wireless communication: _____.

    a. mobile wireless connection   b. microwave

    c. satellite   d. broadcast radio

10. Infrared connection is used for _____.

    a. long distance communication

    b. short distance communication

    c. line-of-sight communication

    d. transferring data between a notebook computer and a desktop computer in one room

11. Using broadcast radio connection, we should use _____.

    a. Wi-Fi standard   b. Web-enabled devices

    c. mobile telephones   d. handheld computers

12. Bluetooth _____.

    a. is one type of microwave communication

    b. requires line-of-sight communication

    c. can pass through nonmetal barriers

    d. can be used to transmit data over short distance of up to approximately 10 meters

13. Satellite communication has _____.

    a. an uplink and a downlink

    b. a limitation of its application as bad weather can interrupt the flow of data

    c. an interesting application which is GPS

    d. the capability for sending and receiving large volumes of data

14. Mobile wireless connection _____.

    a. should be used to connect fixed devices

    b. should be used to connect mobile devices

    c. can be used to access Internet

    d. allows the device to be moved from place to place

PART II

INTERNET APPLICATIONS

CHAPTER 4
TRADITIONAL INTERNET APPLICATIONS

4.1 OVERVIEW OF WORLD WIDE WEB (WWW)

4.1.1 ABOUT WWW

WWW or Web is a large network of Internet servers providing hypertext and other services to terminals running client applications such as a browser[1].

WWW enables users to search, access, and download information from a worldwide series of networked servers where information is dynamically interlinked. A Web client passes a user's request for information to a server, usually by way of a Web browser. The server and client communicate through a transfer protocol, usually the HyperText Transfer Protocol (HTTP) [2]. The server then accesses a Web page using a Uniform Resource Locator (URL). Search engines are available to simplify access by enabling users to enter search criteria on a topic and have several URLs returned for Web pages that pertain to the desired information. [3]

1. Providers

The most common way to access the Internet is through an Internet service provider (ISP). The providers are already connected to the Internet and provide a path or connection for individuals to access the Internet. Your college or university most likely provides you with free access to the Internet either through its local area networks or through a dial-up or telephone connection[4]. There are also some companies that offer free Internet access.

The most widely used commercial Internet service providers are national providers (e.g. America Online (AOL)) and wireless providers.

2. Browsers

Browsers are programs that provide access to Web resources. This software connects you to remote computers, opens and transfers files, displays text and images, and provides in one tool an uncomplicated interface to the Internet and Web documents. Browsers allow you to explore, or to surf the Web by easily moving from one Web site to another[5]. Three well-known browsers are Mozilla Firefox, Netscape Communications, and Microsoft Internet Explorer (See Figure 4-1).

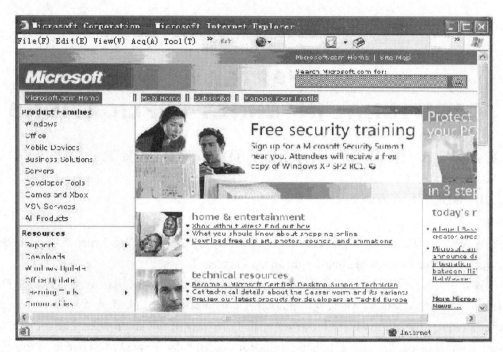

Figure 4-1 Internet Explorer

Browser is a GUI-based hypertext client application, used to access hypertext documents and other services located on innumerable remote servers throughout the WWW and Internet.

As you see from Figure 4-2, what goes on when you click a hyperlink is a pretty significant series of events, involving not only your Web browser software, but also a Web server somewhere, and the transactions involved rely heavily on the HTML language[6].

Figure 4-2 Web browser/server communication

3. Personal Web Sites

Do you have something to share with the world? Would you like a personal Web site, but don't want to deal with learning HTML? Creating your own home on the Internet is easy, and there are many services available to get you started[7].

A service site on the Web provides access to tools to create personal Web pages[8]. After registering with the site, you create your Web pages using the tools provided. Once completed, the service site acts as a host for your personal Web site and others are free to visit it from anywhere in the world.

NOTES

[1] terminals 是指连入因特网的各种终端。

[2] usually 后面是同位语。HTTP(HyperText Transfer Protocol)即超文本传输协议,是互联网上应用最为广泛的一种网络协议。所有的 WWW 文件都必须遵守这个标准。

[3] 由 and 连接的是两个并列成分,主语为 Search engines; search criteria 是指搜索判据(关键字)。

[4] most likely 意为"很可能";either through…or through…方式状语。dial-up or telephone connection 是指用 modem 或 ADSL 方式连接,有关这方面的内容请参阅 3.3 节。

[5] explore 是探索,surf 是冲浪,即在因特网上查看资料的行为。

[6] 本段就是一句话。As 引导的是方式状语从句。主句中 what 引导的是主语从句,involving…现在分词短语做状语,其结构为 not only…but also…。

[7] home 是指主页 homepage。

[8] access to 此处为"使用"。

KEYWORDS

WWW (World Wide Web)	万维网,环球信息网
HyperText	超文本
client	客户
browser	浏览器
search	搜索
access	访问
download	下载
server	服务器
protocol	协议
HTTP (HyperText Transfer Protocol)	超文本传输协议
web page	网页
URL (Uniform Resource Locator)	统一资源定位符(地址)
search engine	搜索引擎
search criteria	搜索条件
ISP (Internet Service Provider)	因特网服务提供商
path	通路,路径
AOL(America On Line)	美国在线
explore	探索
surf	冲浪

personal Web site	个人网站
register	注册
host	宿主机，主机

EXERCISES

Multiple Choices

1. Web_____.
   a. provides hypertext services
   b. allows the user's terminals running server programs to be used
   c. allows the user's terminals running client programs to be used
   d. allows the user's terminals running browser to be used

2. A Web client_____.
   a. passes a user's request for information to another client
   b. passes a user's request for information to a server
   c. uses HTTP to transfer its request
   d. can use search engines

3. A Web server_____.
   a. accesses a Web page using URL
   b. can communicate with client via HTTP
   c. has several search engines to be used
   d. can send a search result to a client

4. ISP_____.
   a. is the abbreviation of Internet Service Provider
   b. can provide a connection for individuals to access the Internet
   c. has always to connect the Internet
   d. usually offers you with free access to the Internet

5. Browsers_____.
   a. are programs that provide access to Web resources
   b. provide a simple interface to the Web documents
   c. allow you to surf the Web
   d. connect you to remote computers

6. Creating our own Website_____.
   a. is difficult
   b. is easier
   c. can use some creating tools
   d. can ask some service sites to give helps

7. A service site on the Web_____.
   a. offers some tools to create personal Web pages
   b. should ask you to register to this site before using the tools of creating personal Web pages

CHAPTER 4 TRADITIONAL INTERNET APPLICATIONS

   c. can act as a host for your personal Web site

   d. allows other sites to visit the personal Web site freely from anywhere in the world

8. Browser is _____.

   a. a GUI-based hypertext client application

   b. a GUI-based hypertext server application

   c. used to access hypertext documents

   d. used to access other services located on innumerable remote servers throughout the Web

9. To create a personal Web site, we _____.

   a. must learn HTML   b. don't need to learn HTML

   c. need some tools for creating the Web site   d. should choose a server site to be a host

10. A Web browser wants to communicate with a server via Internet _____.

    a. the browser should send a http request

    b. the browser should send a html request

    c. the server should return a http reply

    d. the server should return a html reply

4.1.2 SEARCH ENGINES

1. Overview

Imagine walking into a library where books are piled up and strewn about without any order[1]. Finding what you are looking for would be next to impossible [2]. As the World Wide Web rapidly grew, it is necessary to categorize message of Web page and to keep track of "what's what" and "what's where". In the initial stages of the Web, it was hard to locate useful information.

In the early days of the World Wide Web, two graduate students at Stanford University, Jerry Yang and David Filo, came up with a way to organize hyperlinks by category and they found it useful. In late 1993 this way was known as "Jerry Yang's Guide to the WWW." The name was soon changed to Yahoo! and the first search tool was born.

Today there is quite a collection of search tools available that allows us to find required information on the Web quickly and easily. The collection of search tools is constantly evolving, with new ones coming on the scene and others disappearing. Rather than report on how each of the popular search tools works, we will explain a few of them and suggest some Web presentations that provide reviews of all of the current search tools available [3].

2. How search engines work

A search engine operates, in the following order

(1) Web crawling

(2) Indexing [4]

(3) Searching

Web search engines work by storing information about many web pages, which they retrieve from the html itself. These pages are retrieved by a Web crawler (sometimes also known as a spider) — an automated Web browser which follows every link on the site[5]. The contents of each page are then analyzed to determine how it should be indexed (for example, words are extracted from the titles, headings, or special fields called meta tags)[6]. Data about web pages are stored in an index database for use in later queries. A query can be a single word. The purpose of an index is to allow information to be found as quickly as possible. Some search engines, such as Google, store all or part of the source page (referred to as a cache) as well as information about the web pages, whereas others, such as AltaVista, store every word of every page they find. This cached page always holds the actual search text since it is the one that was actually indexed, so it can be very useful when the content of the current page has been updated and the search terms are no longer in it[7].

When a user enters a query into a search engine (typically by using key words), the engine examines its index and provides a listing of best-matching web pages according to its criteria, usually with a short summary containing the document's title and sometimes parts of the text[8].

3. About Google Chrome

Google Chrome is a freeware web browser developed by Google. It used the WebKit layout engine until version 27 and, with the exception of its iOS releases, from version 28 and beyond uses the WebKit fork Blink[9].

It was first released as a beta version for Microsoft Windows on September 2, 2008, and as a stable public release on December 11, 2008[10].

As of January 2015, StatCounter estimates that Google Chrome has a 51% worldwide usage share of web browsers, indicating that it is the most widely used web browser in the world[11].

Google releases the majority of Chrome's source code as an open-source project Chromium[12]. A notable component that is not open source is the built-in Adobe Flash Player.

Google Chrome features a minimalistic user interface, with its user-interface principles later being implemented into other browsers. For example, the merging of the address bar and search bar into the omnibox[13]. Chrome also has a reputation for strong browser performance.

NOTES

[1] where 引导的定语从句修饰 library；pile up 堆积。strewn 是 strew 的过去分词，意为"撒布"，about 此处为"到处"之意。

[2] what 引导的是宾语从句； next to 用于否定句之前，意为"几乎"。

[3] Rather than 引导方式状语从句，表示"而不是……"；how 引导的是宾语从句；that 引导的是定语从句，修饰 presentations，that 在从句中做主语。

[4] Indexing，搜索引擎单独控制它们的索引和排名算法。

[5] crawler 网上浏览器；破折号后面为同位语； which 引导的是定语从句。

[6] title 和 heading 均是标题； meta tag，元标签，是用于插入到网页首部的 HTML 代码。

[7] since 引导的是原因状语从句，it 代表主句中的 cached page，one 是指 page；that 引导的定语从句，修饰 one；so 引导的是结果状语从句，最后的 it 仍然是 cached page。

[8] 长句。When 引导的是时间状语；usually with…为状语；containing…分词短语做定语，修饰 summary。

[9] iOS 原为 iPhone OS，是苹果公司开发的移动操作系统，而且是由苹果硬件独享的。用在该公司的很多设备，如 iPhone、iPad、iPod touch 上。WebKit 是一个开源浏览器引擎，也是苹果 Mac OX 系统引擎。Blink 是基于 WebKit 的 forkWeb 渲染引擎。

[10] 有关 beta version 请见 2.3 节 NOTES [1]。

[11] StatCounter 是万维网流量分析工具。用它分析网络基本业务是免费的，分析高级业务，每月收费 5~119 美元。

[12] Chromium 是一个开源 Web 浏览器课题，从这里 Google Chrome 给出它的开源码。

[13] omnibox，多功能框，地址栏，又兼作搜索插件（search plug in）栏，允许渐进式搜索（incremental search）。

KEYWORDS

search tool	搜索工具
spider	网络蜘蛛，万维网查询工具
index	索引
index database	索引数据库
cache	高速缓存
query	查询
key word	关键字
crawler	爬行者，浏览器
freeware	免费（自由）软件，免费（自由）件
built-in	内置的
majority	大多数，大部分
minimalistic	最小的，极小的
merge	合并，融合，融入
address bar	地址栏

EXERCISES

True/False

1. _____ As the WWW rapidly grew, it is hard to find useful information.
2. _____ Google is the first search tool in the world.
3. _____ Collection of search tools is constantly evolving, with new ones coming on the scene and others disappearing.
4. _____ In order to operate a search engine, we should follow five operations.
5. _____ Search engines retrieve with http itself.
6. _____ Web crawler is an automated Web browser.
7. _____ Index database stores data concerning Web pages.
8. _____ Google stores every word of every page it finds.
9. _____ Cached page always holds the actual search text.
10. _____ A personal Web page is very useful when the content of its current page has been updated.
11. _____ When we enter a query into a search engine, the engine examples its index and provides a listing of best-matching Web pages.
12. _____ Purpose of an index is to allow information to be found as quickly as possible.
13. _____ Google Chrome almost uses the WebKit fork Blink engine.
14. _____ Google Chrome is the most widely used Web browser in the world.
15. _____ Chromium is an open-source project.
16. _____ Omnibox is a multiple function box, an address bar, and a search plug-in engine.

4.2 E-MAIL

E-mail or electronic mail is the transmission of electronic messages over the Internet. At one time, e-mail consisted only of basic text messages. Now E-mail routinely includes graphics, photos, and many different types of file attachments. People all over the world send e-mail to each other. You can E-mail your family, your co-workers. All you need to send and receive E-mail is an E-mail account, access to the Internet, and an E-mail program[1]. Two of the most widely used E-mail programs are Microsoft's Outlook Express and Mozilla Thunderbird.

A typical e-mail message has three basic elements: header, message, and signature (See Figure 4-3). The header appears first and typically includes the following information:

Business Email Sample
To: "Anna Jones" <annajones@buzzle.com<
Cc: All Staff
From: "James Brown"
Subject: Welcome to our Hive!

Figure 4-3 Basic elements of an E-mail message

(1) **Addresses**: Addresses of the persons sending, receiving, and, optionally, anyone else who is to receive copies. E-mail addresses have two basic parts (See Figure 4-4). The first part is the user's name and the second part is the domain name, which includes the top-level domain. In our example e-mail, dcoats is Dan's user name. The server providing e-mail service for Dan is usc.edu. The top-level domain indicates that the provider is an educational institution[2].

Figure 4-4   Two parts of an e-mail address

(2) **Subject**: A one-line description, used to present the topic of the message. Subject lines typically are displayed when a person checks his or her mailbox.

(3) **Attachments**: Many e-mail programs allow you to attach files such as documents and worksheets. If a message has an attachment, the file name appears on the attachment line.

The letter or message comes next. It is typically short and to the point. Finally, the signature line provides additional information about the sender. Typically, this information includes the sender's name, address, and telephone number.

E-mail can be a valuable asset in your personal and professional life. However, like many other valuable technologies, there are drawbacks too. Americans receive billions of unwanted and unsolicited e-mails every year. This unwelcome mail is called spam. While spam is indeed a distraction and nuisance, it also can be dangerous. For example, computer viruses or destructive programs are often attached to unsolicited e-mail.

In an attempt to control spam, anti-spam laws have been added to our legal system. For example, the recently enacted CAN-SPAM Act requires that every marketing-related e-mail provide an opt-out option[3]. When the option is selected, the recipient's e-mail address is to be removed from future mailing lists. This approach, however, has minimal impact since over 50 percent of all spam originates from servers outside the United States[4]. A more effective approach has been the development and use of spam blockers (See Figure 4-5). These programs use a variety of different approaches to identify and eliminate spam.

Spam Blocker	Site
InBoxer	www.inboxer.com
OnlyMyEmailPersonal	www.onlymymail.com
Qurb	www.qurb.com
Vanquish vqME	www.vanquish.com

Figure 4-5   Spam blockers

NOTES

[1] you need…定语从句，修饰 All；is 后面有 3 个并列的表语。

[2] that 引导的是宾语从句。

[3] opt-out，退出。

[4] since 此处做副词用，为"其后""从那时起"之意。

KEYWORDS

E-mail	电子邮件
electronic message	电子消息（报文，电文）
attachment	附件
account	账户（号）
header	头部
signature	签名
address	地址
user's name	用户名
domain name	域名
subject	主题
spam	垃圾邮件
mailing list	邮件地址表，邮件清单
spam blocker	垃圾邮件拦截器

EXERCISES

Fill in the blanks with using terms, words or phrases found behind this exercise.

1. E-mail is the transmission of electronic messages over the _____.
2. Now E-mail routinely includes _____.
3. An E-mail account, access to the Internet, and an E-mail program are _____.
4. A typical E-mail message has three basic elements: _____.
5. The header of an E-mail message includes the following information: _____.
6. E-mail addresses include two basic parts: _____.
7. _____ is used to present the topic of a message.
8. Documents and worksheets can be used to be _____.
9. E-mail can be a _____ in your personal and professional life.
10. The unwanted and unsolicited E-mail are called _____.
11. _____ are often attached to unwelcome E-mail.
12. In our legal system _____ have been added.
13. If we want to remove our E-mail address from future mailing lists, we should select an _____.

14. _____ use a variety of different ways to identify and eliminate spam.

    a. spam

    b. addresses, subject and attachments

    c. Internet

    d. Subject

    e. Spam blockers

    f. all you need to send and receive E-mail

    g. valuable asset

    h. anti-spam laws

    i. graphics, photos, and many different types of file attachment

    j. user's name and domain name

    k. opt-out option

    l. Computer viruses or destructive programs

    m. header, message, and signature

    n. attachments

4.3 ELECTRONIC COMMERCE AND THE INTERNET OF THINGS

4.3.1 ELECTRONIC COMMERCE AND SOCIAL COMMERCE

1. What is Electronic Commerce

Electronic commerce is a system that includes not only those transactions that center on buying and selling goods and services to directly generate revenue, but also those transactions that support revenue generation, such as generating demand for those goods and services, offering sales support and customer service (See Figure 4-6), or facilitating communications between business partners[1].

Electronic commerce builds on the advantages and structures of traditional commerce by adding the flexibilities offered by electronic networks.

Electronic commerce enables new forms of business, as well as new ways of doing business. Amazon.com, for example, is a bookseller based in Seattle, Washington. The company has not physical stores, sells all their books via the Internet, and coordinates deliveries directly with the publishers so they do not have to maintain any inventory[2]. Companies such as Kantara and software.net take this a step further.

Because all of their products (commercial software packages) are electronic, and can be stored on the same computers that they use for processing orders and serving the Web, their inventory is totally digital[3]. As another example, AMP Inc. is offering its clients the opportunity to purchase electronic connectors and related components directly from its Web-based catalog, bypassing the need for EDI-based purchase orders and confirmations [4].

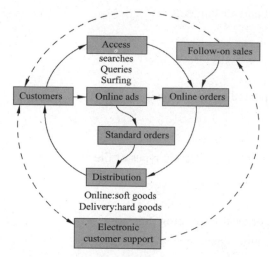

Figure 4-6　Cycle of electronic commerce

2. E-Commerce Business Models

A company's policies, operations, and technology define its business model. In essence, a company's business model describes how the company generates revenue. There are a number of standard e-commerce business models, including business to consumer (B2C), business to business (B2B), consumer to consumer (C2C), and business to government (B2G). Some of the most commonly used models are discussed next.

（1）**Business to Consumer (B2C)**

With the business-to-consumer (B2C) model, businesses sell goods or services to individual consumers. The B2C model was one of the first major types of e-commerce business models to be defined and implemented using the Web. Some examples of B2C businesses include Amazon.com, L.L. Bean, Walmart.com, and Polo.com (See Figure 4-7). These businesses can be Internet-only stores, or click-and-mortar stores with both online and traditional storefronts[5].

（2）**Business to Business (B2B)**

Business-to-business (B2B) applications include any type of e-commerce transaction taking place between two businesses. B2B revenue is increasing and expected to continue to grow tremendously in the next few years.

（3）**Consumer to Consumer (C2C)**

The consumer-to-consumer (C2C)—sometimes referred to as the person-to-person or P2P—business model almost solely consists of consumer auctions, where consumers sell products to other consumers[6]. With millions of products for sale every day, eBay is one of the largest C2C e-commerce business today.

Figure 4-7   Business to consumer

(4) Business to Government (B2G)

With U.S. government spending exceeding $500 billion per year with no signs of slowing down, business-to-government (B2G) organizations are becoming more prominent[7]. These organizations (See Figure 4-8) sell products and services to local, state, and federal government buyers. In general, the government sector has been slower to embrace online buying than the private sector.

Related B2G activities include some states allowing citizens to make payments online, such as paying taxes, renewing driver's licenses, and so forth[8]. This is sometimes referred to as C2G or customer-to-government e-commerce. The federal government's new Pay.gov Web site shown in Figure 4-8 is expected to allow transactions at the federal level.

3. Applications of E-Commerce

(1) Online Shopping

Most people equate e-commerce with online shopping, which is made possible by the security and encryption features built into popular browsers [9]. According to a recent survey, fewer than half of the Web's users have bought something online, but this percentage is growing each year.

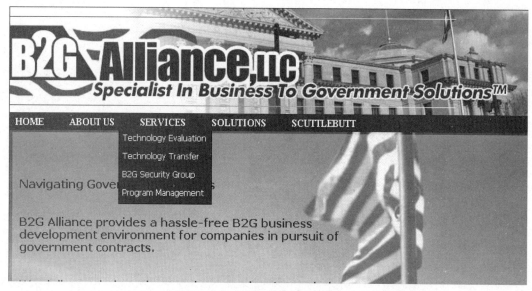

Figure 4-8 Business to government

One of the tremendous advantages of online shopping is the low capital investment needed to set up shop. For less than $2000, a startup can open a Web storefront and start selling product online. For example, Amazon.com, the "world's largest bookstore", as it describes itself, is nothing more than a few offices in a Seattle office building.

If one word could sum up what all these sites share, it's disintermediation. All of these online success stories take intermediaries out of the picture[10]. These sites put customers in direct contact with incredibly-rich resources of information. They enable customers to make their own choices, without being restricted to stock-on-hand (or a salesperson's interference).

(2) **Online Banking**

Banks implement online banking in different ways. One method makes use of checkbook programs such as Microsoft Money or Intuit's Quicken, enabling customers to balance their checkbooks automatically. The drawback, however, is that you have to access your online banking account from the computer that has all the Money or Quicken data, and this data could be examined by anyone with access to your computer[11]. An advantage is that Money and Quicken offer powerful features for budgeting your spending and analyzing your spending habits.

Newer Web-based systems that require only program (a Web browser) are easier to use. All the data is stored on the bank's computer, not your own, which means you can access your account anywhere. Web-based online banking is also much easier to use. However, the Web-based systems don't offer advanced features such as budgeting and spending analysis. What's in it for the banks? Plenty: those bank branches and tellers cost a lot of money[12]!

4. Social Commerce

Social commerce is a subset of electronic commerce that involves social media, online media that supports social interaction, and user contributions to assist online buying and selling of products and services.

More succinctly, social commerce is the use of social network(s) in the context of e-commerce transactions[13].

The term social commerce was introduced by Yahoo! in November 2005 which describes a set of online collaborative shopping tools such as shared pick lists, user ratings and other user-generated content-sharing of online product information and advice[14].

The concept of social commerce was developed by David Beisel to denote user-generated advertorial content on e-commerce sites, and by Steve Rubel to include collaborative e-commerce tools that enable shoppers "to get advice from trusted individuals, find goods and services and then purchase them"[15]. The social networks that spread this advice have been found to increase the customer's trust in one retailer over another.

Social commerce aims to assist companies in achieving the following purposes. Firstly, social commerce helps companies engage customers with their brands according to the customers' social behaviors. Secondly, it provides an incentive for customers to return to their website. Thirdly, it provides customers with a platform to talk about their brand on their website. Fourthly, it provides all the information customers need to research, compare, and ultimately choose you over your competitor, thus purchasing from you and not others.

Today, the range of social commerce has been expanded to include social media tools and content used in the context of e-commerce, especially in the fashion industry. Examples of social commerce include customer ratings and reviews, user recommendations and referrals, social shopping tools (sharing the act of shopping online), forums and communities, social media optimization, social applications and social advertising. Technologies such as Augmented Reality have also been integrated with social commerce, allowing shoppers to visualize apparel items on themselves and solicit feedback through social media tools[16].

Some academics have sought to distinguish "social commerce" from "social shopping", with the former being referred to as collaborative networks of online vendors; the latter, the collaborative activity of online shoppers.

NOTES

[1] 这一段就是一句话，其中第一个 that 引出了 not only…，but also…结构的定语从句，在此从句中又有两个 that 引导的定语从句，分别修饰各自的 transactions；such as 后面为同位语，此同位语由 3 个分词短语构成。

[2] 本句有三个谓语：has、sells 和 coordinates； so 引导的是结果状语从句。

[3] 长句。前面大部分是 Because 引导的原因状语从句，仅最后的 their…为主句。

[4] bypassing，此处可译为"无须……"。

[5] click-and-mortar，混合经营的。

[6] where 引导的是非限定性定语从句。

[7] with 引出的短语作状语。

[8] allowing citizens…现在分词短语作定语。

[9] which 代表 online shopping，它引出的是非限定性定语从句。

[10] out of the picture 意为"消失""退出"。

[11] 长句。that 引导的是表语从句，从句中又由 that 引导出定语从句。

[12] Plenty，副词：的确，此句主语为 those bank branches and tellers，谓语为 cost。

[13] in the context of…，此处可译为"在……中"，下同。

[14] 长句。句中 which 代表 Yahoo!；user rating，用户信誉度。

[15] 长句，句子结构为：The concept…was developed by…to…，and by…to…。

[16] Augmented Reality(AR)，强化现实，是一个具体的现实世界环境的直接或间接的实时观看，但环境中的元素是由计算机产生的，像声音、视频、图形或 GPS 数据那样的感官信息，是追加进去的。

KEYWORDS

electronic commerce	电子商务
B2B(B to B, Business-to-Business)	商业对商业
business mode	企业模式，业务模式
B2C (Business-to-Consumer)	企业对消费者
C2C (Consumer-to-Consumer)	消费者对消费者
B2G (Business-to-Government)	企业对政府
P2P (Person-to-Person)	个人对个人
C2G (Customer-to-Government)	客户对政府
supply chain	供应链
click-and-mortar	混合经营的
online shopping	在线购物
encryption	加密
out of the picture	不合适，不相干的
disintermediation	<美>非居间化（指由储蓄银行存款转为直接的证券投资）
stock-on-hand	现有存货
online banking	在线银行
checkbook	支票簿
social commerce	社交电子商务，社区化商务
subset	子集，子集合
interaction	交互作用，互动
social network	社交网络

e-commerce site	电子商务网站
customer	顾客，客户，买主
retailer	零售商
platform	平台
website	万维网网站
user ratings	用户信誉度
customer ratings	客户信誉度
optimization	最优化
feedback	反馈

EXERCISES
Multiple Choices

1. Electronic Commerce supports _____.
   a. buying goods
   b. selling goods
   c. services of directly generating revenue
   d. other transactions
2. Electronic Commerce is _____.
   a. an EDI-based system only
   b. built on the structures of traditional commerce
   c. used with electronic networks
   d. built on the advantages of traditional commerce
3. The company that runs electronic stores has _____.
   a. not physical store   b. one physical store at least
   c. a digital inventory   d. a Web site
4. Electronic Commerce _____.
   a. provides new forms of doing business
   b. does not have to maintain any inventory
   c. is a system that offers customer service
   d. provides new ways of doing business
5. Business model can be defined by a company's _____.
   a. policies   b. operations   c. technology   d. revenues
6. B2C means _____.
   a. businesses sell goods to individual consumers
   b. business-to-consumer
   c. consumers sell service to businesses
   d. individuals buy goods from businesses
7. C2C means _____.

a. consumer sells auctions to other consumers

   b. consumer sells auctions to individuals

   c. consumer-to-consumer

   d. P2P

8. With B2G, citizens can make _____ online.

   a. paying taxes   b. watching TV

   c. renewing driver's licenses   d. payments

9. Online shopping _____.

   a. has the tremendous advantages

   b. is not possible without the security feature

   c. is possible with encryption built into popular browsers

   d. is one of E-commerce applications

10. Amazon.com _____.

    a. is the largest bookstore in the world   b. has a few offices only

    c. is bookseller based in Washington DC   d. sells all their books via the Internet

11. Web-based on line banking _____.

    a. has all the data stored on your own computer

    b. has all the data stored on the bank's computer

    c. is much easier to use

    d. does not offer budgeting and spending analysis

12. Standard E-commerce business models include _____.

    a. B2C   b. B2B   c. C2C   d. B2G

13. Social commerce includes _____.

    a. social media   b. online media   c. user contributions   d. all of the above

14. Social commerce _____.

    a. is the use of social networks in e-commerce transactions

    b. has a set of online collaborative shopping tools

    c. is equal to social shopping

    d. is a subset of electronic commerce

15. The spread of social commerce _____.

    a. decreases the customer's trust in one retailer over another

    b. helps companies engage customers according to the customer's social behaviours

    c. gives customers with a platform to talk about their brand on their website

    d. provides an incentive for customer to return to their website

16. Applications of social commerce include _____.

    a. context of e-commerce   b. fashion industry

    c. customer ratings   d. sharing the act of shopping online

4.3.2 INTERNET OF THINGS

1. Overview of The Internet of Things

The Internet of Things (IoT) is the network of physical objects or "things" embedded with electronics, software, sensors and connectivity to enable it to achieve greater value and service by exchanging data with the manufacturer, operator and/or other connected devices[1]. Each thing is uniquely identifiable through its embedded computing system but is able to interoperate within the existing Internet infrastructure.

Typically, IoT is expected to offer advanced connectivity of devices, systems, and services that goes beyond machine-to-machine communications (M2M) and covers a variety of protocols, domains, and applications. The interconnection of these embedded devices (including smart objects), is expected to usher in automation in nearly all fields, while also enabling advanced applications like a Smart Grid[2].

Things, in the IoT, can refer to a wide variety of devices such as heart monitoring implants, biochip transponders on farm animals, electric clams in coastal waters, automobiles with built-in sensors, or field operation devices that assist fire-fighters in search and rescue. These devices collect useful data with the help of various existing technologies and then autonomously flow the data between other devices. Current market examples include smart thermostat systems and washer/dryers that utilize wifi for remote monitoring.

Besides the plethora of new application areas for Internet connected automation to expand into, IoT is also expected to generate large amounts of data from diverse locations that is aggregated at a very high-velocity, thereby increasing the need to better index, store and process such data[3].

2. Architecture of The Internet of Things

The IoT system is likely to have an event-driven architecture. In Figure 4-9, IoT development is shown with a three-layer architecture. The top layer is formed by driven applications. The application space of the IoT is huge. The bottom layers represent various types of sensing devices: namely RFID tags, ZigBee or other types of sensors, and road-mapping GPS navigators [4]. Signals or information collected at these sensing devices are linked to the applications through the could computing platform at the middle layer.

The signal processing clouds are built over the mobile networks, the Internet backbone, and various information networks at the middle layer. In the IoT, the meaning of a sensing event does not follow a deterministic or syntactic model. In fact, the service-oriented architecture (SOA) model is adoptable here[5]. A large number of sensors and filters are used to collect the raw data. Various compute and storage clouds and grids are used to process the data and transform it into information and knowledge formats. The sensed information is used to put together a decision-making system for intelligence applications. The middle layer is also considered as a Semantic Web[6]. Some actors (service, components, avatars) are self-referenced[7].

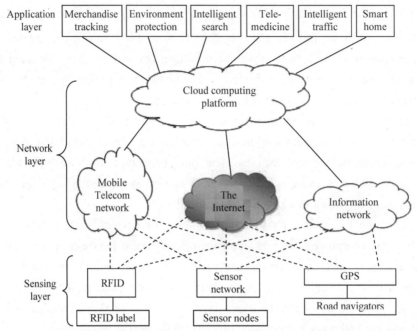

Figure 4-9　The architecture of an IoT consisting of sensing devices that are connected to various applications via mobile networks, the Internet, and processing clouds

3. Applications

According to Gartner, Inc. (a technology research and advisory corporation), there will be nearly 26 billion devices on the Internet of Things by 2020. ABI Research estimates that more than 30 billion devices will be wirelessly connected to the Internet of Things (Internet of Everything) by 2020[8]. As per a recent survey and study done by Pew Research Internet Project, a large majority of the technology experts and engaged Internet users who responded—83 percent—agreed with the notion that the Internet/Cloud of Things, embedded and wearable computing (and the corresponding dynamic systems) will have widespread and beneficial effects by 2025[9]. It is, as such, clear that the IoT will consist of a very large number of devices being connected to the Internet.

The ability to network embedded devices with limited CPU, memory and power resources means that IoT finds applications in nearly every field. Such systems could be in charge of collecting information in settings ranging from natural ecosystems to buildings and factories, thereby finding applications in fields of environmental sensing and urban planning. On the other hand, IoT systems could also be responsible for performing actions, not just sensing things. Intelligent shopping systems, for example, could monitor specific users' purchasing habits in a store by tracking their specific mobile phones. These users could then be provided with special offers on their favorite products, or even location of items that they need. Additional examples of sensing and actuating are reflected in applications that deal with heat, electricity and energy management, as well as cruise-assisting transportation systems.

However, the application of the IoT is not only restricted to these areas. Other specialized use cases of the IoT may also exist. An overview of some of the most prominent application areas is provided here. Based on the application domain, IoT products can be classified broadly into five different categories: smart wearable, smart home, smart city, smart environment, and smart enterprise. The IoT products and solutions in each of these markets have different characteristics.

NOTES

[1] 长句。embedded with…，过去分词短语作定语，修饰 objects or "things"。句中 to enable it…是目的状语，it 仍代表 objects or "things"。

[2] Smart Grid，智能电网，即电网智能化。

[3] 长句，主句是 IoT is also…，句中 that 引导的定语从句，修饰 data，thereby…结果状语。

[4] RFID（Radio Frequency Identification）即射频识别，又称无线射频识别，是通过无线电信号识别特定目标并读写相关数据的一种技术。在识别中，无须在识别系统与特定目标之间建立机械或光学接触。ZigBee 是一种近距离、低复杂度、低功耗、低成本的双向无线通信技术。GPS（Global Positioning System）即全球定位系统，起源于1958年美国军方的一个项目。到1994年，全球覆盖率达98％的 GPS 卫星星座已布设完成。

[5] SOA（Service Oriented Architecture） model 即面向服务的体系结构模型，它将应用程序的不同功能单元（称为服务），通过定义良好的接口和协议联系起来，使得各种系统的服务，可以使用一个统一的和通用的方式进行交互。

[6] Semantic Web，语义网，是1998年就提出来的一个概念，其核心是：通过给全球信息网上的文档（如 HTML）添加能够被计算机所理解的语义（元数据），从而使整个互联网成为一个通用的信息交换媒体。

[7] avatar，阿凡达，指在虚拟实境中互动地呈现一个人，即所谓的计算机化身。

[8] ABI Research，是一个技术市场智能公司，它收集所有公司的研究新闻，且每周都进行公布。Internet of Everything (IoE)，称为万物互联。

[9] 长句。As per a recent survey…Project，为方式状语。主句为 a large majority of…agreed…，其中 that 引导的是宾语从句。

KEYWORDS

  Internet of Things (IoT)   物联网
  object   目标，对象，客体

embedded	嵌入（式）的
sensor	传感器
connectivity	连通性，连通度
AR (Augmented Reality)	强化现实
identifiable	可辨识的，可视为同一的，可确认身份的
interoperate	（相）互操作
infrastructure	基础设施，基础结构
M2M (Machine-to-Machine)	机器对（到）机器
protocol	协议
domain	域，定义域，范畴
interconnection	互连，互接
biochip	生物芯片
transponder	发射应答机，转发器，询问机
thermostat	自动调温器
aggregate	集合，聚集，集合的，合计的
event-driven	事件驱动
RFID(Radio Frequency Identification)	射频识别，无线射频识别
GPS navigator	全球定位导航器
backbone	主干，骨干
syntactic	句法的
SOA (Service Oriented Architecture)	面向服务的体系结构
decision-making system	决策支持系统
avatar	具体化，具体表现，阿凡达
self-reference	自参考，自引用
wearable computing	可穿戴计算技术
ecosystem	生态系统

EXERCISES

True/False

1. _____ The things in IoT are embedded with electronics, software, and sensors.
2. _____ M2M communications go beyond IoT in offering advanced connectivity of devices systems, and services.
3. _____ Physical objects in IoT can refer to a wide variety of devices.
4. _____ Automobiles with built-in sensors are the devices that connect into Internet of Things.
5. _____ We do not need to better index, store and process volume data in IoT recently.
6. _____ IoT system has an event-driven architecture.
7. _____ IoT development has a four-layer architecture.
8. _____ The signal processing clouds are at the middle layer in the architecture of an IoT.

9. _____ In IoT the meaning of a sensing event should follow a deterministic or syntactic model.
10. _____ Various compute and storage clouds are used to process the raw data.
11. _____ By 2025 wearable computing is a widespread industry only.
12. _____ The ability to network embedded devices means that IoT finds applications in nearly every fields.
13. _____ The major task of IoT system is sensing only.
14. _____ Through tracking users' mobile phones the intelligent shopping systems could monitor their habits.
15. _____ IoT products can be classified into five categories.
16. _____ The Network layer in Figure 4-9 can be considered as a Semantic Web.

4.4 COMPUTER AND NETWORK SECURITY

4.4.1 COMPUTER CRIMINALS AND CRIME

1. Computer Criminals

A computer crime is an illegal action in which the perpetrator uses special knowledge of computer technology[1]. Typically, computer criminals are either employees, outside users, hackers, crackers, organized crime members, or terrorists.

What are hackers and crackers? Some people think of these two groups as being the same, but they are not. Hackers are people who gain unauthorized access to a computer system for the fun and challenge of it[2]. Crackers do the same thing but malicious purposes. They may intend to steal technical information or to introduce what they call a bomb—a destructive computer program—into the system[3].

2. Computer Crime

The FBI estimates that businesses lose over $1.5 trillion a year from computer crimes. The number of these crimes has tripled in the past two years. Computer crime can take various forms including the creation of malicious programs, denial of service attacks, Internet scams, theft, and data manipulation.

(1) Malicious Programs

Hackers and crackers are notorious for creating and distributing malicious programs. These programs are called malware, which is short for malicious software[4]. They are specifically designed to damage or disrupt a computer system. The three most common types of malware are viruses, worms, and Trojan horses.

1) Viruses

① What Are Viruses?

A virus--or more formally, a computer virus, is a computer program that is able to make a copy of itself without you knowing what is happening. A virus may copy itself from one part of your hard disk to another, or it may copy itself from one computer to another.

Most viruses do more than make copies of themselves. Some of them cause real damage, say, by deleting files on your hard disk. Others are merely annoying. They may display a message on your monitor or cause something strange to happen as you are working. All viruses are malevolent in that they do their work without your knowing what is happening, and they can cause problems merely by spreading uncontrollably[5].

② Types of Viruses

Although thousands of viruses roam machines and networks all over the world, most have common traits that can be categorized into one of the following groups[6]:

- Boot sector viruses
- File infector viruses
- Macro viruses

③ How Do You Get a Virus?

Most viruses today are transmitted through the Internet. The macro virus is the fastest spreading virus because it is usually transferred from machine to machine via e-mail attachments. However, viruses can also be transmitted through floppy disks, file transfers, and Web downloads. In addition, macro viruses infect all open files of a specific application.

Therefore, if your machine has a Microsoft Word macro virus, it will infect all Word documents that are open on your machine. If you save those infected files to a floppy disk, you will transfer the infection along with the file when you copy it to another hard drive, or open it on another machine using the same application [7].

In addition, using the Web to download files is another common way to get a virus. Sites that allow users to share information over the Internet, such as Napster and Gnutella, are a potential breeding ground for computer viruses, much like real-world viruses breed in high traffic areas such as malls and schools[8].

2) Worms

Worms are a special type of virus that does not attach itself to programs and databases.

(2) Denial of Service

Like a worm a denial of service (DoS) attack attempts to slow down or stop a computer system or network. Unlike a worm that self-replicates, a DoS attack floods a computer or network with requests for information and data[9]. The targets of these attacks are usually Internet service providers (ISP) and specific Web sites. Once under attack, the servers at the ISP or the Web site become overwhelmed with these requests for service and are unable to respond to legitimate users. As a result, the ISP or Web site is effectively shut down.

(3) Internet Scams

A scam is a fraudulent or deceptive act or operation designed to trick individuals into spending their time and money for little or no return. An Internet scam is simply a scam using the Internet. Internet scams are becoming a serious problem and have created financial and legal problems for thousands of people. Almost all of the scams are initiated by a mass mailing to unsuspecting individuals.

(4) Theft

Theft can take many forms—of hardware, of software, of data, of computer time. Thieves steal equipment and programs, of course, but there are also white-collar crimes. These crimes include the theft of data in the form of confidential information such as preferred-client lists. Another common crime is the use (theft) of a company's computer time by an employee to run another business.

(5) Data Manipulation

Finding entry into someone's computer network and leaving a prankster's message may seem like fun, which is why hackers do it[10]. It is still against the law. Moreover, even if the manipulation seems harmless, it may cause a great deal of anxiety and wasted time among network users.

NOTES

[1] in which…定语从句，修饰 action。

[2] who 引导的定语从句，修饰 people； for…，目的状语；it 代表 computer system。

[3] what 引导的是宾语从句；破折号中间的词组为同位语。

[4] which 引导的是非限定性定语从句；malware 一词取自 malicious 的前三个字母和 software 的后四个字母。

[5] in that 引导的是两个并列的状语从句。

[6] Although 引导让步状语从句，that 引导的定语从句，修饰 common traits。categorize 意为"使列入……的范畴，将……分类"。

[7] If 引导的是条件状语从句，主句是 you will…，其中 when 引导的是时间状语从句，从句中两个 it 均代表 infected files。

[8] 长句，句子结构为 Sites…are…，其中 that 引导的定语从句，修饰 sites；句中有两个 such as…均为同位语； much like…为状语。 Napster 是为两个热点音乐在线服务起的名字，它是共享因特网服务的对等式文件的先驱，主要是共享 MP3 编码格式的音乐文件。 Gnutella 在写本文时，是同类网络中，第一个非集中式对等网，后来，其他网络也采取了这种模型。

[9] flood 原意为洪水泛滥，此处是指信息和数据在计算机或网络中到处传递，引起计算机或网络瘫痪。

[10] Finding…and leaving…现在分词短语作主语；which 引导的是非限定性定语从句。

KEYWORDS

computer criminal	计算机罪犯
computer crime	计算机犯罪
hacker	黑客
cracker	破坏者
unauthorized access	非授权访问
steal	窃取
FBI (Federal Bureau of Investigation)	（美）联邦调查局
trillion	（美）1万亿，10^{12}
malicious program	恶意程序
attack	攻击
scam	欺诈
data manipulation	数据操纵
malware	不良件
virus	病毒
worm	蠕虫
Trojan horse	特洛伊木马
confidential information	机密信息

EXERCISES

Multiple Choices

1. Computer criminals can be _____.
   a. employees   b. outside users   c. hackers   d. organized crime members
2. A computer crime _____.
   a. is a legal action
   b. needs special knowledge of computer technology
   c. is an illegal action
   d. does not need special knowledge of computer technology
3. Hackers _____.
   a. are the same as crackers
   b. are different from crackers
   c. access to a computer for the fun and challenge of it
   d. access to a computer for malicious purposes
4. Hackers and crackers may intend to _____.
   a. steal technical information
   b. guard computers against an attack
   c. introduce a destructive program into a computer
   d. gain unauthorized access to a computer

5. Computer crimes include _____.
   a. creation of malicious programs
   b. denial of service
   c. Internet scams
   d. data manipulation
6. Malware is _____.
   a. a malicious program
   b. an Internet scam
   c. short for malicious software
   d. specifically designed to damage a computer
7. There are several types of malware, they are _____.
   a. viruses
   b. worms
   c. Trojan horses
   d. all above
8. A computer virus _____.
   a. is a computer program
   b. can make copy itself from one part of your hard disk to another
   c. can make a copy of itself with you knowing what is happening
   d. is a malicious program
9. Viruses can _____.
   a. cause real damage to the computer
   b. delete files on your hard disk
   c. be merely annoying
   d. transfer on Internet
10. Ways to transmit viruses are through _____.
    a. hard disk
    b. floppy disk
    c. E-mail
    d. Web downloads
11. Worms _____.
    a. are a special type of virus
    b. like a Denial of Service
    c. attach itself to databases
    d. do not attach itself to programs
12. Denial of Service _____.
    a. attempts to slow down or stop a computer or network
    b. floods a computer on network with request for information and data
    c. attacks Internet Service Providers
    d. attacks specific Web sites
13. Internet scams _____.
    a. are a fraudulent action
    b. trick individuals with big return
    c. are becoming a serious problem
    d. have created financial and legal problems
14. Computer thieves should steal _____.
    a. hardware
    b. software
    c. data
    d. computer time
15. Data manipulation _____.
    a. is still against the law
    b. seems harmless
    c. seems harm
    d. may cause a great deal of anxiety
16. We can categorize viruses into one of the following groups _____.
    a. Boot sector viruses
    b. Macro viruses
    c. File infector viruses
    d. Worms

4.4.2 PROTECT COMPUTER SECURITY

Security is concerned with protecting information, hardware, and software from unauthorized use as well as from damage from intrusions, sabotage, and natural disasters. Considering the numerous ways in which computer systems and data can be compromised, we can see why security is a growing field[1]. Some of the principal measures to protect computer security are encryption, restricting access, anticipating disasters, backing up data, and firewall.

1. Encrypting Messages

Whenever information is sent over a network, the possibility of unauthorized access exists. The longer the distance the message has to travel, the higher the security risk is. For example, an e-mail message on a LAN meets a limited number of users operating in controlled environments such as offices. An e-mail message traveling across the country on the Internet affords greater opportunities for the message to be intercepted.

Businesses have been encrypting, or coding, messages for years. One of the most widely used personal encryption programs is Pretty Good Privacy[2].

2. Restricting Access

Security experts are constantly devising ways to protect computer systems from access by unauthorized persons. Sometimes security is a matter of putting guards on company computer rooms and checking the identification of everyone admitted. Other times it is using biometric. Scanning devices such as fingerprint and iris scanners.

Oftentimes it is a matter of being careful about assigning passwords to people and of changing the passwords when people leave a company[3]. Passwords are secret words or numbers that must be keyed into a computer system to gain access.

3. Backing Up Data

Equipment can always be replaced. A company's data, however, may be irreplaceable. Most companies have ways of trying to keep software and data from being tampered with in the first place[4]. They include guarding of passwords, and auditing of data and programs from time to time. An essential procedure, however, is to make frequent backups of data and to store them in safe remote locations.

4. Firewalls

The purpose of a network firewall is to provide a shell around the network which will protect the systems connected to the network from various threats.

A firewall can reduce risks to network systems by filtering out inherently insecure network services. Network File System (NFS) services, for example, could be prevented from being used from outside of a network by blocking all NFS traffic to or from the network. This protects the individual hosts while still allowing the service, which is useful in a LAN environment, on the internal network[5]. Instead what is needed is a way to filter access to the network while still allowing users access to the "outside world"[6]. A typical network firewall can be depicted as shown in Figure 4-10.

Figure 4-10　Typical network Firewall configuration

In this configuration, the internal network is separated from external networks by a firewall gateway. A gateway is normally used to perform relay services between two networks[7]. In the case of a firewall gateway, it also provides a filtering service which limits the types of information that can be passed to or from hosts located on the internal network[8]. There are three basic techniques used for firewalls: packet filtering, circuit gateway, and application gateways. Often, more than one of these is used to provide the complete firewall service.

NOTES

　　[1] Considering…现在分词短语作原因状语；in which…定语从句，修饰 ways；主句中 why 引导的是宾语从句。

　　[2] Pretty Good Privacy (PGP)，智能加密，为了加密 e-mail，以便安全通信而产生的一种方案。

　　[3] 句中 of being…and of changing…为介词短语作定语，修饰 a matter。

　　[4] in the first place，当初。

　　[5] while 引导的是状语从句，句中省略了 it is；which 引导的是非限定性定语从句。

　　[6] instead，副词；what is needed，主语从句；while 的用法和上一句一样。

　　[7] gateway，网关，比较通用的称呼是路由器。

　　[8] 句中有 which 和 that 两个关系代词引导的定语从句,分别修饰 a filtering service 和 information。

KEYWORDS

　　　　computer security　　　　　　计算机安全
　　　　intrusion　　　　　　　　　　侵入窃密，闯入，打扰
　　　　encryption　　　　　　　　　 加密

restricting access	限制访问
anticipating disaster	灾难预测
backing up	备份
firewall	防火墙
security risk	安全风险
intercept	截取（断）
identification	身份证，识别，鉴定
biometric scanning	生物扫描
password	口令
gateway	网关

EXERCISES

Fill in the blanks with using terms, words or phrases found behind this exercise.

1. Security is concerned with protecting information, hardware, and software from _____.
2. Principal measures to protect computer security are _____.
3. Whenever information is sent over a network, the possibility of _____ exists.
4. An E-mail message traveling on the Internet can be _____.
5. One of the most widely used personal encryption program is _____.
6. _____ is a matter of security.
7. In restricting access we can use biometric scanning devices such as _____.
8. _____ are secret words or numbers that must be keyed into a computer system to gain access.
9. The ways of trying to keep software and data from being tampered with in the first place include _____.
10. There are three basic techniques used for firewall, they are_____.
11. A firewall can reduce risks to network systems by_____.
12. A gateway is normally used to perform_____between two networks.

    a. unauthorized access

    b. guarding of passwords, and auditing of data and programs

    c. filtering and inherently insecure network services

    d. encryption, restricting access, anticipating disasters, backing up data, and firewall

    e. Passwords

    f. intercepted

    g. checking the identification of everyone admitted

    h. Pretty Good Privacy

    i. relay services

    j. unauthorized use, damage, intrusion, sabotage, and natural disasters

    k. packet filtering, circuit gateway, application gateway

    l. fingerprint and iris scanners

CHAPTER 5   NEW INTERNET APPLICATIONS

5.1   INSTANT MESSAGING (IM)

5.1.1   OVERVIEW OF INSTANT MESSAGING (IM)

Instant messaging (IM) is an extension of e-mail that allows two or more people to contact each other via direct, live communication. To use instant messaging, you specify a list of friends and register with an instant messaging server. Whenever you connect to the Internet, special software informs your messaging server that you are online[1]. In response, the server will notify you if any of your contacts are online. At the same time, it notifies your friends that you are online. You can then send messages directly back and forth to one another. Most instant messaging programs also include video conferencing features, file sharing, and remote assistance. Many businesses routinely use these instant messaging features.

Instant messaging is a popular variation of chat in which you are informed when someone on your buddy list—a list containing the names of friends and associates that you specify—is on line; when they are on line, you can send them a message that immediately appears on their screen (See Figure 5-1).[2] You can then have a real-time typed conversation.

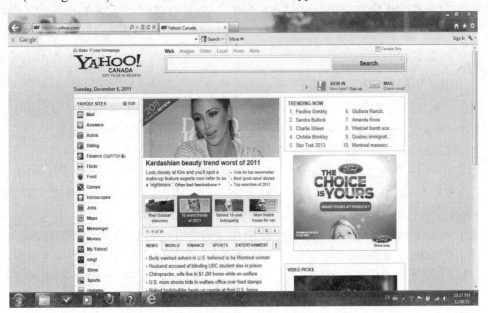

Figure 5-1   Instant Messaging

As of 2010, social networking providers often offer IM abilities. Facebook Chat is a form of instant messaging, and Twitter can be thought of as a Web 2.0 instant messaging system[3]. Similar server-side chat features are part of most dating websites, such as OKCupid or Plenty of Fish[4]. The spread of smartphones and similar devices in the late 2000s also caused increased competition with conventional instant messaging, by making text messaging services still more ubiquitous.

Many instant messaging services offer video calling features, voice over IP and web conferencing services[5]. Web conferencing services can integrate both video calling and instant messaging abilities. Some instant messaging companies are also offering desktop sharing, IP radio, and IPTV to the voice and video features[6].

Each modern IM service generally provides its own client, either a separately installed piece of software, or a browser-based client. These usually only work with the supplier company's service, although some allow limited function with other services. Third party client software applications exist, that will connect with most of the major IM services. Adium, Empathy, Miranda IM, Pidgin, Qnext and Trillian are a few of the common ones[7].

Standard complementary instant messaging applications offer functions like file transfer, contact list(s), the ability to hold several simultaneous conversations, etc. These may be all the functions that a small business needs, but larger organizations will require more sophisticated applications that can work together[8]. The solution to finding applications capable of this is to use enterprise versions of instant messaging applications. These include titles like XMPP, Lotus Sametime, Microsoft Office Communicator, etc., which are often integrated with other enterprise applications such as workflow systems[9]. These enterprise applications, or enterprise application integration (EAI), are built to certain constraints, namely storing data in a common format.

NOTES

[1] that 引导的是宾语从句。

[2] 本句是用分号隔开的两个句子，前一个句子中，in which 引导的是定语从句，此从句中 when 引导的又是时间状语从句；两个破折号中间的句子是同位语。

[3] Facebook Chat，脸谱聊天。有关 Facebook，请见 5.1.3 节。

Twitter，推特。有关 Twitter，请见 5.1.4 节。

[4] OKCupid，免费的在线约会（online dating），是一个朋友和社交网站，其特点是成员之间可以猜谜和讨论问题。Plenty of Fish，"很多鱼"网站。

[5] voice over IP，即 IP 电话，有关 IP 电话的内容，请见 3.1.3 节 NOTES[6]。

[6] IP radio，即 Radio over IP。IP TV，即交互式网络电视，是一种利用宽带网，集互联网、多媒体、通信等技术于一体，向家庭用户提供包括数字电视在内的多种交互式服务的崭新技术。它能够很好地适应当今网络技术的飞速发展。

[7] Adium，是用于 OS X 操作系统的自由和开源即时消息客户端软件，支持

Windows Live Messenger、Yahoo! Messenger、Google Talk、ICQ、XMPP 等多个 IM 网。Empathy，是 IM 和 VoIP 客户端软件，支持文本、语音、视频、文件传输和各种 IM 协议上的应用程序之间的通信。 Miranda IM，为微软 Windows 设计的开源多协议即时消息应用软件。 Pidgin，（以前的名字是 Gaim），是一个开源多平台即时消息客户软件。Qnext，是一个安全远程文件访问的应用软件，可使用计算机和移动设备，所访问的文件可放在多个存储设备或云服务设备上。Trillian，是为微软 Windows、苹果 Mac OS X、Linux、Android、iOS 等系统设计的专利多协议即时消息应用软件。

[8] 并列句，两个句子中都有 that 引导的定语从句。

[9] which 引导的是非限定性定语从句。XMPP（Extensible Messaging and Presence Protocol），可扩展的消息和展示协议，是一种基于 XML 的面向消息的中间件的通信协议。Lotus Sametime，以前是 IBM Lotus Sametime，现为 IBM Sametime，是为企业提供实时统一通信和合作的客户-服务器应用和中间件的平台。其功能有消息展示、企业即时消息、Web 会议、团体合作以及通话和集成等。 Microsoft Office Communicator，现在是 Skype for Business，是与 Microsoft Lync Server 一起使用的即时消息客户端软件。workflow system，即 WfMS（Workflow Management System），是用来建立、执行和监视科学工作流的基础软件结构。

KEYWORDS

IM (Instant Messaging)	即时消息
live communication	实时通信
online	在线，联机
video conferencing	视频会议
file sharing	文件共享
remote assistance	远程辅助
chat	聊天，闲谈
buddy list	好友通讯录
social network	社交网
smartphone	智能电话
browser-based client	基于浏览器的客户（端）
file transfer	文件传输
enterprise version	企业版
workflow system	文件流系统
EAI(Enterprise Application Integration)	企业应用软件集成

EXERCISES

True/False

1. _____ IM is an extension of E-mail.
2. _____ To use instant messaging, you should register with an IM server.

3. _____ When you connect to the Internet, a special program informs you that you are online.
4. _____ In the process of using IM, an instant messaging server will notify you if any of your contacts are online.
5. _____ Using IM, you can send messages directly back and forth to one another.
6. _____ Only a few of instant messaging can use remote assistance.
7. _____ Instant messaging is a special variation of chat.
8. _____ In instant messaging you are informed when someone on your buddy list is online.
9. _____ A list containing the names of friends and associates is the buddy list.
10. _____ Using IM, you can send your friends a message that will appear on their screen for several minutes.
11. _____ Twitter can be thought as a Web 2.0 IM system.
12. _____ Web conferences integrated video calling with IM abilities.
13. _____ IPTV is an interactive network television.
14. _____ Today most IM services provide their own browser-based client only.
15. _____ Original standard of IM offers file transfer, contact lists and several simultaneous conversations.
16. _____ XMPP is an abbreviation for Extensible Messaging and Presence Protocol.

5.1.2 QQ

1. Overview of the QQ

Tencent QQ, generally referred to as QQ, is the most popular free instant messaging computer program in Mainland China. As of September 30, 2010, the active QQ user accounts for QQ IM amounted to 636.6 million, possibly making it the world's largest online community[1]. The number of simultaneous online QQ accounts exceeded 100 million. In February 2011, QQ.com ranked 10th overall in Alexa's internet rankings just behind Twitter ranked 9th[2]. The program is maintained by Tencent Holdings Limited (HKEX: 0700), owned in part by Naspers[3]. Since its entrance into Chinese households QQ quickly emerged as a modern cultural phenomenon, now being portrayed in popular culture[4]. Aside from the chat program, QQ has also developed many sub-features including games, virtual pets, ringtone downloads, music, shopping, blogs, microblogging, and group and voice chat etc.

The current version of QQ is QQ2010 beta2. Tencent periodically releases special versions of QQ to coincide with events such as the Olympics or Chinese New Year.

The official client runs on Microsoft Windows and a beta public version was launched for Mac OS X version 10.4.9 or newer[5]. The Web versions, WebQQ (full version) and WebQQ Mini (Lite version), which makes use of Ajax, are currently available[6].

As of January 2015, there are 829 million active QQ accounts, with a peak of 176.4 million simultaneous online QQ users.

2. QQ International

(1) Windows

In 2009 QQ began to expand its services internationally with its QQ International client for Windows distributed through a dedicated English-language portal.

QQ International offers non-Mandarin speakers the opportunity to use all the features of its Chinese counterpart to get in touch with other QQ users via chat and videocalls, it provides a non-Mandarin interface to access Qzone, Tencent's social network[7]. The client supports English, French, Spanish, German, Korean, Japanese and traditional Chinese. A wealth of third-party applications is bundled with QQ International and it is mainly aimed at making cross-cultural communications in and out of China more convenient.

One of the main features of QQ International is the optional and automatic machine translation of all chats.

(2) Android

An Android version of QQ International was released in September 2013. The client's interface is in English, French, Spanish, German, Korean, Japanese and traditional Chinese. In addition to text messaging, users can send each other images, video, and audio media messages. Moreover, users can share multimedia content with all contacts through the client's Qzone interface.

The live translation feature is available for all incoming messages and supports up to 18 languages.

(3) iOS / iPhone

QQ International for iPhone and iOS devices was released at the end of 2013, fully equivalent to its Android counterpart.

3. Web QQ

Tencent launched its web-based QQ formally on 15 September 2009, the latest version of which being 3.0. Rather than solely a web-based IM, WebQQ 3.0 functions more like its own operating system, with a desktop in which web applications can be added.

4. Open source and cross-platform clients

Using reverse engineering, open source communities have come to understand the QQ protocol better and have attempted to implement client core libraries compatible with more user-friendly clients[8]. Most of these clients are cross-platform, so they are usable on operating systems which the official client does not support. However, these implementations had only a subset of the functions of the official client and therefore were more limited in features.

NOTES

[1] making 现在分词短语做状语；it 代表 QQ, the world's…为宾语补语。

[2] Alexa，是一家专门发布网站世界排名的网站，同时提供网站的流量、网站访问量等。Twitter（非官方称为"推特"），是一个社交网在线服务网站，有关它的内容请见 5.1.4 节。

[3] Tencent Holdings Limited，是中国的一家国有公司，其子公司提供因特网和移动电话的增值服务和在线广告服务。HKEX（Hong Kong Stock Exchange）为香港股市交易。Naspers 是南非的一家跨国媒体公司，主营电子媒体（包括付费电视、因特网和即时消息用户平台，以及其他相关技术）和印刷媒体（包括出版、发行和杂志，报纸和书籍的印刷，也提供私人教育服务）。

[4] Since 引导的是时间状语从句；portray，描绘。

[5] Mac OS X 是苹果电脑的操作系统，详情请参阅 2.5 节。

[6] Ajax 是一组交叉式 Web 应用的开发方法，用于在客户端创建异步式 Web 应用。

[7] 长句，前半句中，第一个动词不定式 to use…为定语，修饰 opportunity。后一个动词不定式 to get in…做状语。Qzone 是腾讯计算机系统有限公司（港交所 00700)2005 年推出的一个网志系统，目前仍活跃在中国大陆。

[8] reverse engineering 即逆向工程（又称反向工程），是一种技术过程，即对一项目标产品进行逆向分析及研究，从而演绎并得出该产品的处理流程、组织结构、功能性能规格等设计要素，以制作出功能相近，但又不完全一样的产品。

KEYWORDS

blog	博客
social network	社交网
microblogging	微博
voice chat	语音聊天
portal	门户
video call	视频呼叫
multimedia	多媒体
counterpart	副本，复本，对应物
open source	开源
cross platform	跨平台
reverse engineering	逆（反）向工程

EXERCISES

Multiple Choices

1. QQ is the most popular free instant messaging in _____.
   a. Japan b. P.R.C c. Singapore d. Mainland China
2. QQ Instant Messaging _____.

a. has user accounts that exceed 600 million

b. is the world's largest online community

c. has more than 100 million of user simultaneous on line

d. has entered into Chinese households

3. QQ has the following sub-features: _____.
   a. chat   b. virtual pets   c. games   d. blogs
4. Special versions of QQ are used for coinciding with _____.
   a. Christmas   b. Thanksgiving Day
   c. Chinese New Year   d. Olympics
5. Beta public version of QQ was launched for Mac OS X version _____.
   a. newer   b. older   c. Web QQ   d. 10.4.9
6. QQ International for Windows offers _____ speakers to use all the features of its Chinese counterpart.
   a. English   b. Korean   c. non-Mandarin   d. Mandarin
7. QQ International for Android _____.
   a. has a server's interface in Classic Chinese
   b. can be used to send text messaging only
   c. can be used to share multimedia content with all contacts
   d. can support up to 18 languages in the live translation
8. The client core libraries are _____.
   a. compatible with user-friendly clients
   b. open source communities
   c. cross-platforms
   d. usable on operating systems which the official client does not support
9. Tencent Holdings Limited _____.
   a. is a Chinese national company
   b. has a branch company which offers Internet added services
   c. has a branch company which offers mobile added services
   d. has a branch company which offers online advertising services
10. Until now QQ International has _____.
    a. a Windows version   b. an Android version
    c. versions for iPhone and iOS   d. all the above

5.1.3 FACEBOOK

1. Features

Facebook is a social networking service and website launched in February 2004, operated and privately owned by Facebook Inc. As of June 2014, Facebook has more than 1.3 billion active users. Users may create a personal profile with photo, lists of personal interests, contact information, and other personal information, add other users as friends, and exchange messages, including automatic notifications when they update their profile[1]. Additionally, users may join common-interest user groups, organized by workplace, school or college, or other characteristics.

Facebook has a number of features with which users may interact:

(1) Wall, a space on every user's profile page that allows friends to post messages for the user to see[2].

(2) Pokes, which allows users to send a virtual "poke" to each other (a notification then tells a user that they have been poked)[3].

(3) Photos, where users can upload albums and photos.

(4) Status, which allows users to inform their friends of their whereabouts and actions.

Depending on privacy settings, anyone who can see a user's profile can also view that user's Wall.

On February 23, 2010, Facebook was granted a patent on certain aspects of its News Feed. The patent covers News Feed in which links are provided so that one user can participate in the same activity of another user. The patent may encourage Facebook to pursue action against websites that violate its patent, which may potentially include websites such as Twitter[4].

One of the most popular applications on Facebook is the Photos application, where users can upload albums and photos. Facebook allows users to upload an unlimited number of photos, compared with other image hosting services such as Photobucket and Flickr, which apply limits to the number of photos that a user is allowed to upload[5]. During the first years, Facebook users were limited to 60 photos per album. As of May 2009, this limit has been increased to 200 photos per album.

2. Technical aspects

Facebook is built in PHP which is compiled with HipHop for PHP, a 'source code transformer' built by Facebook engineers that turns PHP into C++ [6]. The deployment of HipHop reportedly reduced average CPU consumption on Facebook servers by 50%.

Facebook used a combination platform based on HBase to store data across distributed machines [7]. Using a tailing architecture, new events are stored in log files, and the logs are tailed. The system rolls these events up and writes them into storage. The User Interface then pulls the data out and displays it to users. Facebook handles requests as AJAX behavior [8]. These requests are written to a log file using Scribe (developed by Facebook) [9].

NOTES

[1] 长句，主句为 Users may create…，add…，and exchange…，有 3 个并列的谓语；

including…分词短语做状语。

[2] Wall 在 Facebook 中是指一块可供大家在上面留言的墙。

[3] poke，原意是"存入，插入"。此处是用手指戳戳的意思，是用来"烦"别人的。

[4] 句中有 that 和 which，分别引导了定语从句和非限定性定语从句。Twitter 的解释请见 5.1.4 节。

[5] 长句。compared with…过去分词短语做状语；which 引导的是非限定性定语从句；that 引导的是定语从句。Photobucket 是美国一个影像寄存、视频寄存、幻灯片制作与照片分享的网站。Flickr，雅虎旗下图片分享网站，提供免费及付费数码照片存储、分享方案之线上服务，也是网络社群服务的平台。

[6] PHP（Hypertext Preprocessor）即超文本预处理器，是一种通用开源脚本语言，详见 6.4 节。HipHop for PHP 是 Facebook 的一个项目，它由一个 PHP 到 C++的转换程序，一个重新实现的 PHP 运行库和许多常用 PHP 扩展的重写版本构成，目的旨在加速和优化 PHP。

[7] HBase – Hadoop Database 是一个高可靠性、高性能、面向列、可伸缩的分布式存储系统，利用 HBase 技术可在廉价 PC Server 上搭建起大规模结构化存储集群。

[8] AJAX（Asynchronous JavaScript And XML）即异步 JavaScript 和 XML，是指一种创建交互式网页应用的网页开发技术。

[9] Scribe 是 Facebook 开源的日志收集系统，在 Facebook 内部已经得到大量的应用。它能够从各种日志源上收集日志，存储到一个中央存储系统（可以是 NFS，分布式文件系统等）上，以便于进行集中统计分析处理。它为日志的"分布式收集，统一处理"提供了一个可扩展的、高容错的方案。

KEYWORDS

profile	配置文件，概貌，轮廓
patent	专利
hosting	托管
upload	加载，上载，（向上）装入
compile	编译
consumption	消耗，耗散，消费
monolithic	单块（片）的
distributed machines	分布式（计算）机
log file	日志文件
aggregate	集合，聚集，集合的，合计的

EXERCISES

True/False

1. _____ Facebook is a social networking service.
2. _____ Users of Facebook can create a personal profile.

3. _____ Wall in Facebook is like a firewall.
4. _____ Poke in Facebook is a notification.
5. _____ With Photos of Facebook, we can download albums and photos.
6. _____ With Status of Facebook, users can locate their friends.
7. _____ Certain aspects of Facebook's NewsFeed are patent.
8. _____ Flickr allows users to upload unlimited number of photos.
9. _____ Facebook allows users to upload limited number of photos.
10. _____ Before 2005 Facebook users were limited to 60 photos per album.
11. _____ HipHop for PHP is a source code transformer.
12. _____ A report shows that the use of HipHop reduced average CPU consumption on Facebook servers by 40%.
13. _____ Using a tailing architecture, new events are stored in log files.
14. _____ HBase is a high reliable, high performance and resize-able distributed storage system.

5.1.4 TWITTER

Twitter is an online social networking service that enables users to send and read short 140-character messages called "tweets" [1].

Registered users can read and post tweets, but unregistered users can only read them. Users access Twitter through the website interface, SMS, or mobile device app. Twitter Inc. is based in San Francisco and has more than 25 offices around the world.

Twitter was created in March 2006. The service rapidly gained worldwide popularity, with more than 100 million users who in 2012 posted 340 million tweets per day. The service also handled 1.6 billion search queries per day. In 2013 Twitter was one of the ten most-visited websites, and has been described as "the SMS of the Internet." As of December 2014, Twitter has more than 500 million users, out of which more than 284 million are active users [2].

1. Features

Tweets are publicly visible by default, but senders can restrict message delivery to just their followers. Users can tweet via the Twitter website, compatible external applications (such as for smartphones), or by Short Message Service (SMS) available in certain countries[3]. Retweeting is when a tweet is forwarded via Twitter by users. Both tweets and retweets can be tracked to see which ones are most popular. While the service is free, accessing it through SMS may incur phone service provider fees.

Twitter allows users to update their profile via their mobile phone either by text messaging or by apps released for certain smartphones and tablets.

As a social network, Twitter revolves around the principle of followers. When you choose to follow another Twitter user, that user's tweets appear in reverse chronological order on your main Twitter page. If you follow 20 people, you'll see a mix of tweets scrolling down the page: breakfast-cereal updates, interesting new links, music recommendations, even musings on the future of education.

2. Implementation

Twitter places great reliance on open-source software. The Twitter Web interface uses the Ruby on Rails framework, deployed on a performance enhanced Ruby Enterprise Edition implementation of Ruby[4].

As of April 6, 2011, Twitter engineers confirmed they had switched away from their Ruby on Rails search stack, to a Java server they call Blender[5].

The service's application programming interface (API) allows other web services and applications to integrate with Twitter[6].

NOTES

[1] tweet，推文，tweet 原意是小鸟啾啾地叫，此处为短小消息。

[2] out of which，原意是出自……，which 是指 5 亿用户。

[3] tweet，此处为动词。 SMS(Short Message Service)，短信服务。

[4] Ruby on Rails 是一个可以使开发、部署、维护 Web 应用程序变得简单的框架。Ruby Enterprise Edition (REE)是标准 Ruby（一种为简单快捷的面向对象编程而开发的脚本语言）解析器的改进版本,号称能够让 Rails（是一个更符合实际需要而且更高效的 Web 开发框架）应用节约 1/3 的内存使用量，并且能更好地提高性能。

[5] confirmed 后面省略了 that，这是一个宾语从句。Blender 是一款开源的跨平台全能三维动画制作软件，提供从建模、动画、材质、渲染、到音频处理、视频剪辑等一系列动画短片制作解决方案。

[6] 此处 service 是指 Blender。

KEYWORDS

register	注册，寄存器
post	邮政，邮寄，记入，位置
SMS(Short Message Service)	短消息服务
tweet	推文
retweet	转推文
profile	配置文件，简表，概貌，轮廓
stack	栈，堆栈
Java	一种程序设计语言

EXERCISES

Fill in the blanks with appropriate words or phrases found behind this exercise.

1. Twitter allows users to send and receive short _____ called "tweets".
2. Unregistered Twitter users can only read _____.
3. By way of _____ users can access Twitter.
4. Senders in Twitter can _____ message delivery to just their followers.
5. In order to see which persons are most popular, we can _____ tweets and retweets.
6. Twitter places great reliance on _____.
7. Twitter revolves around the principle of _____.
8. _____ allows other web services and applications to integrate with Twitter.
9. Ruby is an object-oriented _____ language.
10. Twitter allows users to update their _____ via their mobile phone.

  a. scripting
  b. API
  c. 140-character messages
  d. open-source software
  e. website, interface, SMS, or mobile device app
  f. profile
  g. tweet
  h. track
  i. restrict
  j. followers

5.1.5 MICROSOFT SERVICE NETWORK (MSN)

MSN (originally The Microsoft Network) is a collection of Internet sites and services provided by Microsoft. The Microsoft Network debuted as an online service and Internet service provider on August 24, 1995, to coincide with the release of the Windows 95 operating system[1].

In addition to its original MSN Dial-up service, Microsoft has used the 'MSN' brand name for a wide variety of products and services over the years, notably Hotmail (which is now Outlook.com), Messenger (which was once synonymous with 'MSN' in Internet slang and has now been replaced by Skype), and its search engine (which is now Bing), as well as several other rebranded and discontinued services[2].

The current website and suite of apps offered by MSN was first introduced by Microsoft in 2014 as part of a complete redesign and relaunch. MSN is based in the United States and offers international versions of its portal for dozens of countries around the world.

In 2012, MSN announced on its blog that it would be unveiling a new version of the

MSN.com home page on October 26, exclusively for Windows 8, saying that the new version would be "clean, simple, and built for touch"[3]. Microsoft said it would be more app-like due to the speed of Internet Explorer 10. More new features included 'Flip Ahead', which allowed users to swipe from one article to the next[4].

Microsoft launched a completely rewritten and redesigned MSN website, making use of the company's modern design language, on September 30, 2014. The new MSN portal features a new version of the logo that follows a style similar to other current Microsoft products. The website no longer offers original content, instead employing editors to repurpose existing content from partners at popular and trusted organizations. Much of the existing content on MSN was eliminated as the website was simplified into a new home page and categories, most of which have corresponding apps[5]:

- News
- Weather
- Entertainment
- Sports
- Money
- Lifestyle
- Health & Fitness
- Food & Drink
- Travel
- Autos
- Video

The top of the home page provides access to Microsoft services Bing, Outlook.com, Skype, Office Online, OneNote, OneDrive, Bing Maps, and Xbox Music, as well as popular social media services Facebook and Twitter[6]. Signing into MSN with a Microsoft account allows for personalized content to appear and to be synchronized across devices on the website and in the corresponding apps. The redesign of the website led to the closure of MSN's longtime personalized home page service 'My MSN', which was made up of customized RSS feeds, as the new website no longer supports user-specified RSS content[7]. However, it added some customizability, allowing each category on the home page to be reordered or hidden.

With the 2014 relaunch, MSN now supports responsive design and eliminates the need for a separate mobile website. The redesign of MSN proved positive and helped increase traffic with an additional 10 million daily visitors after two months.

Along with the 2014 redesign of the MSN web portal, Microsoft relaunched many of the Bing apps that originally shipped with Windows and Windows Phone as a suite of MSN apps. In December 2014, the new apps became available across all the other major mobile device platforms as well: iOS, Android, and Fire OS[8].

NOTES

[1] debut，首次露面；to coincide…动词不定式短语做状语。

[2] Hotmail 是互联网免费电子邮件提供商之一，世界上的任何人都可以通过网页浏览器对其进行读取，收发电子邮件。MSN Messenger 是微软公司推出的即时通信软件，凭借该软件自身的优秀的性能，目前在国内已经拥有了大量的用户群。使用 MSN Messenger 可以与他人进行文字聊天、语音对话、视频会议等即时交流，还可以通过此软件来查看联系人是否联机。Skype 是一款即时通信软件，其具备 IM 所需的功能，比如视频聊天、多人语音会议、多人聊天、传送文件、文字聊天等功能。Bing，微软必应，

是微软公司于 2009 年 5 月 28 日推出，用以取代 Live Search 的全新搜索引擎服务。为符合中国用户使用习惯，Bing 中文品牌名为"必应"。

[3] 第一个 that 引导的是宾语从句；saying that…现在分词短语做状语，其中 that 引导的是一个宾语从句。

[4] which 引导的是非限定性定语从句。Flip Ahead，向前翻转，是 微软 IE10 才有的新功能，是 IE 中的一个选项。

[5] as 引导的是原因状语从句。

[6] Office Online，网上办公，可以在任何地方与任何人一起实时处理 Office 文件并且是免费的。可以在线创建、存储和共享文档、电子表格、演示文稿和笔记本。OneNote 是一款同时支持手写输入和键盘输入的记事本软件，它记录的内容可以利用索引按科目进行分类整理，还可以进行语音输入，具有时间图章的功能。OneDrive，原名 SkyDrive，2014 年 1 月，美国微软公司正式宣布 SkyDrive（云存储服务）更名为 OneDrive。Bing Maps 是微软公司推出的 Bing 服务线上地图服务。使用它，可以在网络浏览器中观察到世界上的每一个角落。它的原名叫做"MSN Virtual Earth"。Xbox Music 是一个为 Xbox 360、Windows Phone 及 Windows 8 提供跨平台的音乐服务。

[7] which 引导的是非限定性定语从句。My MSN，是一个门户网站，其前身是 Windows Live Personalized Experience（即 My.Live.com）。RSS（Rich Site Summary），常称为 Really Simple Syndication，采用标准的 Web feed 格式系列去发表经常更新的消息：博客条目、新闻标题、音频和视频网摘。 RSS feed（又称 Web feed 或通道），就是 RSS 文档。

[8] Android 请见 2.6 节。 Fire OS，是亚马逊为它的 Fire Phone 和 Kindle Fire 开发的、基于 Linux 核的移动操作系统。

KEYWORDS

debut	初次登台
coincide	同时发生，一致，相同，巧合
dial-up	（电话）拨号
portal	门户网站
home page	主页
swipe	猛击
logo	标识
eliminate	除去，删除，排除
headline	标题，书眉
personalized	个人化的
customizability	可定制性
reorder	重排序，再订货
hidden	隐藏，隐式

CHAPTER 5 NEW INTERNET APPLICATIONS

EXERCISES

Multiple choices

1. MSN_____.
   a. is a collection of Internet sites
   b. is a collection of Internet services
   c. has an original name the Microsoft Network
   d. debuted on 9. 24. 1995

2. Hotmail is _____.
   a. a product of Microsoft   b. replaced by Outlook.com now
   c. a search engine   d. an email provider

3. New version of the MSN.com home page _____.
   a. can be used for Windows 7 and Windows 8
   b. can be used for Windows 8 only
   c. would be unveiling on October 26. 2012
   d. would be built for touch

4. Flip Ahead _____.
   a. is a new feature provided by IE10
   b. allowed users to swipe from one article to the next
   c. cab be used to speed up IE10
   d. is an IE option

5. A new MSN portal on 2014 _____.
   a. is a completely rewritten and redesigned MSN website
   b. has a new version of the logo
   c. can offer existing content
   d. uses the company's modern design language

6. The new MSN portal has many apps, they are _____.
   a. News   b. Weather   c. Entertainment   d. about eleven apps

7. Using Office Online we can _____.
   a. make a cloud storage service   b. process office files on real time
   c. create documents online   d. present PowerPoint presentations

8. Signing into MSN _____.
   a. should use a Microsoft account
   b. should use a public account
   c. we can synchronize across devices on the website
   d. we can be allowed for personalized content to appear

9. The 2014 redesign of the MSN web portal _____.
   a. provides many of the Bing apps
   b. makes the new apps to be available across all the other major mobile device platforms

        c. can be used for Fire OS

        d. can be used for Android

10. MSN now _____.

        a. supports 'My MSN'

        b. supports responsive design

        c. eliminates the need for a separate mobile website

        d. helps increase traffic

5.1.6 WECHAT

   WeChat is a mobile text and voice messaging communication service developed by Tencent in China, first released in January 2011. It is the largest standalone messaging app by monthly active users.

   The app is available on Android, iPhone, BlackBerry, Windows Phone and Symbian phones, and there are also Web-based and OS X clients but these require the user to have the app installed on a supported mobile phone for authentication[1]. As of August 2014, WeChat has 438 million active users; with 70 million outside of China.

   Users can register WeChat with Facebook account or sign up with phone number. WeChat currently supports phone numbers of more than 100 countries to register. Registration cannot be done directly through Tencent QQ[2]. But users can connect their WeChat account with Tencent QQ account after registering through phone number.

   WeChat provides text messaging, hold-to-talk voice messaging, broadcast (one-to-many) messaging, sharing of photographs and videos, and location sharing. It can exchange contacts with people nearby via Bluetooth, as well as providing various features for contacting people at random if desired and integration with social networking services such as those run by Facebook and Tencent QQ[3]. Photographs may also be embellished with filters and captions, and a machine translation service is available.

   WeChat supports users to register as a public account, which enables them to push feeds to subscribers, interact with subscribers and provide them with service[4]. By the end of 2014, number of Wechat public accounts had reached 8 million.

   In China, Wechat public accounts have become a common service or promotion platform for government, news media and companies. Specific public account subscribers use the platform for services like hospital pre-registration, visa renewal or credit card service.

   On 30 September 2014, WeChat 6.0 was launched with new features and functions include Sight capture and share.

   According to GlobalWebIndex, WeChat is the fifth most used smartphone app worldwide and in August 2013, following Google Maps, Facebook, YouTube and Google+[5]. WeChat claimed it had 100 million registered international users which is achieved in only 3 months from 50 million registered international users. It also claimed 300 million registered Chinese users.

According to Xinhua, WeChat total users reached 600 millions users worldwide in October 2013. In addition, approximately 30 percent of the total WeChat are users abroad.

NOTES

[1] 由 and 连接的并列句，前一句中的 The app 是指前一段中的 the largest standalone messaging app，即 WeChat，后一句中 there are…省略了 to be used by the app。Android 见 2.6 节，iPhone 见 2.5 节。 BlackBerry，加拿大 RIM(Research In Motion Ltd) 公司生产的黑莓手机。Windows Phone（WP）是微软发布的一款手机操作系统，它将微软旗下的 Xbox Live 游戏、Xbox Music 音乐与独特的视频体验集成至手机中。OS X 见 2.5 节。

Symbian 系统是塞班公司为手机而设计的操作系统。

[2] Tencent QQ，腾讯 QQ，详见 5.1.2 节。

[3] 长句。It 代表 WeChat。句子最后的 run by…过去分词短语，修饰 those。Facebook，请见 5.1.3 节。

[4] push feeds 在 SNS 和微博中都要用到 feed，在微信中 feed 也指每条微信里的新鲜事物。在 feed 架构中要用到推（push）拉（pull）模式。其中 push 模式是把一篇微信推送给所有关注的人。

[5] Global WebIndex 是一个市场研究公司，其业务主要是向广告行业提供客户数据。Google Maps（谷歌地图）是谷歌提供的电子地图服务，包括局部详细的卫星照片。此款服务可以提供含有行政区和交通以及商业信息的矢量地图、不同分辨率的卫星照片和可以用来显示地形和等高线地形的视图。You Tube 是一个共享视频的网站，该网站 2006 年 11 月被谷歌公司收购，现为谷歌公司的一个子公司。 Google+即 Google Plus，一个社交网络和为谷歌服务的社区。

KEYWORDS

WeChat	微信
message	消息，信息，电文
register	注册，寄存器
sign up	签约，参加，入队
text messaging	文本消息
broadcast	广播
photograph	照片，相片
video	视频
Bluetooth	蓝牙
random	随机的
integration	集成，整合

social networking service	社交网络服务
embellish	美化，装饰，修饰
filter	过滤器，滤波器，筛选程序
caption	标题，图片说明，字幕
subscriber	用户

EXERCISES

Fill in the blanks with appropriate words or phrases found behind this exercise.

1. WeChat is a _____.

2. The app of WeChat should install on a _____ for authentication.

3. We can register WeChat with _____.

4. User can't register WeChat directly with _____.

5. Using WeChat users can share _____.

6. With WeChat we can contact _____ at random if desired.

7. If we registered WeChat as a public account, we can interact with _____.

8. In China, WeChat _____ have become a promotion platform for government.

9. _____ possesses new features such as Sight capture and share.

10. Approximately _____ of the total WeChat are users abroad.

11. WeChat was developed by _____.

12. _____ was produced by RIM company in Canada.

  a. public accounts

  b. photographs and videos

  c. mobile text and voice messaging communication service

  d. BlackBerry

  e. WeChat6.0

  f. subscribers

  g. supported mobile phone

  h. 30%

  i. people

  j. Tencent QQ

  k. Tencent in China

  l. Facebook

5.2　SOCIAL NETWORKING SERVICE (SNS)

5.2.1　OVERVIEW OF SOCIAL NETWORKING SERVICE (SNS)

  A social networking service (also social networking site or SNS) is a platform to build social networks or social relations among people who share interests, activities, backgrounds or real-life connections[1]. A social network service consists of a representation of each user (often a profile), his or her social links, and a variety of additional services. Social network sites are web-based services that allow individuals to create a public profile, to create a list of users with whom to share connections, and view and cross the connections within the system. Most social network services provide means for users to interact over the Internet, such as e-mail and instant messaging. Social network sites are varied and they incorporate new information and communication tools such as mobile connectivity, photo/video/sharing and blogging. Online community services are sometimes considered a social network service, though in a broader sense, social network service usually means an individual-centered service whereas online community services are group-centered[2]. Social networking sites allow users to share ideas, pictures, posts, activities, events, interests with people in their network.

The main types of social networking services are those that contain category places (such as former school year or classmates), means to connect with friends (usually with self-description pages), and a recommendation system linked to trust. Popular methods now combine many of these, with American-based services such as Facebook, Google+, LinkedIn, Instagram, Reddit, Pinterest, Vine, Tumblr, and Twitter widely used worldwide[3].

One of the fastest-growing uses of the Internet is social networking, or connecting individuals to one another. There are three basic categories of social networking sites: reuniting, friend-of-a-friend, and common interest.

1. Reuniting sites

Reuniting sites are designed to connect people who have known one another but have lost; for example, an old high school friend that you have not seen for several years. You join a social network by connecting to a reuniting site and providing profile information such as your age, gender, name of high school, and so forth. This information is added to the reuniting site's member database. Members are able to search the database to locate individuals. Many of the sites will even notify you whenever a new individual joins that matches some parts of your profile (such as high school class)[4]. Two of the best-known reuniting sites are Classmates Online and Facebook[5].

2. Friend-of-a-friend sites

Friend-of-a-friend sites are designed to bring together two people who do not know one another but share a common friend. The theory is that, if you share a common friend, then it is likely that you would become friends[6]. For example, a network could be started by one of your acquaintances by providing profile information on him-or herself and a list of friends. You could visit your acquaintance's site to connect to a friend(s) of your acquaintance. You could even join the list of friends provided at the site. Two well-known friend-of-a-friend sites are Friendster and MySpace(See Figure 5-2).

Screenshot of the current Myspace homepage

Figure 5-2　Friend-of-a-friend site

3. Common interest sites

Common interest sites bring together individuals that share common interests or hobbies. You select a networking site based on a particular interest. For example, if you wanted to share images, you might join Flickr[7]. If you are looking for business contact, you might join LinkedIn. If wanted to locate or create a special interest group, you might join Meetup.

Before providing any information to a social networking site or to any individual, consider carefully what you are disclosing. Do not provide inappropriate or overly personal information.

NOTES

[1] who 引导的是定语从句，修饰 people。

[2] 长句，though 引导的是让步状语从句。

[3] LinkedIn（领英）创建于 2002 年，致力于向全球职场人士提供沟通平台。作为全球最大的职业社交网站，LinkedIn 会员人数在世界范围内已超过 3 亿。Instagram 是一款运行在 iOS 平台上的移动应用。Reddit 是一个社交新闻站点，用户确认能够浏览并且可以提交因特网上内容的链接或发布自己的原创或有关用户提交文本的帖子。Pinterest 采用的是瀑布流的形式展现图片内容，无须用户翻页，新的图片不断自动加载在页面底端，让用户不断地发现新图片，堪称图片版的 Twitter。Vine 是微软公司开发基于地理位置的 SNS 系统，类似于 Twitter 服务。Tumblr（汤博乐）是一种介于传统博客和微博之间的全新媒体形态，是当前最受年轻人欢迎的社交网站之一。

[4] whenever…是宾语从句；that 引导的是定语从句，修饰 individual。

[5] 有关 Facebook 的内容，请见本章 5.1.3 节。

[6] that 引导的是表语从句，此从句的开始部分是条件状语从句。

[7] Flickr 见 5.1.3 节 NOTES[5]。

KEYWORDS

reuniting site	再重聚网站
database	数据库
profile	简表，概貌，配置文件
SNS (Social Networking Service)	社交网络服务
platform	平台
background	背景，桌面背景，网页背景，后台
representation	表示，代表
online community	在线社区，在线团体

EXERCISES

Fill in the blanks with terms, words or phrases found behind this exercise.

1. Social networking sites have three basic categories: _____.
2. Reuniting sites are used to connect people who have known one another but have _____.
3. Using reuniting site, your personal information must store into the reuniting site's _____.
4. Two of the best-known reuniting sites are _____.
5. Friend-of-a-friend sites are designed to bring together two people who do not know one another but _____.
6. A network containing the Friend-of-a-friend sites could be started by one of your _____.
7. Common interest sites bring together individuals that _____.
8. If you are looking for business contact, you might join _____.
9. Social networking service is also known as _____.
10. Social networking service consists of a _____.
11. Social networking sites are _____.
12. Most Social network services provide services of _____ over the Internet.
13. Social network sites incorporate _____.
14. In social network sites we can find _____ and so forth.

    a. lost
    b. LinkedIn
    c. acquaintances
    d. reuniting, friend-of-a-friend, and common interest
    e. share a common friend
    f. share common interests or hobbies
    g. Classmates Online and Facebook
    h. member database
    i. category place, means to connect with friends

j. social networking site

k. e-mail and instant messaging

l. web-based services

m. mobile connectivity, photo/video/sharing and blogging

n. profile, user social links and a variety of additional services

5.2.2 WIKI

A wiki is a website that allows the creation and editing of any number of interlinked web pages via a web browser using a simplified markup language or a WYSIWYG text editor.[1] Wikis are typically powered by wiki software and are often used collaboratively by multiple users. Examples include community websites, corporate intranets, knowledge management systems, and note services[2]. The software can also be used for personal notetaking.

Wikis serve different purposes. Some permit control over different functions (levels of access). For example, editing rights may permit changing, adding or removing material. Others may permit access without enforcing access control.

The essence of the wiki concept is as follows:

(1) A wiki invites all users to edit any page or to create new pages within the wiki Web site, using only a plain-vanilla Web browser without any extra add-ons[3].

(2) Wiki promotes meaningful topic associations between different pages by making page link creation almost intuitively easy and showing whether an intended target page exists or not[4].

(3) A wiki is not a carefully crafted site for casual visitors. Instead, it seeks to involve the visitor in an ongoing process of creation and collaboration that constantly changes the Web site landscape.

A single page in a wiki website is referred to as a "wiki page", while the entire collection of pages, which are usually well interconnected by hyperlinks, is "the wiki"[5]. A wiki is essentially a database for creating, browsing, and searching through information.

A defining characteristic of wiki technology is the ease with which pages can be created and updated. Generally, there is no review before modifications are accepted. Many wikis are open to alteration by the general public without requiring them to register user accounts. Sometimes logging in for a session is recommended, to create a "wiki-signature" cookie for signing edits automatically. Many edits, however, can be made in real-time and appear almost instantly online. This can facilitate abuse of the system. Private wiki servers require user authentication to edit pages, and sometimes even to read them.

A wiki is an application also, typically a web application, which allows collaborative modification, extension, or deletion of its content and structure. In a typical wiki, text is written using a simplified markup language (known as "wiki markup") or a rich-text editor.

While a wiki is a type of content management system, it differs from a blog or most other such systems in that the content is created without any defined owner or leader, and wikis have little implicit structure, allowing structure to emerge according to the needs of the users [6].

The encyclopedia project Wikipedia is the most popular wiki on the public web in terms of page views, but there are many sites running many different kinds of wiki software. Wikis can serve many different purposes both public and private, including knowledge management, notetaking, community websites and intranets. Some permit control over different functions (levels of access). For example, editing rights may permit changing, adding or removing material. Others may permit access without enforcing access control. Other rules may also be imposed to organize content.

NOTES

[1] 长句。that 引导定语从句，修饰 website；via…using…均为状语。WYSIWYG 的原文是"What You See Is What You Get"，即所见即所得。

[2] intranet 为因特网内联网，或称内部网、内联网、内网，是一个使用与因特网同样技术的计算机网络，它通常建立在一个企业或组织的内部，并为其成员提供信息的共享和交流等服务，例如万维网服务、文件传输、电子邮件等，是一种企业内部各个部门相互连接而形成的内部网。

[3] add-ons，附件，附加。

[4] by making…and showing…，方式状语。

[5] which 引导的是非限定性定语从句。

[6] 长句。While 引导的是让步状语从句，表示虽然……。in that 后面的结果状语从句是并列的，由 and 连接。

KEYWORDS

interlink	相互链接
community website	团体网站
intranet	内联网
add-on	附件
landscape	景色，风景
hyperlink	超链接
database	数据库
searching	搜索
user account	用户账户
log	登录
cookie	网络跟踪器，"小甜饼"
authentication	验证，鉴别
markup language	标记（置标）语言

text editor	文本编辑程序，文本编辑器
implicit structure	隐式结构
emerge	出现，露出，显现，发生
encyclopedia	百科全书
access	访问，接入，存取

EXERCISES

True/False

1. _____ With a markup language we can create and edit a lot of interlinked Web pages within the Wiki site.

2. _____ Wikis serve different purpose.

3. _____ Levels of access in wiki mean different functions.

4. _____ In wiki users who edit or create Web pages must use extra add-ons.

5. _____ Wiki can make decision to determine whether an intended target page exists or not.

6. _____ A wiki is a carefully crafted site for casual visitors.

7. _____ A single page in a wiki Web site is referred to as a "wiki page".

8. _____ Term "the wiki" means that the entire collection of pages are usually well interconnected by hyperlinks.

9. _____ When altering an opened wiki, we must register user accounts.

10. _____ Private wiki servers require user authentication to edit pages.

11. _____ When a wiki is used as a content management system, it is the same as a blog.

12. _____ Wiki allows collaborative modification, extension, or deletion of its content and structure.

13. _____ Wiki markup language is a simplified markup one.

14. _____ When we use wiki to create a file, we need to define the file's owner or leader.

15. _____ Wikis have been allowed its structure to be emerged.

16. _____ Wikipedia is an encyclopedia project.

17. _____ Wikis can serve many different purposes in public only.

18. _____ Some of wikis should permit to change the editing rights.

5.2.3　BLOG AND MICROBLOG

1. Blog

A blog (a blend of the term web and log) is a type of website or part of a website. Blogs are usually maintained by an individual with regular entries of commentary, descriptions of events, or other material such as graphics or video. Entries are commonly displayed in reverse-chronological order. Blog can also be used as a verb, meaning to maintain or add content to a blog.

Most blogs are interactive, allowing visitors to leave comments and even message each other via widgets on the blogs and it is this interactivity that distinguishes them from other static websites[1].

Many blogs provide commentary or news on a particular subject; others function as more personal online diaries[2]. A typical blog combines text, images, and links to other blogs, Web pages, and other media related to its topic. The ability of readers to leave comments in an interactive format is an important part of many blogs. Most blogs are primarily textual, although some focus on art (art blog), photographs (photoblog), videos (video blogging), music (MP3 blog), and audio (podcasting)[3]. Microblogging is another type of blogging, featuring very short posts.

As of 16 February 2011, there were over 156 million public blogs in existence. On 20 February 2014, there were around 172 million Tumblr and 75.8 million WordPress blogs in existence worldwide [4]. According to critics and other bloggers, Blogger is the most popular blogging service used today, however Blogger does not offer public statistics. Technorati has 1.3 million blogs as of February 22, 2014 [5].

There are many different types of blogs, differing not only in the type of content, but also in the way that content is delivered or written.

2. Microblog

Microblog is a broadcast medium in the form of blog. A microblog differs from a traditional blog in that its content is typically smaller in actual file size[6]. Microblogs "allow users to exchange small elements of content such as short sentences, individual images, or video links".

As with traditional blog, microbloggers post about topics ranging from the simple, such as "what I'm doing right now," to the thematic, such as "sports cars"[7]. Commercial microblogs also exist, to promote websites, services and/or products, and to promote collaboration within an organization.

Some microblog services offer features such as privacy settings, which allow users to control who can read their microblogs, or alternative ways of publishing entries besides the web-based interface[8]. These may include text messaging, instant messaging, E-mail, or digital audio.

Microblog services have revolutionized the way information is consumed. It has empowered citizens themselves to act as sensors or sources of data which could lead to important pieces of information. People now share what they observe in their surroundings, information about events, and what their opinions are about certain topics, for example, government policies in healthcare[9].

Moreover, these services store various metadata from these posts, such as location and time. Aggregated analysis of this data includes different dimensions like space, time, theme, sentiment, network structure etc., and gives researchers an opportunity to understand social perceptions of people in the context of certain events of interest [10]. Microblogging also promotes authorship. On the microblogging platform Tumblr, the reblogging feature links the post back to the original creator [11].

Microblogging has the potential to become a new, informal communication medium, especially for collaborative work within organizations. Over the last few years, communication patterns have shifted primarily from face-to-face to online in email, IM, text messaging, and other tools. However, some argue that email is now a slow and inefficient way to communicate. For instance, time-consuming "email chains" can develop, whereby two or more people are involved in lengthy communications for simple matters, such as arranging a meeting [12]. The one-to-many broadcasting offered by microblogs is thought to increase productivity by circumventing this.

Users and organizations can set up their own microblog service: free and open source software is available for this purpose. Hosted microblog platforms are also available for commercial and organizational use.

NOTES

[1] 并列长句。allowing…分词短语做状语；后一句中 that 引导的是定语从句，修饰 interactivity。

[2] function 此处为动词，意为"尽职责，起作用"；others 代表另外一些博客。

[3] art blog，艺博；photoblog，影博；video blogging，有时简称 vlogging 或 vidding 或 vidblogging，这种微博的媒体形式是视频或网络电视。

podcasting 是一种数字媒体广播技术（非流式网络广播），而 podcast 这种数字媒体文件按插曲出售，通常也可以通过万维网联合组织（web syndication）下载。

[4] Tumblr（汤博乐）成立于 2007 年，是目前全球最大的轻博客网站，也是轻博客网站的始祖。Tumblr 是一种介于传统博客和微博之间的全新媒体形态，既注重表达，又注重社交，而且注重个性化设置，成为当前最受年轻人欢迎的社交网站之一。雅虎公司董事会 2013 年 5 月 19 日决定，以 11 亿美元收购 Tumblr。WordPress 是一种使用 PHP 语言开发的博客平台，用户可以在支持 PHP 和 MySQL 数据库的服务器上架设属于自己的网站。也可以把 WordPress 当作一个内容管理系统（CMS）来使用。

[5] Technorati 是一个著名的博客搜索引擎，截至 2007 年 4 月，Technorati 已经索引了超过 7000 万个博客站点。可以说，Technorati 已经成为世界上最重要的博客搜索引擎之一。

[6] in that 引导的是状语从句。

[7] As 引导的是比较状语，主句中 ranging…为现在分词短语做定语；句中有两个

such as 构成的同位语。sports cars，跑车，赛车。

　　[8] such as privacy settings，…or alternative ways…为同位语； which 引导的是非限定性定语从句，who 引导的是宾语从句。

　　[9] 两个 what 引导的都是宾语从句。

　　[10] 由 and 连接的长句，主语是 Aggregated analysis of…。

　　[11] reblogging，是博客中的一种机制，它允许用户转发另一个用户博客的内容，并指出那个用户是信息源。

　　[12] whereby，此处的用法相当于 by which。

KEYWORDS

entry	条目，词条，入口，进入
commentary	评论，注释
reverse-chronological order	反时间顺序
interactive	交互式的
widget	小装置（窗口），窗件
art blog	艺博
photoblog	影博
video blogging	视博
podcasting	Pod 广播技术
blogger	博客版主
statistics	统计，统计数字，统计表，统计学
microblog	微博
aggregate file	聚合文件
post	邮件，邮箱
microblogger	微博版主
open source software	开源软件
instant messaging	即时消息
digital audio	数字音频
authorship	著述业，著者，来源
circumventing	阻遏，设法规避，欺诈，绕行
hosted	托管的

EXERCISES

Multiple Choices

1. Blog is _____.
　　a. a type of website　　　　　　　b. maintained usually by an individual
　　c. a blend of the term web and log　d. a part of a website

2. A blog consists of _____.
   a. regular entries of commentary
   b. descriptions of events
   c. E-mail
   d. graphics or video
3. Most blogs _____.
   a. are similar to other static websites
   b. allow visitors to leave comments
   c. are interactive
   d. use widgets on the blogs
4. A typical blog links to _____.
   a. other blogs
   b. text
   c. web pages
   d. other media related to its topic
5. The following is belong to special blog _____.
   a. art blog
   b. photo blog
   c. video blogging
   d. MP3 blog
6. Difference among blogs is in _____.
   a. the type of content
   b. the way that content is delivered
   c. the way that content is written
   d. the way of linking
7. Microblog _____.
   a. is a broadcast medium in the form of blog
   b. allows user to exchange short sentences
   c. allows user to exchange longer sentences
   d. allows user to exchange individual images
8. Commercial microblogs can _____.
   a. replace traditional blog
   b. promote websites
   c. promote collaboration within an organization
   d. promote services
9. Entries published by microblog include _____.
   a. text
   b. instant messaging
   c. E-mail
   d. digital audio
10. Microblog services _____.
    a. have empowered citizens themselves to act as sensors of data
    b. have revolutionized the way information is consumed
    c. can be set up by organizations
    d. can be the hosted microblog platforms
11. Metadata gathered from Microblog services includes _____.
    a. space
    b. time
    c. theme
    d. network structure
12. Communication patterns have shifted to _____.
    a. face-to-face
    b. online e-mail
    c. Instant Messaging
    d. text messaging

5.3 CLOUD COMPUTING

Cloud computing is the delivery of computing as a service rather than a product, whereby shared resources, software, and information are provided to computers and other devices as a utility (like the electricity grid) over a network (typically the Internet)[1].

1. Overview

Cloud computing is a marketing term for technologies that provide computation, software, data access, and storage services that do not require end-user knowledge of the physical location and configuration of the system that delivers the services[2]. A parallel to this concept can be drawn with the electricity grid, wherein end-users consume power without needing to understand the component devices or infrastructure required to provide the service[3].

Cloud computing describes a new supplement, consumption, and delivery model for IT services based on Internet protocols, and it typically involves provisioning of dynamically scalable and often virtualised resources[4]. It is a byproduct and consequence of the ease-of-access to remote computing sites provided by the Internet. This may take the form of web-based tools or applications that users can access and use through a web browser as if the programs were installed locally on their own computers[5].

Cloud computing providers deliver applications via the internet, which are accessed from web browsers and desktop and mobile apps, while the business software and data are stored on servers at a remote location. In some cases, legacy applications (line of business applications that until now have been prevalent in thin client Windows computing) are delivered via a screen-sharing technology, while the computing resources are consolidated at a remote data center location; in other cases, entire business applications have been coded using web-based technologies such as AJAX[6].

At the foundation of cloud computing is the broader concept of infrastructure convergence (or Converged Infrastructure) and shared services[7]. This type of data center environment allows enterprises to get their applications up and running faster, with easier manageability and less maintenance, and enables IT to more rapidly adjust IT resources (such as servers, storage, and networking) to meet fluctuating and unpredictable business demand[8].

Most cloud computing infrastructures consist of services delivered through shared data-centers and appearing as a single point of access for consumers' computing needs[9].

The tremendous impact of cloud computing on business has prompted the federal United States government to look to the cloud as a means to reorganize their IT infrastructure and decrease their spending budgets. With the advent of the top government official mandating cloud adoption, many agencies already have at least one or more cloud systems online.

A cloud computing logical diagram shows in Figure 5-3.

Figure 5-3　Cloud computing logical diagram

2. Public, Private, and Hybrid Clouds

A public cloud is built over the Internet and can be accessed by any user who has paid for the service. Public clouds are owned by service providers and are accessible through a subscription. The callout box in top of Figure 5-4 shows the architecture of a typical public cloud. Many public clouds are available, including Google App Engine (GAE), Amazon Web Services (AWS), Microsoft Azure, IBM Blue Cloud, and Salesforce.com's Force.com[10]. The providers of the aforementioned clouds are commercial providers that offer a publicly accessible remote interface for creating and managing VM instances within their proprietary infrastructure[11]. A public cloud delivers a selected set of business processes. The application and infrastructure services are offered on a flexible price-per-use basis.

A private cloud is built within the domain of an intranet owned by a single organization[12]. Therefore, it is client owned and managed, and its access is limited to the owning clients and their partners. Its deployment was not meant to sell capacity over the Internet through publicly accessible interfaces. Private clouds give local users a flexible and agile private infrastructure to run service workloads within their administrative domains. A private cloud is supposed to deliver more efficient and convenient cloud services. It may impact the cloud standardization, while retaining greater customization and organizational control. Intranet-based private clouds are linked to public clouds to get additional resources.

A hybrid cloud is built with both public and private clouds, as shown at the lower-left corner of Figure 5-4. Private clouds can also support a hybrid cloud model by supplementing local infrastructure with computing capacity from an external public cloud. For example, the Research Compute Cloud (RC2) is a private cloud, built by IBM, that interconnects the computing and IT resources at eight IBM Research Centers scattered throughout the United States, Europe, and Asia[13]. A hybrid cloud provides access to clients, the partner network, and third parties.

Figure 5-4   public, private, and hybrid clouds illustrated by functional architecture and connectivity of representative clouds

In summary, public clouds promote standardization, preserve capital investment, and offer application flexibility. Private clouds attempt to achieve customization and offer higher efficiency, resiliency, security, and privacy. Hybrid clouds operate in the middle, with many compromises in terms of resource sharing.

3. Characteristics

Cloud computing exhibits the following key characteristics:
(1) Empowerment of end-users of computing resources by putting the provisioning of

those resources in their own control, as opposed to the control of a centralized IT service.

(2) Agility improves with users' ability to re-provision technological infrastructure resources.

(3) Application programming interface (API) accessibility to software that enables machines to interact with cloud software in the same way the user interface facilitates interaction between humans and computers[14]. Cloud computing systems typically use REST-based APIs[15].

(4) Device and location independence enable users to access systems using a web browser regardless of their location or what device they are using (e.g., PC, mobile phone). As infrastructure is off-site (typically provided by a third-party) and accessed via the Internet, users can connect from anywhere.

(5) Multi-tenancy enables sharing of resources and costs across a large pool of users.

(6) Reliability is improved if multiple redundant sites are used, which makes well-designed cloud computing suitable for business continuity and disaster recovery.

(7) Performance is monitored, and consistent and loosely coupled architectures are constructed using web services as the system interface.

(8) Security could improve due to centralization of data, increased security-focused resources, etc.

(9) Maintenance of cloud computing applications is easier, because they do not need to be installed on each user's computer.

NOTES

[1] 长句，whereby 引导的是非限定性定语从句，whereby 的作用相当于 by which；as 引导的是方式状语。

[2] 长句，句中有三个 that 引导的定语从句。

[3] wherein 引导的是非限定性定语从句，wherein 的作用相当于 in which。

[4] 长句，由 and 连接的并列句；后一句中 it 代表 cloud computing；Internet protocols 实际上是指云计算是在因特网上实现的。

[5] This 代表 cloud computing；that 引导的定语从句，修饰 applications；as if…条件状语。

[6] 并列长句，中间由 while 连接，括号中 that 引导的定语从句，修饰 applications。thin client，瘦客户（lean client or slim client）是一种特别依赖于其他计算机（服务器）去完成通常的计算任务的计算机或程序。与其相对应的是胖客户（fat client），胖客户计算机则主要靠自己去完成计算任务。screen-sharing，屏幕共享，是苹果公司开发的，并作为 VNC 客户机纳入 Mac OS X 10.5 版，只要激活屏幕共享程序即可远程观察和控制任何在局域网上的 Mac 机。AJAX 的解释请见 5.1.3 节 NOTES[8]。

[7] 此句为倒装句，At the foundation of cloud computing 是表语；主语是 the broader concept of…。infrastructure convergence，基础会聚或称会聚的基础（converged infrastructure），是 IT 部门用来集中管理 IT 资源，联合各个系统，提高资源利用率和降

低成本的。

[8] 长句。句子结构为 This type of…allows…, and enables…; and 连接的前后两个句子中，分别有 with…和 to meet…两个状语。

[9] delivered…过去分词短语做定语，修饰 services; appearing…现在分词短语做状语。

[10] Google App Engine(GAE)，是一种可以在谷歌的基础架构上运行的网络应用程序。Google App Engine 应用程序易于构建和维护，并可根据访问量和数据存储需要的增长轻松扩展。Amazon Web Services(AWS)，是亚马逊提供的一种云计算服务。通过这个平台，可以将多余的资源提供给有需要的人。Microsoft Azure 是微软基于云计算的操作系统，它和 Azure Services Platform 一样，是微软"软件和服务"技术的名称。Windows Azure 的主要目标是为开发者提供一个平台，帮助开发可运行在云服务器、数据中心、Web 和 PC 上的应用程序。IBM Blue Cloud 是 IBM 推出的"蓝云"计划，它为客户带来即可使用的云计算。Salesforce 是创建于 1999 年 3 月的一家客户关系管理(CRM) 软件服务提供商，总部设于美国旧金山，可提供随需应用的客户关系管理平台。

[11] VM(Virtual Machine)即虚拟机，指通过软件模拟的、具有完整硬件系统功能的、运行在一个完全隔离环境中的完整计算机系统。

[12] intranet，内联网，有关解释，见 5.2.2 节 NOTES [2]。

[13] 长句。中间插入 built by IBM, that 引导的是定语从句，修饰 private cloud。IBM Research Center 是 IBM 公司的研究和开发部门，是世界上最大的研究机构之一。

[14] 长句,句中 that 引导的定语从句修饰 accessibility; 后面的 in the same way the user interface…句中 the user interface 前省略了 by which。

[15] REST 的原文是 Representational State Transfer，它是分布式超媒体系统（如万维网）的一种软件体系结构格式。

KEYWORDS

cloud computing	云计算
delivery	交付，投递
electricity grid	电网
utility	实用程序
end-user	最终用户，直接用户
thin client	瘦客户
data center	数据中心
manageability	可管理性
maintenance	维护
infrastructure	基础设施，基础结构
infrastructure convergence	基础会聚
public cloud	公共云
subscription	订购，订阅

accessible	可访问的，可接入的
callout	标注
architecture	体系结构，架构
VM(Virtual Machine)	虚拟机
private cloud	私有云
client	客户
hybrid cloud	混合云
RC2(Research Compute Cloud)	研究计算云
scatter	撒播，分散
pool of users	用户群
redundant	冗余的

EXERCISE

Multiple Choices

1. Cloud computing is _____.
    a. a marketing term for technologies   b. the delivery of computing
    c. a product   d. a new supplement for IT services
2. With cloud computing we can share _____.
    a. resources   b. software   c. information   d. services
3. Using cloud computing we should _____.
    a. know the physical location of the system
    b. not know the physical location of the system
    c. connect to Internet
    d. know the configuration of the system
4. Cloud computing deliveries _____.
    a. virtualized resources
    b. dynamically scalable resources
    c. the form of Web-based tools
    d. a delivery model for IT services
5. The applications delivered by cloud computing providers can be accessed from _____.
    a. a person who has not any of electronic devices
    b. mobile apps
    c. desktop computer
    d. Web browser
6. In cloud computing we can encounter _____.
    a. legacy applications   b. screen-sharing technology
    c. Web-based technologies   d. data-centers
7. One type of data center environment allows enterprises to _____.

a. get their applications up
   b. run applications slower
   c. manage applications easier
   d. adjust IT resources more rapidly
8. The tremendous impact of cloud computing on business has prompted _____ in this text.
   a. British
   b. federal U.S. government
   c. many agencies
   d. state government in U.S.
9. A public cloud _____.
   a. can be accessed free by any user
   b. can be accessed by any user who has paid for the service
   c. can be accessed through a subscription
   d. is owned by service provider
10. Right now we can access the following public clouds _____.
    a. IBM Blue Cloud
    b. Salesforce
    c. Microsoft Azure
    d. Amazon Web Services
11. A commercial public cloud provides _____.
    a. an infrastructure service on a flexible price-per-use basis
    b. an infrastructure service on a fixed price-per-use basis
    c. a public accessible remote interface
    d. a selected set of business processes
12. A private cloud is _____.
    a. client owned and managed
    b. server owned and managed
    c. owned by a single organization
    d. owned by multiple organizations
13. A private cloud can be accessed by _____.
    a. any public user
    b. the owning clients
    c. their partners
    d. users within their administrative domains
14. A private cloud _____.
    a. may impact the cloud standardization
    b. retain greater customization and organizational control
    c. is Internet-based
    d. is Intranet-based
15. A hybrid cloud _____.
    a. is built with both public and private clouds
    b. can be supported by private clouds
    c. provides access to clients
    d. provides access to third parties
16. In cloud computing the computing resources can be _____.
    a. controlled by users themselves
    b. controlled by centralized IT service
    c. re-provided to consumers
    d. accessed by APIs
17. Accessing cloud computing systems, we should not concern _____.

    a. the location of required devices   b. what device is used
    c. if the Internet is connected or not   d. if the used browser is online or not

18. In cloud computing the following features are improved _____.
    a. security   b. reliability   c. agility   d. maintenance

5.4 BIG DATA

Big data is a broad term for data sets so large or complex that traditional data processing applications are inadequate[1]. Challenges include analysis, capture, curation, search, sharing, storage, transfer, visualization, and information privacy[2]. The term often refers simply to the use of predictive analytics or other certain advanced methods to extract value from data, and seldom to a particular size of data set.

Analysis of data sets can find new correlations, to "spot business trends, prevent diseases, combat crime and so on." Scientists, Practitioners of Media and Advertising and Governments alike regularly encounter limitations due to large data sets in many areas. The limitations affect Internet search, finance and business informatics.

Scientists, for example, encounter limitations in e-Science work, including meteorology, genomics, connectomics, complex physics simulations, and biological and environmental research[3].

Data sets grow in size in part because they are increasingly being gathered by cheap and numerous mobile devices, aerial sensory technologies (remote sensing), software logs, cameras, microphones, radio-frequency identification (RFID) readers, and wireless sensor networks[4]. The world's technological per-capita capacity to store information has roughly doubled every 40 months since the 1980s; as of 2012, every day 2.5 exabytes (2.5×10^{18}) of data were created; the challenge for large enterprises is determining who should own big data initiatives that straddle the entire organization[5].

Figure 5-5 shows the growth of global information storage capacity.

1. Definition

Big data is a popular term used to describe the exponential growth, availability and use of information, both structured and unstructured. According to IDG, it is imperative that organizations and IT leaders focus on the ever-increasing volume, variety, and velocity of information that forms big data[6].

(1) Volume

Many factors contribute to the increase in data volume-transaction-based data stored through the years, text data constantly streaming in from social media, increasing amounts of sensor data being collected, etc. In the past, excessive data volume created a storage issue. But with today's decreasing storage costs, other issues emerge, including how to determine relevance amidst the large volumes of data and how to create value from that is relevant.

Figure 5-5 Growth and Digitization of Global Information Storage Capacity Source

(2) Variety

Data today comes in all types of formats—from traditional databases to hierarchical data stores created by end users and OLAP systems, to text documents, E-mail, meter-collected data, video, audio, stock ticker data and financial transactions[7]. By some estimates, 80 percent of an organization's data is not numeric! But it still must be included in analyses and decision making.

(3) Velocity

According to Gartner, velocity "means both how fast data is being produced and how fast the data must be processed to meet demand". RFID tags and smart metering are driving an increasing need to deal with torrents of data in near-real time.

2. Uses for big data

So the real issue is not that you are acquiring large amounts of data (because we are clearly already in the era of big data). It's what you do with your big data that matters[8]. The hopeful vision for big data is that organizations will be able to harness relevant data and use it to make the best decisions.

Technologies today not only support the collection and storage of large amounts of data, they provide the ability to understand and take advantage of its full value, which helps organizations run more efficiently and profitably[9]. For instance, with big data and big data analytics, it is possible to:

- Analyze millions of SKUs to determine optimal prices that maximize profit and clear inventory[10].
- Recalculate entire risk portfolios in minutes and understand future possibilities to mitigate risk.
- Quickly identify customers who matter the most.
- Generate retail coupons at the point of sale based on the customer's current and past purchases, ensuring a higher redemption[11].
- Send tailored recommendations to mobile devices at just the right time, while customers are in the right location to take advantage of offers.
- Analyze data from social media to detect new market trends in demand.
- Use clickstream analysis and data mining to detect fraudulent behavior[12].
- Determine root causes of failures, issues and defects by investigating user sessions, network logs and machine sensors.

Examples of big data:

- RFID systems generate up to 1000 times the data of conventional bar code systems.
- 10000 payment card transactions are made every second around the world.
- Walmart handles more than 1 million customer transactions an hour.
- 340 million tweets are sent per day. That's nearly 4000 tweets per second[13].
- Facebook has more than 901 million active users generating social interaction data.
- More than 5 billion people are calling, texting, tweeting and browsing websites on mobile phones.

3. Technologies

A number of recent technology advancements are enabling organizations to make the most of big data and big data analytics:

- Cheap, abundant storage and server processing capability.
- Faster processors.
- Affordable large-memory capabilities, such as Hadoop[14].
- New storage and processing technologies designed specifically for large data volumes, including unstructured data.
- Parallel processing, clustering, MPP, virtualization, large grid environments, high connectivity and high throughputs[15].
- Cloud computing and other flexible resource allocation arrangements.

Big data technologies not only support the ability to collect large amounts of data, they provide the ability to understand it and take advantage of its value. The goal of all organizations with access to large data collections should be to harness the most relevant data and use it for optimized decision making.

It is very important to understand that not all of your data will be relevant or useful. But how can you find the data points that matter most? It is a problem that is widely acknowledged. "Most businesses have made slow progress in extracting value from big data. And some companies attempt to use traditional data management practices on big data, only to learn that the old rulers no longer apply," says Dan Briody, in the 2011 Economist Intelligence Unit's publication, "Big Data Harnessing a Game-Changing Asset"[16].

NOTES

[1] that 引导的是结果状语从句。

[2] curation，策展，即策划、筛选并展示的意思。早期定义是艺术展览活动中的构思、组织、管理工作。

[3] e-Science，由英国在 2000 年提出，是为了应对当时各学科研究领域所面临问题的空前复杂化，利用新一代网络技术（Internet）和广域分布式高性能计算环境（Grid）建立的一种全新科学研究模式，即在信息化基础设施支持下的科学研究活动。connectomics（连接组学）是近年来一系列生命科学研究中的一个分支。

[4] 长句。because 引导的是原因状语从句。有关 RFID 的内容，请见 4.3.2 节，NOTES[4]。

[5] 用两个分号隔开的长句。最后一句中 who 引导的是宾语从句。

[6] 第一个 that 引导的是主语从句，第二个 that 引导的是定语从句，修饰 information。IDG（International Data Group）即国际数据集团，是全世界最大的信息技术出版、研究、发展与风险投资公司。IDG 公司 2005 年全球营业总收入达到 26.8 亿美元。IDG 集团公司创建于 1964 年，总部设在美国波士顿。

[7] 长句。OLAP(On-Line Analytical Processing)，联机分析处理，是数据仓库系统最主要的应用，专门设计用于支持复杂的分析操作，侧重对决策人员和高层管理人员的决策支持。

[8] what 引导的是表语从句。 that 引导的是定语从句，修饰 big data。matter，此处为关系重大之意。

[9] 长句。which 引导的是非限定性定语从句。

[10] SKU(Stock Keeping Unit)，库存量单位，即库存进出计量的单位，可以以件、盒、托盘等为单位。

[11] redemption，兑换率，回收，偿还。

[12] clickstream，点击流，一个用来跟踪正在你的站点上访问的用户所到达位置细节的工具。它允许你跟踪访问你的站点的"点击流"或者"传输路径"。data mining，数据挖掘，又译为资料探勘、数据采矿。它是数据库知识发现中的一个步骤。数据挖掘一般是指从大量的数据中通过算法搜索隐藏于其中信息的过程。

[13] tweet，推文，有关它的内容请见 5.1.4 节。

[14] Hadoop,由 Apache 基金会开发的分布式系统基础架构,用户可以在不了解分布式底层细节的情况下,开发分布式程序。充分利用群集的威力进行高速运算和存储。

[15] MPP (Massively Parallel Processing)意为大规模并行处理系统,这样的系统是由许多松耦合处理单元组成的。

[16] 此处引用 Dan Briody 说的原话,前后两段分别用引号标明,中间是插入语:says Dan Briody, in…。 Economist Intelligence Unit's publication,经济学人智库出版物。

KEYWORDS

big data	大数据
data set	数据集,数据传输机
capture	捕获,捕捉,截获
curation	策展
visualization	可视化,目视,显像
predictive analytics	预测分析
informatics	信息控制论,信息科学,情报学,资料学
genomics	基因学
connectomics	连接组学
biological	生物学的,生物的
remote sensing	远程传感技术
RFID(Radio Frequency Identification)	射频识别
per-capita	人均的,人均
exabyte(10^{18} byte)	艾字节
exponential	指数,指数的,幂
IDG(International Data Group)	国际数据集团
stream	流,序列
social media	社交媒体
sensor	传感器
database	数据库
hierarchical	层次的,分层的
OLAP(On-Line Analytical Processing)	联机分析处理
smart metering	智能仪表
parallel processing	并行处理
clustering	聚类,分类归并,群集,分群,分组,划分
MPP(Massively Parallel Processing)	大规模并行处理系统
grid	格栅,栅极
throughput	吞吐量,通过量,总处理能力
decision making	决策支撑

EXERCISES
True/False

1. _____ Big data can be used to process very complex data.
2. _____ Usually big data is directed against a particular size of data set.
3. _____ Analysis of data set can find new correlations among events.
4. _____ Scientists alike regularly encounter limitations as large data set.
5. _____ e-Science is a new idea introduced by America.
6. _____ It is important to possess the big data initiatives for large enterprises.
7. _____ The ever-increasing volume, variety, and velocity of information promote big data.
8. _____ Right now excessive data volume creates storage issue.
9. _____ On-line Analytical Processing (OLAP) is the major application of data mining technology.
10. _____ About 80% of an organization's data is numeric.
11. _____ Velocity in big data means how fast data is being produced only.
12. _____ We are clearly already in the era of big data.
13. _____ Organizations can use big data to make the best decisions.
14. _____ Stock Keeping Unit (SKU) is used to determine optimal prices.
15. _____ A point of sale should generate retail coupons only based on the customer's current purchases.
16. _____ In order to detect fraudulent behavior we can use clickstream analysis.
17. _____ We have listed six samples of big data.
18. _____ Cloud computing can help organizations to make the most of big data.
19. _____ Big data technologies only support the ability to collect large amounts of data.
20. _____ Most businesses have made fast progress in extracting value from big data.

5.5 MOOC AND FLIPPED CLASS ROOM

1. MOOC

A massive open online course is an online course aimed at unlimited participation and open access via the web. In addition to traditional course materials such as filmed lectures, readings, and problem sets, many MOOCs provide interactive user forums to support community interactions between students, professors, and teaching assistants (TAs). MOOCs is a recent development in distance education which was first introduced in 2008 and emerged as a popular mode of learning in 2012.

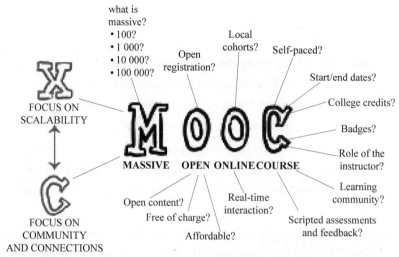

Figure 5-6　Poster, entitled "MOOC, every letter is negotiable," exploring the meaning of the words "Massive Open Online Course"

Early MOOCs often emphasized open-access features, such as open licensing of content, structure and learning goals, to promote the reuse and remixing of resources. Some later MOOCs use closed licenses for their course materials while maintaining free access for students.

Figure 5-6 gives the meaning of the words "Massive Open Online Course".

(1) Technology

Unlike traditional courses, MOOCs require additional skills, provided by videographers, instructional designers, IT specialists and platform specialists. Georgia Tech professor Karen Head reports that 19 people work on their MOOCs and that more are needed [1]. The platforms have availability requirements similar to media/content sharing websites, due to the large number of enrollees. MOOCs typically use cloud computing and are often created with authoring systems. Authoring tools for the creation of MOOCs are specialized packages of educational software like Elicitus, IMC Content Studio and Lectora that are easy-to-use and support e-learning standards like SCORM and AICC [2].

Course delivery involves asynchronous access to videos and other learning material, exams and other assessment, as well as online forums [3]. Before 2013 each MOOC tended to develop its own delivery platform. EdX in April 2013 joined with Stanford University, which previously had its own platform called Class2Go, to work on XBlock SDK, a joint open-source platform [4]. It is available to the public under the Affero GPL open source license, which requires that all improvements to the platform be publicly posted and made available under the same license [5]. Stanford Vice Provost John Mitchell said that the goal was to provide the "Linux of online learning." This is unlike companies such as Coursera that have developed their own platform [6].

(2) Potential benefits

The MOOC Guide lists 12 benefits:

- Appropriate for any setting that has connectivity (Web or Wi-Fi)
- Any language or multiple languages
- Any online tools
- Escape time zones and physical boundaries
- Produce and deliver in short timeframe
- Contextualized content can be shared by all
- Informal setting
- Peer-to-peer contact can trigger serendipitous learning
- Easier to cross disciplines and institutional barriers
- Lower barriers to student entry
- Enhance personal learning environment and/or network by participating
- Improve lifelong learning skills

2. FLIPPED CLASS ROOM

The flipped classroom describes a reversal of traditional teaching where students gain first exposure to new material outside of class, usually via reading or lecture videos, and then class time is used to do the harder work of assimilating that knowledge through strategies such as problem-solving, discussion or debates[7].

In the flipped classroom, the roles and expectations of students and teachers change where:
- students take more responsibility for their own learning and study core content either individually or in groups before class and then apply knowledge and skills to a range of activities using higher order thinking,
- significant learning opportunities can be gained through facilitating active learning, engaging students, guiding learning, correcting misunderstandings and providing timely feedback using a variety of pedagogical strategies[8],
- there is a greater focus on concept exploration, meaning making and demonstration or application of knowledge in the face-to-face setting as shown below (Figure 5-7).

Educational technologies (See Figure 5-8) are an important feature of the flipped classroom as they can be used to:
- capture key content for students to access at their own convenience and to suit their pace of learning (e.g. lecture material, readings, interactive multimedia),
- present learning materials in a variety of formats to suit different learner styles and multimodal learning (e.g. text, videos, audio, multimedia),
- provide opportunities for discourse and interaction in and out of class (e.g. polling tools, discussion tools, content creation tools),
- convey timely information, updates and reminders for students (e.g. micro-blogging, announcement tools),

- provide immediate and anonymous feedback for teachers and students (e.g. quizzes, polls) to signal revision points,
- capture data about students to analyze their progress and identify 'at risk' students (e.g. analytics).

Figure 5-7 Learning opportunities of the flipped classroom (adapted from Gerstein)Educational technologies (See Figure 5-8) are an important feature of the flipped classroom

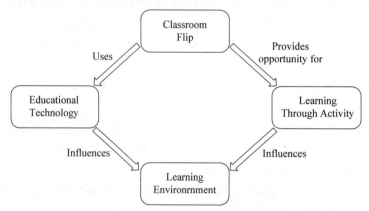

Figure 5-8 Key elements of a flipped classroom (Strayer)

NOTES

[1] 长句。Georgia Tech，即 Georgia Institute of Technology，佐治亚理工学院。句中两个 that 引导的都是宾语从句。

[2] Elicitus 为课程开发者提供了制作导航、模板界面以及几十种界面风格，使课程开发者无须了解 HTML 编程语言就能制作出风格多样、互动性强的电子课件。IMC Content Studio and Lectora，其中 IMC 是 Interactive Multimedia Culture，为交互式多媒体文化，Content Studio 是内容工作室。Lectora 是一款功能强大的课件制作工具。SCORM（Shareable Content Object Reference Model），译为共享内容对象参考模型或共享组件引用模式，由 ADLI（Advanced Distribution Learning Initiative）集成各个学习标准制定而成。AICC 为 Adobe Illustrator CC，是一款矢量图像设计软件，广泛应用于印刷出版、专业插画、多媒体图像处理和互联网页面的制作等。

[3] Course delivery，授课方式。

[4] 长句，which 引导的是非限定性定语从句。EdX 是麻省理工学院和哈佛大学于 2012 年 4 月联手创建的大规模开放在线课堂平台。Class 2Go，是斯坦福大学开发的 MOOC 新平台，共有 16 组课程。XBlock SDK 软件开发工具包（SDK，Software Development Kit）一般都是一些被软件工程师用于为特定的软件包、软件框架、硬件平台、操作系统等建立应用软件的开发工具的集合。

[5] It 代表 EdX，which 引导的是非限定性定语从句，that 引导的是宾语从句，be posted，知晓。Affero GPL，即 Affero General Public License，最初是由 Affero 公司撰写的公众特许条款，被广泛用作自由软件特许条款。

[6] Coursera，是免费大型公开在线课程项目，由美国斯坦福大学的两名计算机科学教授创办。旨在和世界顶尖大学合作，在线提供免费的网络公开课程。

[7] 本段就是一句话，其中 where 引导的是定语从句，then 此处为"而后""接着"的意思。

[8] 长句。through 引导的是方式状语。

KEYWORDS

MOOC(Massive Open Online Course)	慕课，大型开放式网络课程，大规模开放在线课程
participation	参与，共享，分担
lecture	讲演，讲课，教训
forum	论坛，讨论会
interactive	互动的，交互作用的，互相影响的
licensing	许可，认可，发给执照
videographer	视频图形程序
enrollee	会员，入会者
could computing	云计算
authoring system	写作系统，创作系统
specialized package	专业化程序包，专业化插件
asynchronous access	异步访问
time zone	时区
time framc	时间范围，时帧
contextualized	依照上下文的，依照前后关系的
peer-to-peer	一对一，对等
discipline	训练，修养，纪律，学科
micro-blogging	微博
anonymous	匿名的
flipped classroom	翻转课堂
assimilate	吸收，消化，理解，同化
debate	辩论，讨论

constructivist	解释者
pedagogy	教育学，教授法，教育，教学
consolidation	巩固，合并，联合
facilitation	方便，促进，助长，简化
moderation	适度，节制
exploration	探测，探讨
multimodal	多样式，多形态

EXERCISES

Multiple Choices

1. MOOC _____.
   a. is an abbreviation for Massive Open Online Course
   b. is a distant education
   c. was first introduced in 2007
   d. was emerged as a popular mode of learning in 2012

2. Many MOOCs provide _____.
   a. traditional course materials   b. interactive user forums
   c. problem sets online   d. filmed lectures online

3. Early MOOCs _____.
   a. use closed licenses for their course material
   b. use open licenses for their course material
   c. promote the reuse and remixing of resources
   d. can be accessed via the web

4. In MOOCs additional skills can be provided by _____.
   a. IT specialists   b. platform specialists
   c. videographers   d. students

5. In order to create a MOOC platform, we can use _____.
   a. Elicitus   b. Lectora
   c. IMC Content Studio   d. some other authoring system

6. With MOOC we can asynchronous access to _____.
   a. video   b. offline forum
   c. other learning materials   d. online forum

7. Class 2Go _____.
   a. Was developed by MIT   b. was developed by Stanford
   c. is a new MOOC platform   d. has sixteen sets of course

8. In the flipped classroom the students should _____.
   a. first face to new material inside of class
   b. first face to new material outside of class

    c. do the easer work in classroom
    d. do the harder work in classroom
9. In the flipped classroom the students _____.
    a. take more responsibility for their own learning
    b. can apply knowledge and skills to a range of activities
    c. can use higher order thinking
    d. should change their own roles
10. Significant learning opportunities can be gained through _____.
    a. facilitating active learning   b. engaging teachers
    c. guiding learning   d. providing timely feedback
11. Learning opportunities of the flipped classroom include _____.
    a. concept exploration   b. meaning making
    c. demonstration/application   d. all of the above
12. In the flipped classroom we can gate different learning materials, they are:_____.
    a. text   b. video   c. audio   d. multimedia
13. Key elements of a flipped classroom include _____.
    a. classroom flip   b. learning through activity
    c. learning environment   d. educational technology
14. AICC is _____.
    a. Adobe Illustrator CC   b. an image design software
    c. used for printing and press industries d. used for making web pages

PART III

PROGRAM DESIGN LANGUAGES AND DATABASES

CHAPTER 6   PROGRAMMING LANGUAGES

6.1   OVERVIEW OF PROGRAMMING LANGUAGES

A language is a system of communication. A programming language consists of all the symbols, characters, and usage rules that permit people to communicate with computers. Some programming languages are created to serve a special purpose (e.g. controlling a robot), while others are more flexible general purpose tools that are suitable for many types of applications[1]. However, every programming language must accept certain types of written instructions that will enable a computer system to perform a number of familiar operations[2]. That is, every language must have instructions that fall into the following familiar categories:

(1) **Input/output instructions.** Required to permit communication between I/O devices and the central processor, these instructions provide details on the type of input or output operation to be performed and the storage locations to be used during the operation.

(2) **Calculation instructions.** Instructions to permit addition, subtraction, multiplication, and division during processing are, of course, common in all programming languages.

(3) **Logic/comparison instruction**. These instructions are used to transfer program control, and are needed in the selection and loop structures that are followed to prepare programs [3].

(4) **Storage/retrieval and movement instructions.** These instructions are used to store, retrieve, and move data during processing. Data may be copied from one storage location to another and retrieved as needed.

But even though all programming languages have an instruction set that permits these familiar operations to be performed, there's a marked difference to be found in the symbols, characters, and syntax of machine languages, assembly languages, and high-level languages[4].

1. Machine Languages

A computer's machine language consists of strings of binary numbers and is the only one the CPU directly "understands". An instruction prepared in any machine language will have at least two parts. The first part is the command or operation, and it tells the computer what function to perform. Every computer has an operation code or "op code" for each of its functions. The second part of the instruction is the operand, and it tells the computer where to find or store the data or other instructions that are to be manipulated[5]. The number of operands in an instruction varies among computers. In a single-operand machine, the binary equivalent of "ADD 0184" could cause the value in address 0184 to be added to the value stored in a register in the arithmetic-logic unit[6]. And the programmer writing the instruction to "ADD 0184" for an early IBM machine would have written:

00010000000000000000000000101111000

In addition to remembering the dozens of code numbers for the commands in the machine's instruction set, a programmer was also forced to keep track of the storage locations of data and instructions. The initial coding often took months, was therefore quite expensive, and often resulted in error. Checking instructions to locate errors was about as tedious as writing them initially. And if a program had to be modified at a later date, the work involved could take weeks to finish.

2. Assembly Languages

To ease the programmer's burden, mnemonic operation codes and symbolic addresses were developed in the early 1950s. One of the first steps in improving the program preparation process was to substitute letter symbols—mnemonics—for the numeric machine—language operation codes[7]. Each computer now has a mnemonic code, although, of course, the actual symbols vary among makes and models [8]. Machine language is still used by the computer as it processes data, but assembly language software first translates the specified operation code symbol into its machine-language equivalent.

And this improvement sets the stage for further advances. If the computer could translate convenient symbols into basic operations, why couldn't it also perform other clerical coding functions, such as assigning storage addresses to data? [9] Symbolic addressing is the practice of expressing an address not in terms of its absolute numerical location, but rather in terms of symbols convenient to the programmer.

Programmers no longer assign actual address numbers to symbolic data items as they did earlier. Now they merely specify where they want the first location in the program to be, and an assembly language program takes it from there, allocating locations for instructions and data[10].

This assembly program, or assembler, also enables the computer to convert the programmer's assembly language instructions into its own machine code. A program of instructions written by a programmer in an assembly language is called a source program. After this source program has been converted into machine code by an assembler, it's referred to as an object program.

A big drawback of assembly languages is that they are machine-oriented. That is, they are designed for the specific make and model of processor being used. Programs might have to be recoded for a different machine.

3. High Level Languages

The earlier assembly programs produced only one machine instruction for each source

CHAPTER 6 PROGRAMMING LANGUAGES

program instruction. To speed up coding, assembly programs were developed that could produce a variable amount of machine language code for each source program instruction. In other words, a single macroinstruction might produce several lines of machine language code. For example, the programmer might write "READ FILE", and the translating software might then automatically provide a detailed series of previously prepared machine language instructions which would copy a record into primary storage from the file of data being read by the input device[11]. Thus, the programmer was relieved of the task of writing an instruction for every machine operation performed.

The development of mnemonic techniques and macroinstructions led, in turn, to the development of high level languages that are often oriented toward a particular class of processing problems. For example, a number of languages have been designed to process problems of a scientific mathematic nature, and other languages have appeared that emphasize file processing applications[12].

Unlike assembly programs, high level language programs may be used with different makes of computers with little modification. Thus, reprogramming expense may be greatly reduced when new equipment is acquired. Other advantages of high level languages are:

(1) They are easier to learn than assembly languages.

(2) They require less time to write.

(3) They provide better documentation.

(4) They are easier to maintain.

(5) A programmer skilled in writing programs in such a language is not restricted to using a single type of machine.

NOTES

[1] 长句，由 while 连接的两个并列句，后一句中有 that 引导的定语从句。

[2] accept 接受，认可，此处可译为遵守；that 引导的定语从句，修饰 instructions。

[3] that are followed to prepare programs 是定语从句,修饰 structures；其中的 followed 在此作"遵从""采用"讲。

[4] 长句，even though 引导的是让步状语从句；that 引导的定语从句修饰 instruction set; marked difference，明显的区别。

[5] where 引导的是复合宾语，that 引导的是定语从句。

[6] binary equivalent，意指"ADD 0184"的二进制等效形式。因机器不能识别 ADD 这样的符号。

[7] substitute，取代，此动词表示用……取代……，被取代者前面加 for。

[8] makes and models 意指不同的型号，makes 译为"牌号"，model 直译为"模型"。

[9] clerical coding function，事务性编码功能；such as 引导的是同位语。

[10] and 连接的并列句，前一句中 where 引导的是宾语从句；后一句中 it 代表前一句中的 the first location。

[11] 并列长句，后一句中有 which 引导的定语从句，修饰 instructions。

[12] scientific mathematic nature，科学数学性质，科学计算性质；后一句中 that 引导的同位语从句。

KEYWORDS

programming languages	程序设计语言
instruction	指令
operation code	操作码
operand	操作数
mnemonic	助记的
symbolic address	符号地址
assembly languages	汇编语言
assembler	汇编程序，汇编器
source program	源程序
object program	目标程序
high level language	高级语言
macroinstruction	宏指令
compiler	编译程序，编译器
selection structure	选择结构
loop structure	循环结构
machine language	机器语言
addressing	编址，寻址

EXERCISES

True/False

1. _____ Computation instructions are common in all programming languages.
2. _____ To speed up coding we developed the macro-assembly language.
3. _____ The development of mnemonic techniques and macroinstructions led to the development of high-level languages.
4. _____ In assembly languages programmers need assign actual address numbers to symbolic data items.
5. _____ Assembler converts the assembly language instructions into machine code.
6. _____ We have some languages that emphasize file processing applications.
7. _____ Using machine languages we do not need remember a number of binary code.
8. _____ Symbolic addressing is the practice of expressing an address in terms of symbols convenient to the programmer.
9. _____ High-level language programs may be used with the same makes of computers with little modification.

10. _____ One advantage of high level languages is requiring less time to write.

6.2 C, C++, AND C#

C combines the best features of a structured high-level language and an assembly language—that is, it's relatively easy to code (at least compared to assembly language) and it uses computer resources efficiently[1]. Although originally designed as a system programming language (in fact, the first major program written in C was the UNIX operating system), C has proven to be a powerful and flexible language that can be used for a variety of applications[2]. It is used mostly by computer professionals to create software products.

A newer object-oriented version of C is called C++ (See Figure 6-1). C++ includes the basic features of C, making all C programs understandable to C++ compilers, but has additional features for objects, classes, and other components of an OOP. There is also a visual version of the C++ language. All in all, C++ is one of the most popular programming languages for graphical applications.

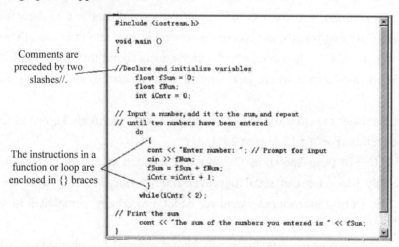

Figure 6-1   Adding-two-numbers program implemented in C++

C++ fully supports object-oriented programming, including the four pillars of object-oriented development: encapsulation, data hiding, inheritance, and polymorphism.

The property of being a self-contained unit is called encapsulation. With encapsulation, we can accomplish data hiding. Data hiding is the highly valued characteristic that an object can be used without the user knowing or caring how it works internally[3]. Just as you can use a refrigerator without knowing how the compressor works, you can use a well-designed object without knowing about its internal data members.

C++ supports the properties of encapsulation and data hiding through the creation of user-defined types, called classes [4]. Once created, a well-defined class acts as a fully encapsulated entity—it is used as a whole unit [5]. The actual inner workings of the class should be hidden. Users of a well-defined class do not need to know how the class works; they just need to know how to use it. When the engineers at Acme Motors want to build a new car, they have two choices: They can start from scratch, or they can modify an existing model. Perhaps their Star model is nearly perfect, but they'd like to add a turbocharger and a six-speed transmission. The chief engineer would prefer not to start from the ground up, but rather to say, "Let's build another Star, but let's add these additional capabilities. We'll call the new model a Quasar [6]." A Quasar is a kind of Star, but one with new features.

C++ supports the idea of reuse through inheritance. A new type, which is an extension of an existing type, can be declared. This new subclass is said to derive from the existing type and is sometimes called a derived type [7]. The Quasar is derived from the Star and thus inherits all its qualities, but can add to them as needed.

C++ supports the idea that different objects do "the right thing" through what is called function polymorphism and class polymorphism[8]. Poly means many, and morph means form. Polymorphism refers to the same name taking many forms.

While it is true that C++ is a superset of C, and that virtually any legal C program is a legal C++ program, the leap from C to C++ is very significant[9]. C++ benefited from its relationship to C for many years, as C programmers could ease into their use of C++. To really get the full benefit of C++, however, many programmers found they had to unlearn much of what they knew and learn a whole new way of conceptualizing and solving programming problems [10].

One of the major revisions of the C++ standard, C++11 (formerly known as C++0x), was approved and released on the 12 August 2011.

In 2014, C++14 (also known as C++1y) was released as a small extension to C++11, featuring mainly bug fixes and small improvements. It aims at doing what C++03 did to C++98[11]. The Draft International Standard ballot procedures completed in mid-August 2014[12].

After C++14, a major revision, informally known as C++17, is planned for 2017.

The newer version of C is C# (pronounced "C sharp"). A hybrid of C and C++, C# is Microsoft's newest programming language developed to compete directly with Sun's Java language. C# is an object-oriented programming language designed to improve productivity in the development of Web applications. The most recent version is C# 5.0, which was released on August 15, 2012.

Microsoft Visual C# is Microsoft's implementation of the C# specification, included in the Microsoft Visual Studio suite of products [13]. It is based on the ECMA/ISO specification of the C# language, which Microsoft also created [14]. While multiple implementations of the specification exist, Visual C# is by far the one most commonly used.

NOTES

　　[1] structured language 是指结构化程序设计语言。code 为动词，写代码，即编写程序。

　　[2] Although 引导的是让步状语从句,主句中有 that 引导的定语从句,修饰 language。

　　[3] that 引导的定语从句修饰 characteristic，without 引导的是条件状语，但此句中的 knowing or caring 为独立结构，其逻辑主语为 user。

　　[4] 类型（type）和类（class）在 C++中都称为类。

　　[5] encapsulated entity…封装的实体，被作为一个整体来使用。

　　[6] Quasar，类星体。

　　[7] derived type，派生类型。

　　[8] that 引导的是同位语从句，through 引导的是介词宾语从句，此处作状语。

　　[9] While 引导的是让步状语从句，由两个 that 引导的主语从句构成。

　　[10] 长句。前面 To really get，为动词不定式短语做目的状语，后面一句中的谓语 found 后面省略了由 that 引出的宾语从句，此宾语从句有两个并列成分：to unlearn…and learn。

　　[11] doing，动名词，用作介词 at 的宾语，what 引导的是宾语从句。

　　[12] The Draft International Standard，是指 C++14 的标准。

　　[13] Microsoft Visual Studio（简称 VS）是美国微软公司的开发工具包系列产品。VS 是一个基本完整的开发工具集，它包括整个软件生命周期中所需要的大部分工具，如 UM 工具、代码管控工具、集成开发环境（IDE）等。所生成的目标代码适用于微软支持的所有平台。

　　[14] ECMA/ISO，ECMA（European Computer Manufacturens Association，欧洲计算机制造联合会）是 1961 年成立的旨在建立统一的电脑操作格式标准——包括程序语言和输入输出的组织。ISO（International Standardization Organization）即国际标准化组织。

KEYWORDS

structured language	结构化语言
OOP (Object-Oriented Programming)	面向对象程序设计
compiler	编译器（程序）
class	类
Java language	一种面向对象的程序设计语言
encapsulation	包装，封装
data hiding	数据隐藏
inheritance	继承性
polymorphism	多态性
portable	可移植的
ANSI	美国国家标准学会
superset	超类

bug	故障，错误，安全漏洞
fix	修正，修正程序
specification	规范，规格，说明书，技术要求
suite	套件，程序集，一套，一组
unqualified	非限定的，不合格的
reference	引用，参考，基准，坐标，标记
context	上下文，前后关系，语言环境，场合

EXERCISES

Multiple choices

1. C language possesses best features of _____.
   a. a structured high-level language
   b. coding relatively easy
   c. using computer resources efficiently
   d. an assembly language

2. C++ include the following pillars: _____.
   a. encapsulation
   b. data hiding
   c. inheritance
   d. polymorphism

3. Encapsulation is _____.
   a. derived from data hiding
   b. a self-contained unit
   c. used to accomplish data hiding
   d. supported by C++

4. Classes in C++ _____.
   a. are user-defined types
   b. can be used to support the properties of encapsulation and data hiding
   c. act as a fully encapsulated entity
   d. like a new car started from scratch

5. A new type _____.
   a. is a subclass
   b. can be derived from an existing type
   c. can't be called a derived type
   d. inherits all features of existing type and has additional features

6. Polymorphism in C++ _____.
   a. means the same name taking many forms
   b. supports different objects do the right thing
   c. has function polymorphism
   d. has class polymorphism

7. C++ is _____.
   a. object-oriented
   b. procedure-oriented
   c. extended from C
   d. a superset of C

8. The actual inner workings of the class _____.

a. should not be hidden b. should be hidden
c. like a "black box" d. don't like a "black box"
9. C# is _____.
   a. a hybrid of C and C++
   b. developed to compete indirectly with Sun's Java language
   c. designed to improve productivity for developing Web applications
   d. the newer version of C
10. Visual C# is _____.
    a. MS's implementation of the C# specification
    b. based on the ECMA/ISO specification of the C# language
    c. by far the one most commonly used
    d. included in the MS Visual Studio suite of products

6.3 JAVA

1. Overview of Java

Java is a programming language originally developed by James Gosling at Sun Microsystems (now part of Oracle Corporation) and released in 1995 as a core component of Sun Microsystems' Java platform[1]. The language derives much of its syntax from C and C++ but has a simpler object model and fewer low-level facilities. Java applications are typically compiled to bytecode (class file) that can run on any Java Virtual Machine (JVM) regardless of computer architecture[2]. Java is a general-purpose, concurrent, class-based, object-oriented language that is specifically designed to have as few implementation dependencies as possible[3]. It is intended to let application developers "write once, run anywhere." (WORA), meaning that compiled Java code can run on all platforms that support Java without the need for recompilation. Java is one of the most popular programming languages in use, particularly for client-server web applications, with a reported 9 million developers.

2. Goals

There were five primary goals in the creation of the Java language:
- It should be "simple, object-oriented and familiar"
- It should be "robust and secure"
- It should be "architecture-neutral and portable"
- It should execute with "high performance"
- It should be "interpreted, threaded, and dynamic"

3. Java Platform

One characteristic of Java is portability, which means that computer programs written in the Java language must run similarly on any hardware/operating-system platform[4]. This is achieved by compiling the Java language code to an intermediate representation called Java bytecode, instead of directly to platform-specific machine code. Java bytecode instructions are analogous to machine code, but are intended to be interpreted by a virtual machine (VM) written specifically for the host hardware. End-users commonly use a Java Runtime Environment (JRE) installed on their own machine for standalone Java applications, or in a Web browser for Java applets[5].

Standardized libraries provide a generic way to access host-specific features such as graphics, threading, and networking.

A major benefit of using bytecode is porting. However, the overhead of interpretation means that interpreted programs almost always run more slowly than programs compiled to native executables world[6]. Just-in-Time compilers were introduced from an early stage that compile bytecodes to machine code during runtime[7].

4. Java applet

A Java applet is an applet delivered to users in the form of Java bytecode. Java applets can run in a Web browser using a Java Virtual Machine (JVM), or in Sun's AppletViewer, a stand-alone tool for testing applets[8].

Java applets run at speeds comparable to, but generally slower than, other compiled languages such as C++, but until approximately 2011 many times faster than JavaScript[9]. In addition they can use 3D hardware acceleration that is available from Java. This makes applets well suited for non trivial, computation intensive visualizations. When browsers have gained support for native hardware accelerated graphics in the form of Canvas and WebGL, as well as Just in Time compiled JavaScript, the speed difference has become less noticeable[10].

Since Java's bytecode is cross-platform or platform independent, Java applets can be executed by browsers for many platforms, including Microsoft Windows, UNIX, Mac OS and Linux. It is also trivial to run a Java applet as an application with very little extra code. This has the advantage of running a Java applet in offline mode without the need for any Internet browser software and also directly from the integrated development environment (IDE)[11].

NOTES

[1] as 引导的为方式状语。Sun Microsystems 曾为著名的微型计算机系统公司，以生产工作站出名，后并入 Oracle（俗称甲骨文）公司。

[2] that 引导的定语从句，修饰 bytecode。 bytecode，Java 字节码，是一种中间语言，通常它把 Java 源代码编译成字节码，也可以把其他语言，如 Ada 语言的源代码编译

成字节码。 JVM(Java Virtual Machine)是在虚拟和非虚拟硬件上，在标准操作系统上实现的软件，它提供一个可运行 Java 字节码的环境。

[3] implementation dependencies，实现依赖，是说 union 里的数据，如果你存储时用一种数据类型，然后用另一种数据类型去访问，那么访问得到的结果是未知的。在 C++ 规范中没有规定应该会有怎么样的结果，而依赖于编译器的实现，在不同的编译器中，不同的优化条件可以有不同的返回结果，这就叫实现依赖。

[4] which 引导的是非限定性定语从句。

[5] 长句，JRE（Java Runtime Environment）（包括 JavaPlug-in）是 Sun 公司的产品，包括两部分：Java Runtime Environment 和 JavaPlug-in。JRE 是可以在其上运行、测试和传输应用程序的 Java 平台。它包括 Java 虚拟机、Java 平台核心类和支持文件。它不包含开发工具——编译器、调试器和其他工具。JRE 需要辅助软件 JavaPlug-in，以便在浏览器中运行 applet。

[6] that 引导的是宾语从句，从句中有 more…than 结构。

[7] that 引导的定语从句，修饰 compilers。

[8] 由 or 连接的并列句，句子主语为 Java applets；后一句中省略 can run，此处的 applets 是指另外的小应用程序；Sun 公司的 AppletViewer（小应用程序观察器）是 JDK（Java Development Kit）自带的工具，用于快速测试小应用程序。

[9] JavaScript 的解释见下一节。

[10] When 引导的是时间状语从句。WebGL(Web-based Graphics Library)是一种扩展了 JavaScript 程序设计语言功能的软件库，它可在任何兼容的 Web 浏览器内，制作交互式 3D 图形。Canvas 是一种 HTML 元件，它含有 WebGL 文本。

[11] 并列句，主语为 This，and 后面省略了 runs。IDE(Integrated Development Environment)即集成开发环境，又称集成设计环境、集成调试环境或交互式开发环境，它为程序员提供了很多复杂功能。

KEYWORDS

Java	一种程序设计语言
Java platform	Java 平台
object model	对象模型
bytecode	字节代码
class file	类文件
JVM(Java Virtual Machine)	Java 虚拟机
concurrent	并发（行）的
object-oriented	面向对象的
portable	可移植的
interpret	解释
JRE(Java Runtime Environment)	Java 运行时环境
library	（程序）库

overhead	开销
compiler	编译程序
recompilation	再次编译
client-server	客户-服务器（方式）
Java applet	Java 小应用程序
visualization	可视化
cross-platform	交叉平台
IDE(Integrated Development Environment)	集成开发环境

EXERCISE

Fill in the blanks with using appropriate words of terms found behind this exercise.

1. Java was originally developed at _____.
2. Java derives much of its syntax from _____.
3. _____ are typically compiled to bytecode.
4. Java is a _____ language.
5. There were _____ in creation of the Java language.
6. _____ should be interpreted.
7. _____ WORA means.
8. _____ is an intermediate representation.
9. _____ provide a generic way to access host-specific features.
10. A major benefit of using bytecode is _____.
11. End-users commonly use a _____ installed on their own machine for standalone Java applications.
12. Just-in-Time compilers can _____ bytecode to machine code during runtime.
13. Running Java applet we can use _____.
14. Java applets run generally slower than _____.
15. _____ runs many times slower than Java applets.
16. _____ makes applets well suited for computation intensive visualization.
17. Java's bytecode is _____.
18. Java applets can be executed for _____.

    a. five primary goals
    b. Java Runtime Environment
    c. Sun Microsystem
    d. JavaScript
    e. Unix, Mac OS, and Linux
    f. C and C++
    g. Standardized libraries
    h. 3D hardware acceleration

i. Java Virtual Machine

j. write once, run anywhere

k. concurrent, class-based and object-oriented

l. porting

m. cross-platform

n. Java applications

o. Java bytecode

p. Java

q. compiled languages

r. compile

6.4 MARKUP AND SCRIPTING LANGUAGES

There are languages other than programming languages that are used in conjunction with application development . The majority of these are Web related, as discussed in the next few sections.

1. HTML and Other Markup Languages

(1) HTML

Most Web pages today are written in a markup language. Markup languages are designed to make it possible to transmit documents over a network using minimal line capacity[1]. Instead of sending exact specifications regarding the appearance of a Web page, markup languages define the structure and layout of a Web page by using a variety of tags[2]. The most common markup language for Web pages is HTML(Hypertext Markup Language). HTML uses HTML tags.

When a Web page is created—using either a word processor, text editor, or a special Web site development program—HTML tags are inserted in the appropriate locations within the Web page's text. Some tags are used along; others are used in pairs. For example, <b> turns bolding on for the text that follows up until a tag </b> is reached, so the following HTML statement

<b>This text is bolded. </b>

would produce the following when viewed with most Web browsers.

This text is bolded

A Web page and its corresponding HTML code are shown in Figure 6-2, with some common HTML tags.

HTML5 is a core technology markup language of the Internet used for structuring and presenting content for the World Wide Web. As of October 2014, this is the final and complete fifth revision of the HTML standard of the World Wide Web Consortium (W3C) [3]. The previous version, HTML 4, was standardized in 1997.

Figure 6-2　HTML

　　Its core aims have been to improve the language with support for the latest multimedia while keeping it easily readable by humans and consistently understood by computers and devices (web browsers, parsers, etc.)[4]. HTML5 is intended to subsume not only HTML4, but also XHTML1 and DOM Level 2 HTML[5].

　　In particular, HTML5 adds many new syntactic features. These include the new <video>, <audio> and <canvas> elements, as well as the integration of scalable vector graphics (SVG) content (replacing generic <object> tags), and MathML for mathematical formulas[6]. These features are designed to make it easy to include and handle multimedia and graphical content on the web without having to resort to proprietary plugins and APIs. Other page structuring new elements, such as <main>, <section>, <article>, <header>, <footer>, <aside>, <nav> and <figure>, are designed to enrich the semantic content of documents. New attributes have been introduced for the same purpose, while some elements and attributes have been removed[7]. Some elements, such as <a>, <cite> and <menu> have been changed, redefined or standardized. The APIs and Document Object Model (DOM) are no longer afterthoughts, but are fundamental parts of the HTML5 specification. HTML5 also defines in some detail the required processing for invalid documents so that syntax errors will be treated uniformly by all conforming browsers and other user agents.

(2) XML

Extensible Markup Language (XML) is a markup language that defines a set of rules for encoding documents in a format which is both human-readable and machine-readable[8]. It is defined by the W3C's XML 1.0 Specification and by several other related specifications, all of which are free open standards.

The design goals of XML emphasize simplicity, generality and usability across the Internet. It is a textual data format with strong support via Unicode for different human languages[9]. Although the design of XML focuses on documents, it is widely used for the representation of arbitrary data structures, such as those used in web services.

Several schema systems exist to aid in the definition of XML-based languages, while many application programming interfaces (APIs) have been developed to aid the processing of XML data.

(3) .NET

Very closely interrelated with XML is Microsoft's .NET strategy to increase the convergence of personal computing with the Web[10]. In a nutshell, .NET is Microsoft's platform to implement XML-based Web services. These services allow applications to communicate and share data over the Internet, regardless of the operating system or programming language being used.

2. Scripting Languages

(1) JavaScript

HTML is principally designed for laying out Web pages that have on moving elements, much as a desktop-publishing program is designed for laying out printed pages. Thus, HTML has minimal tools to create Web pages that change as the user looks at them or to enable users to interact with Web pages on their screens, other than some capabilities with DHTML and recent HTML enhancements[11]. If you want to develop pages with a great deal of dynamic content, scripting languages, such as JavaScript, may be appropriate. Such languages enable you to build program instructions, or scripts, directly into a Web page's code to add dynamic content. For example, JavaScript is commonly used to display submenus or new images when a menu item is pointed to (See Figure 6-3).

JavaScript was originally developed by Netscape to enable Web authors to implement interactive Web sites. Although it shares many of the features and structures of the full Java language, it was developed independently. When using JavaScript, it is important to realize that not all scripting commands work with all browsers[12]. Because of this, make sure that the important features you add to your site with JavaScript are not browser specific.

(2) PHP

PHP is a server-side scripting language designed for web development but also used as a general-purpose programming language. As of January 2013, PHP was installed on more than

240 million websites (39% of those sampled) and 2.1 million web servers.

JavaScript
The JavaScript code needed to drive the interactirity is inserted as necessary into a Web page's HTML.

INTERACTIVITY
The JavaScript code enables the description to change whenever a different item is pointed to.

Figure 6-3　JavaScript

　　PHP code can be simply mixed with HTML code, or it can be used in combination with various templating engines and web frameworks [13]. PHP code is usually processed by a PHP interpreter, which is usually implemented as a web server's native module or a Common Gateway Interface (CGI) executable [14]. After the PHP code is interpreted and executed, the web server sends resulting output to its client, usually in form of a part of the generated web page; for example, PHP code can generate a web page's HTML code, an image, or some other data. PHP has also evolved to include a command-line interface (CLI) capability and can be used in standalone graphical applications[15].

　　The canonical PHP interpreter, powered by the Zend Engine, is free software released under the PHP License[16]. PHP has been widely ported and can be deployed on most web servers on almost every operating system and platform, free of charge.

　　Despite its popularity, no written specification or standard existed for the PHP language until 2014, the canonical PHP interpreter is as a de facto standard[17]. Since 2014, there is ongoing work on creating a formal PHP specification.

　　As of 2014, work is underway on a new major PHP version named PHP 7.

NOTES

　　[1] minimal line capacity，此处指用标记语言写的网页简明，在网上传输时占用的带宽少。

　　[2] 长句，Instead of…引导的介词短语做状语，主句为 markup languages…。

　　[3] W3C（World Wide Web Consortium）即万维网联盟，创建于 1994 年，是 Web

技术领域最具权威和影响力的国际中立性技术标准机构。到目前为止，W3C 已发布了 200 多项影响深远的 Web 技术标准及实施指南，如广为业界采用的超文本标记语言（标准通用标记语言下的一个应用）、可扩展标记语言（标准通用标记语言下的一个子集），以及帮助残障人士有效获得 Web 内容的信息无障碍指南（WCAG）等，有效促进了 Web 技术的互相兼容，对互联网技术的发展和应用起到了基础性和根本性的支撑作用。

[4] 长句。while 连接的 keeping it…为分词短语，做伴随状语。

[5] XHTML1 是 XML 风格的 HTML 4.01。但还没等 XHTML 兴起，它的地位就被 HTML5 取代了。DOM Level2 HTML，其中 HTML DOM(Document Object Model) 定义了一套标准的针对 HTML 文档的对象。DOM level1 模型：将 HTML 文档封装成了对象；DOM level2 模型：在 level1 的基础上，加入了名称空间的功能。

[6] MathML，数学置标语言，是一种基于 XML 的标准，用来在互联网上书写数学符号和公式的置标语言。

[7] while 连接了两个并列句。

[8] that 引导的定语从句中又有 which 引导的定语从句。

[9] Unicode（统一码、万国码、单一码）是一种在计算机上使用的字符编码。Unicode 是为了解决传统的字符编码方案的局限而产生的，它为每种语言中的每个字符设定了统一并且唯一的二进制编码，以满足跨语言、跨平台进行文本转换、处理的要求。

[10] 过去分词 interrelated 做表语，此句强调表语，为倒装句。

[11] 长句，主句中 tools 有两个并列的动词不定式做定语，即 to create…or to enable…。而第一个定语中，又有 that 引导的定语从句修饰 pages。DHTML 为动态超文本标记语言，用于在 Web 页面中增加诸如移动、增长、收缩、隐蔽或出现等动态特性的标记语言。

[12] When 引导的是时间状语，it 是先行代词，为形式主语，实际的主语是 to realize…，that 引导的是宾语从句。

[13] templating engine，模板引擎（这里特指用于 Web 开发的模板引擎），是为了使用户界面与业务数据（内容）分离而产生的，它可以生成特定格式的文档，用于网站的模板引擎会生成一个标准的 HTML 文档。

[14] CGI（Common Gateway Interface）即通用网关接口，是 WWW 技术中最重要的技术之一，有着不可替代的重要地位。CGI 是外部应用程序（CGI 程序）与 Web 服务器之间的接口标准，是在 CGI 程序和 Web 服务器之间传递信息的规程。

[15] CLI（command-line interface）即命令行接口，是指可在用户提示符下输入可执行指令的界面，它通常不支持鼠标，用户通过键盘输入指令，计算机接收到指令后，予以执行。

[16] Zend Engine 是 Zend 引擎，是 PHP 实现的核心，提供了语言实现上的基础。Zend 引擎最主要的特性就是把 PHP 的边解释边执行的运行方式改为先预编译（Compile），然后再执行（Execute）。

[17] 长句。Despite 引导的是让步状语从句。

KEYWORDS

markup language	标记（置标）语言
scripting language	脚本[描述] 语言，过程[编制] 语言
HTML (Hypertext Markup Language)	超文本标记语言
line capacity	线路容量
tag	标记
word processor	字处理器（软件）
text editor	文本编辑器（软件）
W3C(World Wide Web Consortium)	万维网联盟
browser	浏览器
parser	语法分析程序
subsume	包含，包括
syntactic	句法的，句法上的
SVG(Scalable Vector Graphics)	可缩放的矢量图形
plugin	插件
API(Application Program Interface)	应用程序接口
XML(eXtensible Markup Language)	可扩展的标记语言
JavaScript	Java 过程（程序）语言，基于对象的脚本语言
DHTML (Dynamic HTML)	动态超文本标记语言
attribute	性质，特性，本性，属性，象征
human-readable	人可读的
machine-readable	机器可读的
interpreter	解释程序

EXERCISES

True/False

1. _____ Markup languages are common programming languages.
2. _____ Markup language is used to write Web pages.
3. _____ We must send exact specifications regarding the appearance of a Web page written by using a markup language.
4. _____ HTML uses a variety of tags.
5. _____ We must use tags of HTML in pairs.
6. _____ HTML5 is a core technology markup language of the Internet.
7. _____ If we want to increase the convergence of personal computing with the Web, we should use .NET platform.
8. _____ The services provided by .NET platform allow applications to communicate and share data over the Internet.

9. _____ A scripting language is used for developing Web pages with a great deal of dynamic content.
10. _____ It is important to realize that all scripting commands of JavaScript work with all browsers.
11. _____ HTML5 is intended to subsume only HTML4.
12. _____ HTML5 adds many syntactic features.
13. _____ The APIs and DOM in HTML5 are afterthoughts.
14. _____ XML has a free open standard.
15. _____ XML emphasize simplicity, generality and usability across the Internet.
16. _____ PHP can be used in combination with various template engines and web frameworks.
17. _____ PHP is a client-side scripting language as well.
18. _____ PHP code is usually processed by a PHP compiler.
19. _____ After the PHP code is executed, the web server sends resulting output to its client.
20. _____ PHP has been widely ported and can be free of charge.

CHAPTER 7 DATABASE

7.1 DATABASE CONCEPTS

People often need to retrieve large amounts of data rapidly. An airline agent on the phone to a client may need to search quickly to find the lowest-cost flight from Atlanta to Toronto. The registrar of a university may have to scan student records swiftly to find all students who will graduate in June with a grade-point average of 3.5 or higher[1]. A clerk in a video store may need to determine if a particular movie is available for rental. The type of software used for such tasks is a database management system. Computerized database management systems are rapidly replacing the paper-based filing systems that people have had to wade through in the past to find the information their jobs require[2]. The basic features and concepts of PC-based relational database software are discussed next, using Microsoft Access as an example when needed.

1. What Is a Database Program

A database is a collection of data that is stored and organized in a manner enabling information to be retrieved as needed. A database management system (DBMS)—also sometimes called just database software—enables the creation of a database on a computer and provides easy access to data stored within it.

Although not all databases are organized identically, most PC-based databases are organized into fields, records, and files. A field is a single type of data to be stored in a database, such as a person's name or a person's telephone number. A record is a collection of related fields—for example, the ID number, name, address, and major of Phyllis Hoffman (See Figure 7-1). A file—frequently called a table in PC databases—is a collection of related records (such as all student address data, all student grade data, or all student schedule data). The resulting set of related files or tables (such as all student data) comprises the database.

The type of database software found on most PCs is the relational database management system.

2. Creating a Database

A database can contain a variety of objects (See Figure 7-2). The object created first in a new database is the table, then other objects can be created to be used in conjunction with that table as needed.

CHAPTER 7 DATABASE

Figure 7-1 Fields, records, files, and databases. Fields, records, and files (tables) organize the data that are to be part of a database

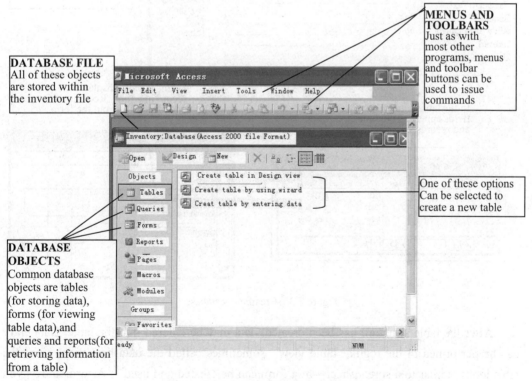

Figure 7-2 A typical database program

When creating a database, the number of tables to be included in the database should be determined. Then the data items to be stored in each table can be identified so that the appropriate field characteristics can be used[3]. For each field, the following should be determined:

(1) Field name (an identifying name unique within the table)

(2) Type of data to be contained in the field (text, numbers, date, etc.)

(3) Field size (how many characters will be needed to store the data)

Once these specifications have been determined, the structure of each table containing the field specifications (See Figure7-3) can be created.

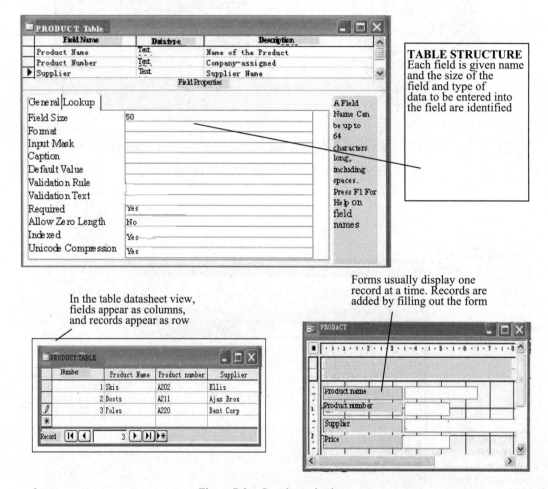

Figure 7-3 Creating a database

After the table structure has been created, data may be entered into the table. Data entry can be performed in the regular table view—sometimes called the datasheet view, since the table looks similar to a spreadsheet—or a form can be created and used[4]. A form allows you to view or edit table content in a more formal manner—usually just working with one record at a time, instead of a full page of records, as in the datasheet view. Figure 7-3 illustrates entering data using both methods once the table structure has been created.

3. Modifying a Database

Once a database table has been created, it may need to be modified. Changes may be made to the table structure or to the data located in the table as needed.

（1）**Modifying the Table Structure**

The table structure needs to be modified only when there are changes to the field properties. For example, a field may need to be widened to accommodate a name that is longer than anticipated, the wrong field type may have been initially selected, or a new field may need to be added.

（2）**Editing, Adding, and Deleting Records**

To make changes to the actual data in a table, the table must first be opened (either using the table datasheet view or a form), and then the necessary changes can be made. To move to a particular record to edit its contents, either the arrows and other keyboard directional keys or the record buttons located at the bottom of the window, can be used (refer again to Figure 7-3)[5]. Since records are typically added to the end of a table, there is usually a New Record button in the group of record buttons that automatically moves you to a blank record at the end of the table[6].

To delete a record, either the Delete key on the keyboard or some type of Delete Record option on the menu bar is used.

4. Queries and Reports

To retrieve information from a database, queries and reports are used. A query is a question, or, in database terms, a request for specific information from the database. A query takes your instructions about what information you want to find and displays just that requested information. A report is a more formal printout of a table or query result.

NOTES

[1] who 引导的定语从句，修饰 students；grade-point，美国学校学生的学习成绩的统计方法，即学绩点。

[2] 长句，句中有两个定语从句，一个是 that 引导的定语从句，修饰 filing systems；另一个是 their jobs require，省略了 that。

[3] so that 引导的是目的状语从句。

[4] 破折号中间的句子是同位语；datasheet，数据单，是一种总结产品、机器、部件、材料、子系统和软件的文档。

[5] To move…目的状语；主句结构为 either…or…can be used。

[6] Since…原因状语从句，主句中 that 引导的定语从句，修饰 New Record button。

KEYWORDS

database	数据库
DBMS (Data Base Management System)	数据库管理系统
relational database	关系型数据库
field	字段
record	记录
file	文件
ID (Identification)	标识，识别，身份证
table	表
data item	数据项
table view	表格视图
data sheet	数据单，数据表
form	表格，表单
query	查询
directional key	定向键
blank record	空记录
menu bar	菜单栏

EXERCISES

Multiple Choices

1. Database management systems can be used to search _____.
   a. the lowest-cost flight tickets   b. student records
   c. video stock   d. another useful data

2. MS Access is _____.
   a. PC-based   b. a DBMS
   c. a hierarchical database   d. a relational database

3. DBMS is _____.
   a. a database software
   b. an abbreviation for database management system
   c. used to create database
   d. used to manage database

4. Most PC-based databases are organized into _____.
   a. notes   b. fields   c. records   d. files

5. The following can be used as a field _____.
   a. user's name   b. user's telephone number
   c. product name   d. product supplier

6. A record in a database can record the _____ of Mr. Clinton.
   a. major   b. birth day   c. ID number   d. address

7. File in a database _____.
   a. is a collection of related records
   b. is called a table in PC database
   c. likes a record
   d. can be used to build a database
8. Objects in a database _____.
   a. have a variety of types
   b. can conjunct with the first created object
   c. can be tables for storing data
   d. can be formed for viewing table data
9. We should determine following specifications for each field _____.
   a. database size
   b. field size
   c. type of data to be contained in the field
   d. field name
10. Data entry can be performed in _____.
    a. a form
    b. the regular table view
    c. a spreadsheet
    d. datasheet view
11. Modifying database means _____.
    a. changing table structure
    b. changing data located in the table
    c. changing field properties
    d. adding new fields
12. To move to a particular record to edit its contents, we can use _____.
    a. a field
    b. the arrows
    c. keyboard directional keys
    d. record buttons located at the bottom of the window

7.2 THE WEB AND DATABASES

Databases are extremely common on the World Wide Web. Virtually all companies that offer products or corporate information, online ordering, or similar activities through a Web site use a database. The most common applications involve client-server database transactions, where the user's browser is the client software[1]. The use of peer-to-peer information exchange, however, is increasing[2].

1. Examples of Web Databases in Use

There are scores of examples of how databases can be used on the Web. Databases facilitate information retrieval and processing, as well as allow more interactive, dynamic content[3]. Following these sections is a discussion and example of how a Web database might work and a brief look at other Web-database-related issues.

(1) Information Retrieval

By their very nature, databases lend themselves to information retrieval on the Web, which is, in essence, a huge storehouse of data waiting to be retrieved. Data is stored in the database, and Web site visitors can request and view it (See Figure 7-4).

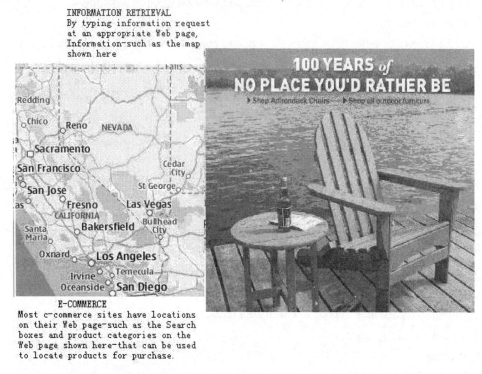

Figure 7-4   Information retrieval and e-commerce

(2) E-Commerce and E-Business

Another widely used database application on the Web is to support and facilitate e-commerce. Catalog information, pricing, customer information, shopping cart contents, and more can be stored in a database to be retrieved on demand using an appropriate script or program to link the database with the Web site[4]. (Refer again to Figure7-4).

(3) Dynamic Web Pages

Static Web pages display the same information for everyone, every time the page is displayed, until the Web page file is modified. In contrast, the appearance and content of dynamic Web pages change based on a user's input. This input can be based on selections specified on a form located on the page or controlled by some other aspect, such as a Java applet, ActiveX control, or the activities that the user has already performed on the site, such as clicking a displayed ad or a product's hyperlink[5].

2. How Web Databases Work

(1) An Example about Database and the Web Work Together

To further illustrate more about how databases and the Web can work together, let's look at an example.

The request to retrieve information from or store data into a Web database is usually initiated by the user. Filling out a Web page form, selecting an option from a menu displayed on a Web page, or clicking an onscreen are common ways database requests are made. The request is received by the Web server, which then converts the request into a database query and passes it on to the database server with the help of intermediary software called middleware. The database server retrieves the appropriate information and returns it to the Web server (again, via middleware) where it is displayed on the user's screen as a Web page. These steps are illustrated in Figure 7-5.

Figure 7-5   A Web database in action

(2) Middleware

Software that connects two otherwise separate applications—such as a Web server and a database management system, as in Figure7-5—is referred to as middleware. The most common types of middleware used to interface between a database and a Web page are CGI and API scripts. A newer scripting language becoming increasingly more popular is PHP and ASP[6].

NOTES

[1] where 引导的是非限定性定语从句。

[2] 与客户-服务器方式不同，peer-to-peer 对等方式是，交换信息的双方不分客户和服务器。

[3] 本句的两个谓语是 facilitate 和 allow，在后一句中 dynamic content 后面省略了 retrieval and processing。

[4] 长句。被动式动词不定式 to be retrieved…为定语修饰 database，不定式 to link 也作为定语，修饰 script or program。

[5] 长句，主句为被动式并列句：This input can be based on…or controlled by…，而后面的句子中又由 or 连接两个并列成分，且都由 such as 引入同位语。Java applet，是以 Java 字节码形式提供给用户的小应用程序，详见 6.3 节。

[6] PHP（Personal Home Page），原来的意思是个人家庭主页，但这个产品很快超出其名字的含义。现在这个缩写词的正式含义是：超文本处理程序，详见 6.4 节。ASP，动态服务器主页，是开发 Web 应用程序的、开放的、免编译的开发应用程序环境，用它可以建立强大的、综合了超文本标记语言（HTML）、脚本以及微软的 ActiveX 技术，以提供动态网站的、基于 Web 的分布式应用程序。

KEYWORDS

online ordering	在线订货
peer-to-peer	对等的
retrieval	检索
script	脚本〔文件〕，稿本，过程
dynamic web page	动态网页
static web page	静态网页
form	表单，表格，窗体
database query	数据库查询
middle ware	中间件
CGI (Common Gateway Interface)	公用网关接口
API (Application Program Interface)	应用程序接口
PHP (Personal Home Page)	个人家庭主页
ASP (Active Server Page)	现用服务器页面，动态服务器主页

EXERCISES

True/False

1. _____All companies offering on line services use the database via Web sites.

2. _____There are two types of Web-based database applications, they are client-server mode and peer-to-peer mode.

3. _____A Web site visitor can require to access a database for retrieving information.

4. _____ To link a database with a Web site we should use an appropriate script.
5. _____ The content of static Web page can be changed by a user's command.
6. _____ We can use Java applet or ActiveX control to change the content of a Web page.
7. _____ There are four common ways to require database.
8. _____ The request of retrieving information is launched by the Web server.
9. _____ In an action of Web database, the database server converts the user's request into a database query.
10. _____ Software that connects a Web server and a DBMS is called middleware.

7.3 MySQL

MySQL is as of July 2013 the world's second most widely used relational database management system (RDBMS) and most widely used open-source RDBMS. It is named after co-founder Michael Widenius's daughter, My[1]. The SQL acronym stands for Structured Query Language.

The MySQL development project has made its source code available under the terms of the GNU General Public License, as well as under a variety of proprietary agreements[2]. MySQL was owned and sponsored by a single for-profit firm, the Swedish company MySQL AB, now owned by Oracle Corporation.

MySQL is a popular choice of database for use in web applications, and is a central component of the widely used LAMP open source web application software stack [3]. LAMP is an acronym for "Linux, Apache, MySQL, Perl/PHP/Python." Free-software-open source projects that require a full-featured database management system often use MySQL.

For proprietary use, several paid editions are available, and offer additional functionality. Applications which use MySQL databases include: TYPO3, MODx, Joomla, WordPress, phpBB, MyBB, Drupal and other software[4]. MySQL is also used in many high-profile, large-scale websites, including Google (though not for searches), Facebook, Twitter, Flickr, and YouTube[5].

1. Interfaces

MySQL is a relational database management system (RDBMS), and ships with no GUI tools to administer MySQL databases or manage data contained within the databases. Users may use the included command line tools, or use MySQL "front-ends", desktop software and web applications that create and manage MySQL databases, build database structures, back up data, inspect status, and work with data records[6]. The official set of MySQL front-end tools, MySQL Workbench is actively developed by Oracle, and is freely available for use.

2. Graphical

The official MySQL Workbench is a free integrated environment developed by MySQL AB, that enables users to graphically administer MySQL databases and visually design database structures[7]. MySQL Workbench replaces the previous package of software, MySQL GUI Tools[8]. Similar to other third-party packages, but still considered the authoritative MySQL front end, MySQL Workbench lets users manage database design & modeling, SQL development (replacing MySQL Query Browser) and Database administration (replacing MySQL Administrator)[9].

MySQL Workbench is available in two editions, the regular free and open source Community Edition which may be downloaded from the MySQL website, and the proprietary Standard Edition which extends and improves the feature set of the Community Edition[10].

Third-party proprietary and free graphical administration applications (or "front ends") are available that integrate with MySQL and enable users to work with database structure and data visually[11].

3. Deployment

MySQL can be built and installed manually from source code, but this can be tedious so it is more commonly installed from a binary package unless special customizations are required. On most Linux distributions the package management system can download and install MySQL with minimal effort, though further configuration is often required to adjust security and optimization settings.

Though MySQL began as a low-end alternative to more powerful proprietary databases, it has gradually evolved to support higher-scale needs as well. It is still most commonly used in small to medium scale single-server deployments, either as a component in a LAMP-based web application or as a standalone database server (See Figure 7-6). Much of MySQL's appeal

Figure 7-6 LAMP software bundle, displayed here together with Squid

originates in its relative simplicity and ease of use, which is enabled by an ecosystem of open source tools such as phpMyAdmin[12]. In the medium range, MySQL can be scaled by deploying it on more powerful hardware, such as a multi-processor server with gigabytes of memory.

There are however limits to how far performance can scale on a single server, so on larger scales, multi-server MySQL deployments are required to provide improved performance and reliability[13]. A typical high-end configuration can include a powerful master database which handles data write operations and multiple slaves that handle all read operations[14]. The master server synchronizes continually with its slaves so in the event of failure a slave can be promoted to become the new master, minimizing downtime. Further improvements in performance can be achieved by caching the results from database queries in memory using memcached, or breaking down a database into smaller chunks called shards which can be spread across a number of distributed server clusters[15].

NOTES

[1] name after…，以……的名字命名。

[2] under the terms of… 在……条件下。

[3] LAMP，Linux+Apache+Mysql/MariaDB+Perl/PHP/Python 一组常用来搭建动态网站或者服务器的开源软件，本身都是各自独立的程序，但是因为常被放在一起使用，拥有了越来越高的兼容度，共同组成了一个强大的 Web 应用程序平台。

[4] Typo3 内容管理系统，是基于 PHP4/PHP5+MySQL 的内容管理系统（框架）（CMS/CMF），兼容 PHP4 和 PHP5 数据库系统。除 MySQL 之外，也能运行于 Oracle、MS-SQL、ODBC、LDAP 等数据库系统，支持 Typo3 的服务器系统有 Apache 或者 IIS 架设的服务器。

MODx 是一个开源的 PHP 应用框架，可以帮助使用者控制自己的网上内容。Joomla!是一套在国外相当知名的内容管理系统。Joomla! 是使用 PHP 语言加上 MySQL 数据库开发的软件系统，可以在 Linux、 Windows、MacOS X 等各种不同的平台上执行。WordPress 的解释请见 5.2.3 节中 NOTES[4]。phpBB 是一个论坛软件，是用 PHP 语言开发的，并开放其原始码。MyBB 是国际上非常优秀的免费论坛软件，最大的特色是简单但功能却很强大。Drupal 是使用 PHP 语言编写的开源内容管理框架（CMF），它由内容管理系统（CMS）和 PHP 开发框架（Framework）共同构成。

[5] 有关 Facebook，请见 5.1.3 节。有关 Twitter，请见 5.1.4 节。Flickr，见 5.1.3 节 NOTES[5]。YouTube 见 5.1.6 节 NOTES[5]。

[6] 长句，句子结构为 Users may use…，or use…。

[7] MySQL AB 是由 MySQL 创始人和主要开发人创办的公司。

[8] MySQL Workbench 是一款专为 MySQL 设计的 ER/数据库建模工具。MySQL GUI Tools，一个可视化界面的 MySQL 数据库管理控制台，提供了 4 个非常好用的图形化应用程序，方便数据库管理和数据查询。

[9] 长句，主句为 MySQL Workbench lets…到句子结束。MySQL 有许多图形化的管理工具，MySQL Administrator 和 MySQL Query Browser 是两个官方公布的工具。MySQL Administrator 是用来管理 MySQL Server 的。而 MySQL Query Browser 可以用来查看数据库内容。

[10] 本段就是一句话，两个 Edition 后面都有 which 引导的定语从句。Community Edition，社区版。关于 MySQL 的版本，从软件的使用授权角度来说，分为 Community 版本和 Commercial 版本，其中 Community 版本就是我们通常可以从 MySQL 官方网站上下载到的社区版，它是基于 GPL 协议的。

[11] that integrate with…是同位语从句，但在 that 前面省略了一个 that。

[12] which 引导的是非限定性定语从句。phpMyAdmin 是一个以 PHP 为基础，以 Web-Base 方式架构在网站主机上的 MySQL 的数据库管理工具，让管理者可用 Web 接口管理 MySQL 数据库。

[13] 长句。how far：到何种程度，so 引导的是结果状语从句。

[14] 句子结构为…can include a powerful master database…and multiple slaves…，两个并列宾语中分别有 which 和 that 引导的定语从句。 slaves 此处为 slave databases。

[15] 长句。句中大部分为 by 引导的方式状语，其结构为 by caching…，or breaking…，而 shards 后面又有 which 引导的定语从句。memcache 即 memory cache，存储器高速缓存。

KEYWORDS

RDBMS(Relational Database Management System)	关系型数据库管理系统
open-source	开源
software stack	软件栈
proprietary	专利的，专卖的，独占的
workbench	工作台，工作架
GUI(Graphic User Interface)	图形用户界面
command line	命令行
front end	前端
desktop software	桌面软件
package	程序包，插件
modeling	模型化，造型，建模，模拟
download	下载
deployment	部署
source code	源代码
master database	主数据库
slave	从属的，随动的
caching	高速缓冲存储
memcache	存储器高速缓存

chunk   程序块，组块，字节片

shard   碎片，破片

cluster   簇，束，聚类，群集

EXERCISES

Fill in the blanks with appropriate words or phrases found behind this exercise.

1. MySQL is as of July 2013 _____ most widely used RDBMS.
2. Structured Query Language can be abbreviated to the _____.
3. We can use MySQL's source code under a variety of _____ .
4. MySQL is _____ for use in web applications.
5. _____ is also used in many high-profile, large-scale website.
6. MODx is an application framework of _____.
7. MySQL Workbench is a _____ tools.
8. If we want to manage MySQL databases graphically, a official _____ needs to buy.
9. MySQL Workbench replaces_____.
10. MySQL Workbench has two editions, they are_____.
11. MySQL can integrate with _____.
12. The better way to install MySQL is from _____.
13. MySQL is still most commonly used in _____.
14. _____ is a MySQL database administrative tool PHP-based and Web-based.
15. A typical high-end configuration of MySQL requires _____.
16. We can improve the performance of MySQL by using_____.

  a. MySQL

  b. small to medium scale single-server deployments

  c. MySQL Workbench

  d. the world's second

  e. memcache

  f. Community Edition and Standard Edition

  g. a powerful master database with multiple slaves

  h. a popular choice of database

  i. GUI

  j. phpAdmin

  k. the third-party proprietary applications

  l. SQL

  m. MySQL GUI Tools

  n. proprietary agreements

  o. a binary package

  p. official set of MySQL front-end

PART IV

APPLICATION SOFTWARE

CHAPTER 8
OFFICE AUTOMATION SOFTWARE

8.1 THE BASICS OF OFFICE AUTOMATION SOFTWARE

When you use any type of application software program, such as a word processor, to type a letter or a tax preparation program to prepare your taxes, there are some basic concepts and functions you need to be familiar with[1]. These include common document-handling tasks, the concept of the software suite, ownership rights for the software you use, and how to get help while you work with the program. These topics and some features are discussed in the next few sections.

1. Basic Concepts of Office Automation Software

(1) Document-Handling Operations[2]

While some document-handling operations are specific for a particular application program, some—such as the concept of opening a document, saving it, and printing it—are fairly universal[3]. A few of the most common document-handling operations are described in Figure 8-1, with examples of the icons used to perform the operations in Windows' applications.

Start a new document	Allows you to create a new blank document, or possibly create a new document from a predefined template. The document will be stored just in RAM until it is saved onto a disk
Open a document	Opens a previously saved document from a disk, usually for editing or printing. Any changes made to the document will be stored just in RAM until the document is saved back onto the disk
Save a document	Saves the current version of the document to a disk
Print a document	Prints the current version of the document onto paper. Some programs give you a variety of print options, such as to print the entire document or just specified pages
Close a document	Removes the document from RAM. Any changes made to the document are lost if the document wasn't saved to disk before it was closed

Figure 8-1   Common document handling tasks

In general, the commands to perform these operations are the same or very similar in all GUI (Graphical User Interface) programs. Because almost everyone needs fast access to document-handling commands, these operations are usually located on easy-to-reach menus or toolbars.

(2) Software Suites

Most office-oriented programs, such as word processors and spreadsheets, are sold bundled together with other related application software in a software suite. The dominant leader in suite sales for office application is Microsoft Office. The high-end edition of this package bundles Word (for word processing), Excel (for spreadsheet work), PowerPoint (for presentation graphics), Access (for database management), together with several other programs such as FrontPage (for Web site development).

One of the biggest advantages to using a software suite is being able to transport or share documents or parts of documents from one program to another. For instance, let's say you are writing a letter in your word processing program, and you want to insert a spreadsheet table. You can launch (start) your spreadsheet program, locate the particular table you want in a stored worksheet, and then copy and paste the table into your letter—all without ever closing the word processing program (See Figure 8-2).

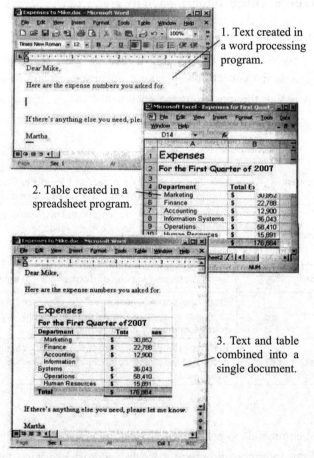

Figure 8-2 Application integration

(3) Online Help

Most people run into problems or need some help with a feature as they work with a software program. To provide help without forcing you to leave your computer screen, most application programs have an online help feature. Programs employ a variety of tools to provide online assistance. Some of the possible configurations are illustrated in Figure 8-3.

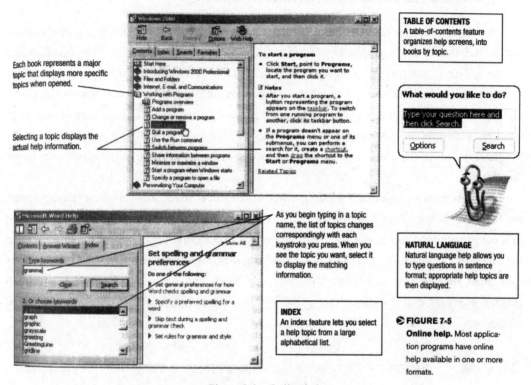

Figure 8-3   Online help

(4) Ownership and Distribution Rights

Sensitive questions sometimes arise about ownership and user rights regarding software products. Typically, a software maker or publisher develops a program, secures a copyright on it, and then retains ownership of all rights to that program[4]. The publisher then dictates who can use, copy, or distribute the program. The various classes of ownership and allowable use are discussed next.

① Proprietary Software

Many of the systems software and application program used today are proprietary software. This means that someone owns the rights to the program, and the owner expects users to buy their own copies.

② Shareware

Some software is available as shareware. While you don't have to pay to install and try out shareware, most shareware specifies that you need to pay to continue to use the software after the trial period—often one month—expires.

③ Freeware

Freeware, or public-domain software, refers to programs that you can use and share with other free of charge.

2. Common Features of Office Automation Software

A user interface is the portion of the application that you work with. Most applications use a GUI that displays graphical elements called icons to represent familiar objects and a mouse[5]. The mouse controls a pointer on the screen that is used to select items such as icons. Another feature is the use of windows to display information. A window is simply a rectangular area that can contain a document, program, or message. (Do not confuse the term window with the various versions of Microsoft's Windows operating systems, which are programs.) More than one window can be opened and displayed on the computer screen at one time.

Most software programs including Microsoft Office 2010, have menus, dialog boxes, toolbars, and buttons (See Figure 8-4). Menus present commands that are typically displayed in a menu bar at the top of the screen. When one of the menu items is selected, an additional list of menu options or a dialog box that provides additional information and requests user input may appear[6]. Toolbars typically are below the menu bar. They contain small graphic elements called buttons that provide shortcuts for quick access to commonly used commands.

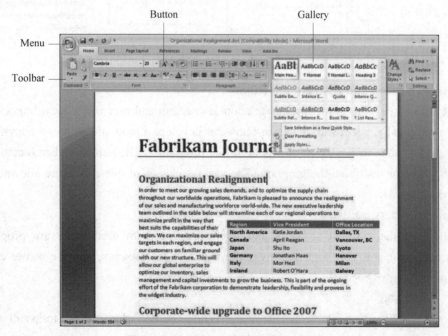

Figure 8-4　Microsoft Office Word 2007

The newest Office version, Microsoft Office 2010, has already a redesigned interface that is intended to make it easier for users to find and use all the features of an application[7]. This new design introduces ribbons, contextual tabs, galleries, and more[8].

(1) Ribbons replace menus and toolbars by organizing commonly used commands into a set of tabs. These tabs display command buttons that are the most relevant to the tasks being performed by the user.

(2) Contextual tabs are tabs that appear automatically. These tabs only appear when they are needed and anticipate the next operations to be pre-formed by the user.

(3) Galleries simplify the process of making a selection from a list of alternatives[9]. This is accomplished by replacing many dialog boxes with visual presentations of potential results.

NOTES

[1] 长句。主句为 there are…。when 引导的是时间状语从句。such as 引出的为两个并列的同位语：a word processor，…or a tax preparation。

[2] document，文档、文件，与 file 同义，但前者多指文档，后者多指文件，本文则将其译为文档较为合适。

[3] While 引导的是让步状语从句。

[4] secure，动词，意为"确保"，此处是指在所开发的程序上保护版权。

[5] that 引导的定语从句修饰 GUI。

[6] When 引导的是时间状语从句，主句中主语为 an additional list… or a dialog box，谓语为 may appear；that 引导的是定语从句。

[7] that 引导的是定语从句，从句中 it 是形式宾语，真正的宾语是 to find and use…。

[8] ribbon，在基于 GUI 的应用软件中，ribbon 是一种接口。在这种接口中，一组工具栏放在一个标记条（tab bar）中。在微软的某些应用软件中，这种格式纳入模块化 ribbon，作为主接口。contextual tabs，在微软 ribbon framework 应用中，它是一个隐藏的标记控件，当在应用工作区中，一个对象，如一幅图像被选中或呈高亮状态时，该控件就会显示在标记行中。

[9] alternatives 为多种选择，此处是指很多图形。

KEYWORDS

template	模（样）板
option	选项
menu	菜单
menu bar	菜单栏
pointer	指针
window	窗口
dialog box	对话框
toolbar	工具栏

button	按钮
list of menu	菜单列表
shortcut	快捷（方式）
ribbon	条形框
tab	选项卡
contextual tab	上下文选项卡
gallery	图库
presentation	显示，表示，呈现，显现

EXERCISES

Multiple Choices

1. Basic concepts and functions of the office automation software include_____.
    a. common document-handling tasks   b. software suite
    c. ownership rights   d. how to get help
2. The following concepts for document-handling are universal_____.
    a. opening a document   b. saving a document
    c. turning on a computer   d. printing a document
3. Microsoft Office includes the following software_____.
    a. Word   b. Excel   c. PowerPoint   d. Access
4. If we want to insert a spreadsheet table into a letter written by using the Word, we should _____.
    a. not be able to use the Excel   b. be able to use the Excel
    c. locate the particular table   d. copy and paste the table into our letter
5. Online help is a tool that_____.
    a. provides online assistance
    b. provides help with forcing you to leave your computer screen
    c. provides help without forcing you to leave your computer screen
    d. has a table-of-contents feature
6. Proprietary software can be_____.
    a. a system software   b. an application program
    c. owned by its maker only   d. owned by its buyer
7. Although shareware is a software that you can install and use, but most shareware specifies that_____.
    a. you need to pay for a trial period   b. you don't need to pay for a trial period
    c. you need to pay for a long term   d. you don't need to pay for a long term
8. Freeware is a_____.
    a. public-domain software
    b. program you can share with others free of charge

c. program you can use after buying it

   d. program you can use free of charge

9. Icons _____.

   a. are graphical elements

   b. are used to represent familiar objects and a mouse

   c. can be selected by a pointer which is controlled by a mouse

   d. can be used to replace commands

10. A window is _____.

    a. simply a rectangular area

    b. one type of Windows operating system

    c. used to contain a document, program, or massage

    d. used to display some objects or elements

11. Microsoft Office 2010 has _____.

    a. menus   b. dialog boxes   c. toolbar   d. buttons

12. The newest Office version introduces _____.

    a. ribbons   b. contextual tabs   c. galleries   d. others

8.2 MICROSOFT OFFICE 2013

Microsoft Office 2013 (formerly Office 15) is a version of Microsoft Office, a productivity suite for Microsoft Windows. It is the successor of Microsoft Office 2010 and includes extended file format support, user interface updates and support for touch among its new features. Office 2013 is suitable for IA-32 and x64 systems and requires Windows 7, Windows Server 2008 R2 or a later version of either [1]. A version of Office 2013 comes included on Windows RT devices [2].

Microsoft Office 2013 comes in twelve different editions, including three editions for retail outlets, two editions for volume licensing channel, five subscription-based editions available through Microsoft Office 365 program, the web application edition known as Office Web Apps and the Office RT edition made for tablets and mobile devices [3].

On February 25, 2014, Microsoft Office 2013 Service Pack 1 (SP1) was released.[4]

1. New features

Office 2013 is more cloud-based than previous versions; a domain login, Office 365 account, or Microsoft account can now be used to sync Office application settings (including recent documents) between devices, and users can also save documents directly to their SkyDrive account[5].

New features include a new read mode in Microsoft Word, a presentation mode in Microsoft PowerPoint and improved touch and inking in all of the Office programs. Microsoft

Word can also insert video and audio from online sources as well as the capability to broadcast documents on the Web. Word and PowerPoint also have bookmark-like features which sync the position of the document between different computers.

The Office Web Apps suite was also updated for Office 2013, introducing additional editing features and interface changes.

Other features of Office 2013 include:
- Flatter look of the Ribbon interface and subtle animations when typing or selecting (Word and Excel)
- A new visualization for scheduled tasks in Microsoft Outlook
- Remodeled start screen
- New graphical options in Word
- Objects such as images can be freely moved; they snap to boundaries such as paragraph edges, document margin and or column boundaries
- Online picture support with content from Office.com, Bing.com and Flickr (by default, only images in public domain) [6]
- Ability to return to the last viewed or edited location in Word and PowerPoint
- New slide designs, animations and transitions in PowerPoint 2013
- Support for Outlook.com and Hotmail.com in Outlook
- Support for integration with Skype, Yammer and SkyDrive[7]
- IMAP special folders support[8]

Figure 8-5 shows the lineup of Microsoft Office 2013 icons.

Figure 8-5　Lineup of Microsoft Office 2013 icons, from left to right: Word, Excel, PowerPoint, Outlook, Access, OneNote, Publisher, Lync and InfoPath

2. Editions

As with previous versions, Office 2013 is made available in several distinct editions aimed towards different markets. All traditional editions of Microsoft Office 2013 contain Word, Excel, PowerPoint and OneNote and are licensed for use on one computer.

Five traditional editions of Office 2013 were released:
- Home & Student: This retail suite includes the core applications Word, Excel, PowerPoint, and OneNote[9].
- Home & Business: This retail suite includes the core applications Word, Excel, PowerPoint, and OneNote plus Outlook[10].
- Standard: This suite, only available through volume licensing channels, includes the core applications Word, Excel, PowerPoint, and OneNote plus Outlook and

Publisher[11].
- Professional: This retail suite includes the core applications Word, Excel, PowerPoint, and OneNote plus Outlook, Publisher and Access[12].
- Professional Plus: This suite, only available through volume licensing channels, includes the core applications Word, Excel, PowerPoint, and OneNote plus Outlook, Publisher, Access, InfoPath and Lync[13].

NOTES

[1] IA-32（Intel Architecture），英特尔体系架构，属于 x86 体系结构的 32 位版本，即具有 32 位内存地址和 32 位数据操作数的处理器体系结构，从 1985 年面世的 80386 直到 Pentium 4，都是使用 IA-32 体系结构的处理器。x64 是指 64 位操作系统，在计算机架构中，64 位整数、内存地址或其他数据单元，是指它们最高达到 64 位（8 字节）宽。Windows Server 2008 R2 是一款服务器操作系统。和 2008 年 1 月发布的 Windows Server 2008 相比，Windows Server 2008 R2 继续提升了虚拟化、系统管理弹性、网络存取方式，以及信息安全等领域的应用，其中有不少功能需要搭配 Windows 7 操作系统。

[2] included on…，过去分词短语做状语。Windows RT 是 Windows 家族的一个新成员，新系统画面与操作方式变化极大，采用全新的 Metro（新 Windows UI）风格用户界面，各种应用程序、快捷方式等能以动态方块的样式呈现在屏幕上，用户可自行将常用的浏览器、社交网络、游戏、操作界面融入。

[3] 长句，介绍 Office 2013 的 12 种版本。Office 365 是微软带给所有企业最佳生产力和高效协同的高端云服务，是微软公司基于云平台的应用套件。Office Web Apps 是由微软推出的基于 Web 端的在线办公工具，它将 Microsoft Office 2010 产品的体验延伸到可支持的浏览器上。Office RT 即 Office 2013 RT 版，是微软为 ARM 设备量身打造的，所有搭载 Windows RT 操作系统的平板等都预装 Office 2013 RT，包括 Surface 平板电脑。

[4] Service Pack1, Service Pack 直译是服务包,操作系统中比较大的而且重要的升级补丁，一般说法是补丁，用途是修补系统、大型软件中的安全漏洞，一般是补丁的集合，简称 SP。

[5] SkyDrive 是微软公司推出的一项云存储服务,用户可以通过自己的 Windows Live 账户进行登录，上传自己的图片、文档等到 SkyDrive 中进行存储。

[6] Bing，微软必应搜索是国际领先的搜索引擎，为中国用户提供网页、图片、视频、词典、翻译、资讯、地图等全球信息搜索服务。Flickr 的注释请见 5.1.3 节的 NOTES[5]。

[7] Skype 是一款即时通信软件，它具备 IM 所需的功能，比如视频聊天、多人语音会议、多人聊天、传送文件、文字聊天等功能。它可以免费高清晰与其他用户语音对话，也可以拨打国内国际电话，无论固定电话、手机、小灵通均可直接拨打，并且可以实现呼叫转移、短信发送等功能。

[8] IMAP（Internet Mail Access Protocol），Internet 邮件访问协议，以前称作交互邮件访问协议（Interactive Mail Access Protocol），它的主要作用是邮件客户端（例如 MS Outlook Express）可以通过这种协议从邮件服务器上获取邮件信息，下载邮件等。

[9] OneNote，是一种数字笔记本，它为用户提供了一个收集笔记和信息的位置，并提供了强大的搜索功能和易用的共享笔记本。

[10] Outlook 是微软办公软件套装的组件之一，它对 Windows 自带的 Outlook express 的功能进行了扩充。Outlook 的功能很多，可以用它来收发电子邮件、管理联系人信息、记日记、安排日程、分配任务。目前最新版为 Outlook 2013。

[11] Publisher 是 Microsoft Office 组件之一。 Publisher 是完整的企业发布和营销材料的解决方案。

[12] Access 是由微软发布的关系数据库管理系统。它结合了 Microsoft Jet Database Engine 和图形用户界面两项特点，是 Microsoft Office 的系统程序之一。

[13] InfoPath 是企业级搜集信息和制作表单的工具，将很多的界面控件集成在该工具中，为企业开发表单搜集系统提供了极大的方便。Lync 是微软带有即时消息、会议和语音功能的真正的统一通信客户端软件。

KEYWORDS

productivity	生产力，生产率
suite	一组，一套，套件，程序序列
touch	触摸，接触
retail outlet	零售商店
licensing	许可，特许
subscription	预定，预约，预约费
cloud-based	基于云的
presentation	显示，表示，呈现，显现
domain	域
login	注册，挂号，登录，进入系统
setting	设置，调整，装置
inking	着墨，给……上油墨
ribbon interface	带状界面
icon	图标，图符，图示
typing	打印，打字
visualization	可视化，目视，显像
remodel	改做，改造，改……之型
snap	抓取，快照，按钮接头，快速移动
animation	动画制作，直观显示
transition	转换，转移，变迁，过渡
folder	文件夹，信息页面

CHAPTER 8 OFFICE AUTOMATION SOFTWARE

EXERCISES

Fill in the blanks with appropriate words or phrases found behind this exercise.

1. Microsoft Office 2013 is the _____ of Microsoft office 2010.
2. Microsoft Office 2013 has the following new features:_____.
3. Microsoft Office 2013 requires using _____.
4. Microsoft Office 2013 has _____ editions.
5. IA (Intel Architecture)-32 belongs to 32-bit edition of _____.
6. Microsoft Office 2013 has three editions for _____.
7. If we want to use the subscription-based editions of Microsoft Office 2013, we should have aid of _____.
8. Office Web Apps is the _____ edition.
9. Office 365 is an application suite of _____.
10. In Office 2013 users can save documents directly to their _____ account.
11. We can insert video and audio from _____ into Microsoft Word.
12. Microsoft Outlook has a new feature of _____ for scheduled tasks.
13. IMAP is an abbreviation for _____.
14. Comparing edition of Home & Student with edition of Home & Business, we found that the successor has one more feature, it is _____.
15. Through volume licensing channels we can use _____ editions.
16. The last edition of outlook is _____.

    a. x86 architecture
    b. Outlook
    c. SkyDrive
    d. successor
    e. web application
    f. Outlook 2013
    g. Windows 7
    h. visualization
    i. Microsoft Office 365 program
    j. Extended file format, user interface updates and touch support
    k. Standard and Professional Plus
    l. on line sources
    m. retail outlets
    n. cloud-based platform
    o. 12
    p. Internet Mail Access Protocol

8.3 MICROSOFT OFFICE MOBILE

Microsoft Office Mobile is a free office suite by Microsoft for mobile devices. It is meant to be compatible with desktop versions of the office suite.

Core applications include Word Mobile, Excel Mobile, and PowerPoint Mobile. Other non-core Office applications, like Lync Mobile, OneNote Mobile, and SharePoint Newsfeed are available for download from mobile app stores[1].

Windows Phone 8 ships with an updated version of the Office Mobile suite, consisting of mobile versions of Word, Excel, PowerPoint, and OneNote[2]. In comparison to their Windows Phone 7 versions, the new versions add an improved Office Hub interface that can sync recently opened and modified documents (including changes to documents stored via Office 365 and SkyDrive), a separated OneNote app with additional features (such as voice notes), and improved document editing and viewing functionality[3].

In June 2013, Microsoft released a version of Office Mobile for iPhone; it is similar to the Windows Phone version, but originally requires an Office 365 subscription to use. A version for Android smartphones was released in July 2013; it, too, originally needed Office 365 for use[4].

Apps for iPad and Android tablet computers were released in March 2014 and January 2015, respectively[5]. These, along with their smartphone equivalents, have been made free for personal use, though certain premium features have been paywalled and require Office 365, which includes licensing of the apps for business use[6].

Windows 10 Mobile, set for release later in 2015, will come with brand new Office apps, and making use of the "universal app" platform pioneered with Windows 10[7].

1. Word

Word Mobile is a word processor that allows creating and editing documents. It supports basic formatting, such as bolding, changing font size, and changing colors. It can add comments, but can't edit documents with tracked changes. It can't open password protected documents, change the typeface, text alignment, or style; create bulleted lists; insert pictures; or undo. Footnotes, endnotes, headers, footers, page breaks, certain indentation of lists, and certain fonts, while not displayed nor able to be inserted while working on a document in Word Mobile, are retained if the original document has them[8]. In addition to the features of the 2013 version, the 2007 version on Windows Mobile also has the ability to save documents in the Rich Text Format and open legacy PSW (Pocket Word) [9]. Furthermore, it includes a spell checker, word count tool, and a "Find and Replace" command.

2. Excel

Excel Mobile is a spreadsheet program that can edit XLSX files[10]. It can edit and format text in cells, calculate formulas, search within the spreadsheet, sort rows and columns, freeze panes, filter the columns, add comments, and create charts[11]. It can't add columns or rows except at the edge of the document, rearrange columns or rows, delete rows or columns, or add spreadsheet tabs. Protection settings, zoom settings, autofilter settings, certain chart formatting, hidden sheets, and other features are not supported on Excel Mobile.

3. PowerPoint

PowerPoint Mobile is included with Windows Mobile 5.0. It is a presentation program capable of reading and editing Microsoft PowerPoint presentations, although authoring abilities are limited to adding notes, editing text, and rearranging slides. It can't create new presentations. Versions of PowerPoint Mobile for Windows Phone 7 can also watch presentation broadcasts streamed from the Internet.

4. OneNote

Microsoft OneNote Mobile was first released in Windows Mobile 6[12]. Microsoft OneNote Mobile is included in the "Office Hub" in Windows Phone 7, but got a standalone application in Windows Phone 8. Still, it was listed in the Microsoft Office Mobile license as a core application. It features integration with Office Lens, an application that can capture whiteboards, blackboards, digital copies and documents on pieces of paper and allow users to edit them in Microsoft OneNote Mobile[13]. Every note is automatically time-stamped and can be filled with words, pictures and recorded audio. It includes features like "to-do lists" but only has a limited option for formatting and layouts and includes no additional fonts[14].

NOTES

[1] Lync，请见上一节注释[13]。OneNote Mobile 可让你随时随地记录和查阅笔记。

[2] Windows Phone 8 是微软公司 2012 年 6 月 22 日发布的一款手机操作系统，是 Windows Phone 系统的第三个大型版本。

[3] 长句，In comparison to…为状语，主句结构为 the new versions add an…, a…, and improved…，共有三个宾语，其中第一个宾语中 that 引导的是定语从句，修饰 interface。Office Hub，办公中心，办公室枢纽，员工在电话上工作的软件。Office 365 见本章 8.2 节的注释[3]，SkyDrive 见 8.2 节的注释[5]。

[4] Android，有关 Android 的内容见 2.6 节。

[5] iPad，苹果公司生产的平板电脑。

[6] though 引导的是让步状语从句，which 引导的是非限定性定语从句。paywall（收费墙）指阻止非付费订阅用户看到网页内容的一个屏蔽系统。

[7] 这一段就是一句话，set for…过去分词短语作状语。

[8] 长句，句子结构为 Footnotes，…，are retained if…中间插入 while not…while working…，第一个 while 引出并列成分，第二个 while 引出时间状语。

[9] Rich Text Format（RTF），富文本格式，能解决在不同的操作系统下的不同的文字处理软件间更换文本文件的问题。

[10] XLSX 是 Office 2007 使用的，是 Microsoft Office Excel 2007 文档的扩展名，是用新的基于 XML 的压缩文件格式取代其目前专有的默认文件格式，在传统的文件扩展名后面添加了字母 x（即 docx 取代 doc、xlsx 取代 xls 等），使其占用空间更小。

[11] in cell，此处 cell 应为 cellular mobile phone，即蜂窝式移动电话。

[12] Windows Mobile 6 使移动办公向前迈进一大步。该系统为用户熟悉的 Outlook Mobile、Office Mobile 和现在用于 Windows Mobile 的 Windows Live™ 等 Microsoft 移动应用程序带来了非常出色的增强功能。

[13] an application 为同位语，that 引导的是定语从句，修饰 application，句子结构为 that can capture…and allow…。Office Lens 可用来修正、增强白板和文档的图片，并将它们发送到 OneNote，需要 Microsoft 账号登录。

[14] to-do lists，任务清单，待办事项清单。

KEYWORDS

suite	一组，一套，套件，程序序列
core application	核心应用程序
hub	中心，中枢，集线器，插孔
voice note	话音说明，话音注记
paywall	收费墙
formatting	格式化，格式编排，格式化操作
bolding	粗体
font	字体，字形，字模
typeface	字体，字样，字形，字面
alignment	校准，定位
bulleted	加重的
undo	还原，取消，废除
footnote	脚注
endnote	结束附注
header	首页，标题，报头
footer	脚注，脚页
PSW(Pocket Word)	袖珍字
spreadsheet	电子表格
pane	显示窗口，区
resolution	分辨率，清晰度，分解，决定

split	分割，划分，分配，分裂，分离
worksheet	工作表
zoom	区域缩放，缩放，伸缩，变焦距
hidden	隐藏，隐蔽，隐式
time-stamped	时间戳
presentation	显示，表示，呈现，显现

EXERCISES
True/False

1. _____ Microsoft Office Mobile is compatible with desktop version of the office suite.
2. _____ Core applications of Office Mobile include Word Mobile, PowerPoint Mobile and Lync Mobile.
3. _____ Windows Phone 8 is an operating system used for smartphones.
4. _____ Windows Phone 7 adds an improved Office Hub interface.
5. _____ Office Mobile for iPhone requires an Office 365 subscription to use.
6. _____ Apps for iPad and Android tablet computers have been made free for personal use.
7. _____ Windows 10 Mobile will have brand new Office apps.
8. _____ Using Word Mobile we can edit documents with tracked changes.
9. _____ With Word Mobile we can open password protected documents.
10. _____ When we work on a document in Word Mobile, its content can be displayed.
11. _____ RTF stands for Rich Text Format.
12. _____ With Excel Mobile we can edit text in smartphone.
13. _____ With Excel Mobile we can add rows and columns in any places of a document.
14. _____ Excel Mobile supports certain chart formatting.
15. _____ With PowerPoint Mobile we can create a new presentation.
16. _____ OneNote Mobile in Windows Phone 8 is a standalone application.
17. _____ OneNote Mobile is a core application in Office Mobile.
18. _____ We can fill into note of the OneNote Mobile with words, pictures and recorded audio.

CHAPTER 9   MULTIMEDIA

9.1   MULTIMEDIA AND ITS MAJOR CHARACTERISTICS

1. Definition

Multimedia refers to content that uses a combination of different content forms. This contrasts with media that use only rudimentary computer displays such as text-only or traditional forms of printed or hand-produced material. Multimedia includes a combination of text, audio, still images, animation, video, or interactive content forms, as shown in Figure 9-1.

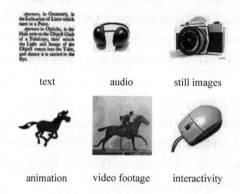

Figure 9-1   Examples of individual content forms combined in multimedia[1]

Multimedia can be recorded and played, displayed, dynamic, interacted with or accessed by information content processing devices, such as computerized and electronic devices, but can also be part of a live performance[2]. Multimedia devices are electronic media devices used to store and experience multimedia content. Multimedia is distinguished from mixed media in fine art; by including audio, for example, it has a broader scope. The term "rich media" is synonymous with interactive multimedia. Hypermedia scales up the amount of media content in multimedia application.

2. Categorization of multimedia

Multimedia may be broadly divided into linear and non-linear categories. Linear active content progresses often without any navigational control for the viewer such as a cinema presentation[3]. Non-linear uses interactivity to control progress as with a video game or self-paced computer based training. Hypermedia is an example of non-linear content.

3. Major characteristics of multimedia

Multimedia presentations may be viewed by person on stage, projected, transmitted, or played locally with a media player. A broadcast may be a live or recorded multimedia presentation. Broadcasts and recordings can be either analog or digital electronic media technology. Digital online multimedia may be downloaded or streamed. Streaming multimedia may be live or on-demand.

Multimedia games and simulations may be used in a physical environment with special effects, with multiple users in an online network, or locally with an offline computer, game system, or simulator[4].

The various formats of technological or digital multimedia may be intended to enhance the users' experience, for example to make it easier and faster to convey information. Or in entertainment or art, to transcend everyday experience[5].

Enhanced levels of interactivity are made possible by combining multiple forms of media content. Online multimedia is increasingly becoming object-oriented and data-driven, enabling applications with collaborative end-user innovation and personalization on multiple forms of content over time[6]. Examples of these ranging from multiple forms of content on Web sites like photo galleries with both images (pictures) and title (text) user-updated, to simulations whose events, illustrations, animations or videos are modifiable, allowing the multimedia "experience" to be altered without reprogramming[7]. In addition to seeing and hearing, haptic technology enables virtual objects to be felt. Emerging technology involving illusions of taste and smell may also enhance the multimedia experience.

NOTES

[1] video footage，视频片段，视频画面，视频录像，录像（见图 9-1）。

[2] 长句，句型为 Multimedia can be…but can also be…，中间插入由 such as 引导的同位语。

[3] progress 此处为动词。

[4] 长句。句中有 3 个 with 引导的方式状语。

[5] 由于紧跟前一句，此句省略了主语和谓语。

[6] 长句。enabling…，现在分词短语作状语。

[7] 长句。主句结构为 Examples of…are modifiable，其中 ranging from…，to…为分词短语作定语，而 allowing…为现在分词短语作状语。

KEYWORDS

rudimentary	基本的，初步的
animation	动画
rich media	富媒体，多元媒体
synonymous	同义的，同义词的

hypermedia	超媒体
convey	传达，运输
collaborative	合作的，协作的
simulation	模拟，仿真
streaming multimedia	流式媒体，流媒体
live	实况，直播
object-oriented	面向对象的
data-driven	数据驱动的
haptic	触觉的

EXERCISES

Multiple choices.

1. Multimedia refers to use _____.
   a. only rudimentary computer displays   b. a combination of different content forms
   c. a single content form only   d. many kinds of digital medium
2. Multimedia includes a combination of _____.
   a. text   b. still images   c. video   d. interactive content forms
3. Multimedia devices are _____ devices.
   a. computerized   b. electronic   c. analog only   d. digital
4. There are several multimedia forms in this text, they are _____ media.
   a. super   b. hyper   c. rich   d. mixed
5. Multimedia can be _____ by information content processing devices.
   a. recorded   b. played   c. displayed   d. accessed
6. The content forms in multimedia can be _____.
   a. mouse   b. video footage   c. animation   d. audio
7. Non-linear multimedia _____.
   a. uses interactivity to control progress
   b. does not need any navigational control when its active content progresses
   c. has an example of hypermedia
   d. has a typical application which is video game
8. Multimedia presentations may be _____.
   a. transmitted   b. projected   c. played   d. viewed
9. Right now emerging technology in multimedia involves _____.
   a. seeing   b. hearing   c. tasting   d. smelling
10. Online multimedia will enable _____.
    a. applications with collaborative end-user innovation
    b. applications with personalization
    c. the multimedia "experience" to be altered without reprogramming

d. virtual objects to be felt

9.2 USAGE/APPLICATION

Multimedia finds its application in various areas including, but not limited to, advertisements, art, education, entertainment, engineering, medicine, mathematics, business, scientific research and spatial temporal applications. Figure 9-2 shows a presentation using PowerPoint. Figure 9-3 shows a virtual reality using multimedia content. Figure 9-4 shows a multimedia-terminal in Dresden (Germany).

Figure 9-2　A presentation using Powerpoint. Corporate presentations may combine all forms of media content

Figure 9-3　Virtual reality uses multimedia content. Applications and delivery platforms of multimedia are virtually limitless

Several examples are as follows:

1. Creative industries

Creative industries use multimedia for a variety of purposes ranging from fine arts, to entertainment, to commercial art, to journalism, to media and software services provided for any of the industries listed below[1]. An individual multimedia designer may cover the spectrum throughout their career. Requests for their skills range from technical, to analytical, to creative.

Figure 9-4   Multimedia-Terminal in Dresden (Germany)

2. Commercial uses

Much of the electronic old and new media used by commercial artists and graphic designers is multimedia. Exciting presentations are used to grab and keep attention in advertising. Business to business, and interoffice communications are often developed by creative services firms for advanced multimedia presentations beyond simple slide shows to sell ideas or liven-up training[2]. Commercial multimedia developers may be hired to design for governmental services and nonprofit services applications as well.

3. Entertainment and fine arts

In addition, multimedia is heavily used in the entertainment industry, especially to develop special effects in movies and animations (VFX, 3D animation, etc.)[3]. Multimedia games are a popular pastime and are software programs available either as CD-ROMs or online. Some video games also use multimedia features. Multimedia applications that allow users to actively participate instead of just sitting by as passive recipients of information are called Interactive Multimedia[4]. In the Arts, there are multimedia artists, whose minds are able to blend techniques using different media that in some way incorporates interaction with the viewer[5].

4. Education

In Education, multimedia is used to produce computer-based training courses (popularly called CBTs) and reference books like encyclopedia and almanacs[6]. A CBT lets the user go through a series of presentations, text about a particular topic, and associated illustrations in various information formats. Edutainment is the combination of education with entertainment, especially multimedia entertainment.

Learning theory in the past decade has expanded dramatically because of the introduction of multimedia. Several lines of research have evolved (e.g. Cognitive load, Multimedia learning, and the list goes on)[7]. The possibilities for learning and instruction are nearly endless.

5. Journalism

Newspaper companies all over are also trying to embrace the new phenomenon by implementing its practices in their work[8]. While some have been slow to come around, other major newspapers like The New York Times, USA Today and The Washington Post are setting the precedent for the positioning of the newspaper industry in a globalized world[9].

News reporting is not limited to traditional media outlets. Freelance journalists can make use of different new media to produce multimedia pieces for their news stories. It engages global audiences and tells stories with technology, which develops new communication techniques for both media producers and consumers.

6. Engineering

Software engineers may use multimedia in Computer Simulations for anything from entertainment to training such as military or industrial training. Multimedia for software interfaces are often done as a collaboration between creative professionals and software engineers.

7. Industry

In the Industrial sector, multimedia is used as a way to help present information to shareholders, superiors and coworkers. Multimedia is also helpful for providing employee training, advertising and selling products all over the world via virtually unlimited web-based technology.

8. Mathematical and scientific research

In mathematical and scientific research, multimedia is mainly used for modeling and simulation. For example, a scientist can look at a molecular model of a particular substance and manipulate it to arrive at a new substance.

9. Medicine

In Medicine, doctors can get trained by looking at a virtual surgery or they can simulate how the human body is affected by diseases spread by viruses and bacteria and then develop techniques to prevent it[10]. Multimedia application like virtual surgeries also help doctors to get practical training.

NOTES

[1] 长句，ranging from…，to…，to…，现在分词短语作定语。fine arts，美艺术（绘画、雕塑、音乐、舞蹈、文学、工艺、建筑等）。

[2] Business to business 即 B2B，电子商务中的商业对商业的运作模式，详见 4.3.1 节。beyond…，介词短语作状语。 liven-up，使有生气，活跃气氛。

[3] VFX（Visual Effects），视觉效果。3D animation，三维动画。

[4] 长句，句子结构为 Multimedia applications…are called…，中间有 that 引导的定语从句以及 instead of…，介词短语，意为"而不是……"。

[5] 长句，主句为 whose minds…，句中 using…现在分词短语作定语，修饰 techniques，而后面的 that 引导的定语从句，修饰 media。

[6] CBT（Computer-Based Training）即计算机辅助训练，广泛应用于培训领域，如驾驶、医疗等。CBT 培训是采用图像、声音、文字、仿真模拟操作等多媒体手段为学员营造直观、真实的培训环境，并提供接近真实的互动操作，使学员在培训过程中有身临其境的感受，增强培训效果。

[7] cognitive load，认知负荷，来源于认知负荷理论（cognitive load theory），是 Sweller 等人在 20 世纪 80 年代提出的，主要从认知资源分配的角度考察学习和问题的解决。the list goes on，意为"还有很多"。

[8] phenomenon，此处是指多媒体这一特殊事务。

[9] 长句。While 引导的是并列分句，主句是 other major newspapers…are setting…。其中 setting the precedent for…为创立……的先例（先机）。

[10] 由 or 连接的并列句，后一句中又是一个由 and 连接的并列句，句子结构为 they can simulate… and then develop…，其中 how 引导的是宾语从句。

KEYWORDS

advertisement	广告，宣传，启事
entertainment	娱乐，招待，表演
spatial	空间的
temporal	时间的
journalism	新闻学（界），报刊编辑
spectrum	光谱，范围
presentation	显示，表示，呈现，显现

slide	幻灯片
animation	动画
illustration	插画，插图，实例
edutainment	教育娱乐，寓教于乐
phenomenon	现象，特殊的事务
pastime	娱乐，消遣
almanacs	年鉴
cognitive	认知的
modeling	建造模型（建模），造型
simulation	模拟

EXERCISES

Fill in the blanks with appropriate words or phrases found behind this exercise.

1. An individual multimedia _____ may cover the spectrum throughout their career.
2. Exciting presentations are used to grab and keep attention in _____.
3. Multimedia games are a popular pastime and are software programs available either as CD-ROMs or _____.
4. Learning theory in the past decade has expanded dramatically because of the introduction of _____.
5. Freelance _____ can make use of different new media to produce multimedia pieces for their news stories.
6. In mathematical and scientific research, multimedia is mainly used for _____ and simulation.
7. Multimedia application like virtual surgeries also help _____ to get practical training.
8. Software engineers may use multimedia in _____.
9. Multimedia in the industries is helpful for providing employee _____.
10. Corporate presentations may combine all forms of _____.
11. Applications and delivery platforms of multimedia are _____.
12. In this text we present _____ kinds of multimedia applications.

    a. journalists
    b. doctors
    c. advertising
    d. computer simulations
    e. multimedia
    f. designer
    g. training, advertising and selling products
    h. nine
    i. online

j. virtually limitless
k. media content
l. modeling

9.3 MULTIMEDIA NETWORKING

Many applications, such as video mail, video conferencing, and collaborative work systems, require networked multimedia. In these applications, the multimedia objects are stored at a server and played back at the client's sites. Such applications might require broadcasting multimedia data to various remote locations or accessing large depositories of multimedia sources[1]. Multimedia networks require a very high transfer rate or bandwidth, even when the data is compressed. Traditional networks are used to provide error-free transmission. However, most multimedia applications can tolerate errors in transmission due to corruption or packet loss without retransmission or correction. In some cases, to meet real-time delivery requirements or to achieve synchronization, some packets are even discarded. As a result, we can apply lightweight transmission protocols to multimedia networks. These protocols cannot accept retransmission, since that might introduce unacceptable delays.

Multimedia networks must provide the low latency required for interactive operation. Since multimedia data must be synchronized when it arrives at the destination site, networks should provide synchronized transmission with low jitter[2].

In multimedia networks, most communications are multipoint as opposed to traditional point-to-point communication. For example, conferences involving more than two participants need to distribute information in different media to each participant. Conference networks use multicasting and bridging distribution methods[3]. Multicasting replicates a single input signal and delivers it to multiple destinations. Bridging combines multiple input signals into one or more output signals, which then deliver to the participants.

In the past, multimedia was difficult to apply on a wide scale because the available technology wasn't sufficient to support it. With today's fast computers and internet connections, multimedia implementation is much more feasible and its use is increasing exponentially. A growing number of Web sites incorporate multimedia, as do computer software and consumer interfaces, such as ATM machines and information kiosks[4]. Multimedia presentations are also commonly used in business.

In the Web-based applications multimedia elements can be added to virtually any Web page to add interest, to better deliver content, or to add functionality. There are some specific types of multimedia applications, however, where multimedia is an integral component of the Web site[5]. The advantages of Web-based applications include:

- Multiple platforms—since Web pages can be viewed by many types of computers and

platforms, multiple versions of the application (such as PC and Macintosh) don't need to be developed.
- Familiar interface—since Web-based applications are accessed with a Web browser, the user should feel somewhat familiar with the application the first time it is accessed.
- Browser and computer incompatibility—Test your site on a variety of browsers, platforms, operating system, and screen resolutions to ensure that your site works on as many different configurations as possible.
- Easily updated—the content of a Web-based application can be updated as frequently as necessary and users always see the most recent version each time the Web site is accessed.

NOTES

[1] broadcasting…or accessing…为动名词作并列宾语。

[2] jitter 抖晃，中华人民共和国新闻出版行业标准 CY/T38-2001 中的标准译名；此句为 since 引导的原因状语从句，此从句中 when 引导的是时间状语从句。

[3] multicasting 又称多播或组播。

[4] as do computer…词序倒装，do 代表主句中的 incorporate multimedia。

[5] where 引导的是地点状语从句。

KEYWORDS

bandwidth	带宽
real-time	实时
synchronization	同步
multicasting	多目标广播，组播
latency	延时，等待
video mail	视频邮件
video conferencing	视频会议
transfer rate	传输速率
multipoint	多点
point-to-point	点对点
ATM(Automated Teller Machine)	自动出纳机
kiosk	信息站，公用信息机，公用电话间

EXERCISES

Fill in the blank with appropriate words or phrases found behind this exercise.

1. Many applications, such as video mail, video conferencing, and collaborative work systems, require _____.

2. Multimedia networks require a very high transfer rate or _____, even when the data is compressed.
3. Multimedia networks must provide the low _____ required for interactive operation.
4. In multimedia networks, most communications are _____ as opposed to traditional point-to-point communication.
5. To add functionality we can put _____ into any Web page.
6. A growing number of _____ incorporate multimedia.
7. _____ replicates a single input signal and delivers it to multiple destinations.
8. Web pages can be viewed by many types of _____.
9. Today we can use _____ to implement multimedia more feasibly.
10. The content of a _____ can be updated easily.

    a. multipoint

    b. Web-based application

    c. Multicasting

    d. latency

    e. computers and platforms

    f. networked multimedia

    g. bandwidth

    h. fast computers and Internet connections

    i. Web sites

    j. multimedia elements

CHAPTER 10
COMPUTER GRAPHICS AND IMAGES

10.1 OVERVIEW

The term computer graphics has been used in a broad sense to describe "almost everything on computers that is not text or sound". Typically, the term computer graphics refers to several different things:
- The representation and manipulation of image data by a computer
- The various technologies used to create and manipulate images

Computer graphics is widespread today. Computer imagery is found on television, in newspapers, for example in weather reports, or for example in all kinds of medical investigation and surgical procedures. A well-constructed graph can present complex statistics in a form that is easier to understand and interpret[1]. In the media "such graphs are used to illustrate papers, reports, thesis", and other presentation material.

There are many ways to get images onto the computer. They can be hand-drawn paintings or photographs that are scanned into the computer. Computer paint programs allow you to draw your images directly in the computer. You can also use the computer to define an imaginary world, with objects placed in this world. These objects can then be rendered by the computer to produce stunningly real-looking images, like those seen in the movie Toy Story[2]. A rendered image is built from the ground up starting with a mathematical description of the objects in the scene, to defining lights and colors to produce the final finished image[3].

Many powerful tools have been developed to visualize data. Computer generated imagery can be categorized into several different types: two dimensional (2D), three dimensional (3D), and animated graphics. As technology has improved, 3D computer graphics have become more common, but 2D computer graphics are still widely used. Computer graphics has emerged as a sub-field of computer science which studies methods for digitally synthesizing and manipulating visual content[4]. Over the past decade, other specialized fields have been developed like information visualization, and scientific visualization more concerned with "the visualization of three dimensional phenomena (architectural, meteorological, medical, biological, etc.), where the emphasis is on realistic renderings of volumes, surfaces, illumination sources, and so forth, perhaps with a dynamic (time) component"[5].

Computer graphics may be used in the following areas:
- Computational biology   • Computational physics   • Computer-aided design

- Computer simulation
- Graphic design
- Rational drug design
- Video games
- Digital art
- Infographics
- Scientific visualization
- Virtual reality
- Education
- Information visualization
- Special Effects for cinema
- Web design

NOTES

[1] that 引导的定语从句，修饰 form。

[2] Toy Story，"玩具总动员"，是 Pixar（皮克斯）1995 年制作的动画系列电影，由华特·迪士尼影片公司和皮克斯动画工作室合作推出，截至 2015 年共制作了三部。

[3] from the ground up，从头，彻底地，完全地。in the scene，在现场，在场景中。

[4] which 引导的定语从句，修饰 sub-field。

[5] 长句。where 引导的是非限定性定语从句。volume 原意是体积，容量，卷，册，此处可译为物体。

KEYWORDS

computer graphics	计算机图形学
presentation	显示，表示，显现，呈现
illustrate	举例说明，作图解
visualization	可视化，目视，显像
render	着色，渲染
visualize	想象，可视化
synthesize	合成，综合
meteorological	气象的，气象学的
realistic	逼真的
illumination	照明
Computer simulation	计算机模拟
Infographics	信息图表
Virtual reality	虚拟现实

EXERCISES

True/False

1. _____ The term computer graphics has been used in a broad sense to describe "almost everything on computers that is not text or sound".

2. _____ A well-constructed graph can present complex statistics in a form that is difficult to understand and interpret.

3. _____ There are many ways to get images onto the computer.

4. _____ A rendered image is built from the ground up starting with a mathematical description of the objects in the scene, to defining lights and colors to produce the final

finished image.

5. _____ As technology has improved, 3D computer graphics have become more common, but 2D computer graphics have disappeared totally.
6. _____ Computer graphics has emerged as a sub-field of computer science which studies methods for digitally synthesizing and manipulating visual content.
7. _____ We have listed sixteen kinds of applications in computer graphics.
8. _____ The computer graphics can refer to the various technologies used to create and manipulate images.
9. _____ Weather forecast on TV is created with computer imagery.
10. _____ Toy Story is a TV play produced by Pixar in America.

10.2 THE VARIOUS COMPUTER GRAPHICS

1. Two-dimensional computer graphics

2D computer graphics are the computer-based generation of digital images—mostly from models, such as digital image, and by techniques specific to them.

2D computer graphics are mainly used in applications that were originally developed upon traditional printing and drawing technologies such as typography. In those applications, the two-dimensional image is not just a representation of a real-world object, but an independent artifact with added semantic value; two-dimensional models are therefore preferred, because they give more direct control of the image than 3D computer graphics, whose approach is more akin to photography than to typography[1].

(1) Pixel art

A large form of digital art being pixel art is created through the use of raster graphics software, where images are edited on the pixel level[2]. Graphics in most old (or relatively limited) computer and video games, graphing calculator games, and many mobile phone games are mostly pixel art.

(2) Sprite graphics

A sprite is a two-dimensional image or animation that is integrated into a larger scene. Initially including just graphical objects handled separately from the memory bitmap of a video display, this now includes various manners of graphical overlays[3].

Originally, sprites were a method of integrating unrelated bitmaps so that they appeared to be part of the normal bitmap on a screen, such as creating an animated character that can be moved on a screen without altering the data defining the overall screen[4]. Such sprites can be created by either electronic circuitry or software. In circuitry, a hardware sprite is a hardware construct that employs custom DMA channels to integrate visual elements with the main screen in that it super-imposes two discrete video sources[5]. Software can simulate this through specialized rendering methods.

(3) Vector graphics

Vector graphics formats are complementary to raster graphics. Raster graphics is the representation of images as an array of pixels and is typically used for the representation of photographic images. Vector graphics consists of encoding information about shapes and colors that comprise the image, which can allow for more flexibility in rendering. There are instances when working with vector tools and formats is best practice, and instances when working with raster tools and formats is best practice[6]. There are times when both formats come together. An understanding of the advantages and limitations of each technology and the relationship between them is most likely to result in efficient and effective use of tools.

Figure 10-1 shows the effect of vector graphics versus raster (bitmap) graphics.

Figure 10-1   Example showing effect of vector graphics versus raster (bitmap) graphics

2. Three-dimensional computer graphics

3D graphics compared to 2D graphics are graphics that use a three-dimensional representation of geometric data. For the purpose of performance this is stored in the computer[7]. This includes images that may be for later display or for real-time viewing.

Despite these differences, 3D computer graphics rely on similar algorithms as 2D computer graphics do in the frame and raster graphics (like in 2D) in the final rendered display[8]. In computer graphics software, the distinction between 2D and 3D is occasionally blurred; 2D applications may use 3D techniques to achieve effects such as lighting, and primarily 3D may use 2D rendering techniques.

3D computer graphics are the same as 3D models. The model is contained within the graphical data file, apart from the rendering. However, there are differences that include the 3D model is the representation of any 3D object. Until visually displayed a model is not graphic. Due to printing, 3D models are not only confined to virtual space. 3D rendering is how a model can be displayed. Also can be used in non-graphical computer simulations and calculations[9].

3. Computer animation

Computer animation is the art of creating moving images via the use of computers. It is a subfield of computer graphics and animation. Increasingly it is created by means of 3D computer graphics, though 2D computer graphics are still widely used for low bandwidth, and faster real-time rendering needs[10]. Sometimes the target of the animation is the computer itself, but sometimes the target is another medium, such as film. It is also referred to as CGI (Computer-generated imagery or computer-generated imaging), especially when used in films.

Figure 10-2 shows an example of computer animation produced using Motion capture.

Figure 10-2   Example of Computer animation produced using Motion capture

Virtual entities may contain and be controlled by assorted attributes, such as transform values (location, orientation, and scale) stored in an object's transformation matrix. Animation is the change of an attribute over time. Multiple methods of achieving animation exist; the rudimentary form is based on the creation and editing of keyframes, each storing a value at a given time, per attribute to be animated. The 2D/3D graphics software will change with each keyframe, creating an editable curve of a value mapped over time, in which results in animation. Other methods of animation include procedural and expression-based techniques: the former consolidates related elements of animated entities into sets of attributes, useful for creating particle effects and crowd simulations; the latter allows an evaluated result returned from a user-defined logical expression, coupled with mathematics, to automate animation in a predictable way (convenient for controlling bone behavior beyond what a hierarchy offers in skeletal system set up)[11].

To create the illusion of movement, an image is displayed on the computer screen then quickly replaced by a new image that is similar to the previous image, but shifted slightly. This technique is identical to the illusion of movement in television and motion pictures.

NOTES

[1] 分号隔开的两个句子，后一句中 because…是原因状语从句，whose approach…为非限定性定语从句。

[2] being pixel art，现在分词短语作定语，修饰 digital art，where 引导的是非限定性

定语从句。

　　[3] 长句。其中 including…现在分词短语作状语。sprite 在国内很多人叫精灵，是一种网页图片应用处理方式。它允许将一个页面涉及的所有零星图片都包含到一张大图中去，这样一来，当访问该页面时，载入的图片就不会像以前那样一幅一幅地慢慢显示出来了。

　　[4] 长句。so that…目的状语从句，such as…为同位语，其中 that 引导的定语从句，修饰 character（角色）。without altering…现在分词短语作状语，而 defining…为现在分词短语作定语，修饰 data。

　　[5] 长句。第一个 that 引导的定语从句，修饰 construct，第二个 in that…为方式状语从句。

　　[6] 句子结构为 There are instances… and instances…。

　　[7] performance，性能，此处意为展现，呈现。this 为 3D 图形数据。

　　[8] as 2D computer graphics do…，方式状语从句。

　　[9] Also 后面省略了主语 3D rendering。

　　[10] 长句，though 引导的是让步状语从句。

　　[11] 长句。分号后面的一句话中有两个状语：分词短语 coupled…和 to automate…。括号中 what… 从句是 beyond 的宾语。

KEYWORDS

photography	摄影，照相
typography	排版
pixel	像素
artifact	人工制品
semantic	语义的
raster graphics	光栅图形（像）
sprite	精灵
bitmap	位图
vector	矢量，向量
encoding	编码
DMA (Direct Memory Access)	直接存储器访问
animation	动画
algorithm	算法
blur	模糊
stylistic	文体的，风格上的，格式上的
frame	帧
matrix	矩阵，阵列
consolidate	联合，合并
hierarchy	层次，体系，谱系

skeletal 骨架
illusion 幻觉，错觉

EXERCISES

Fill in the blanks with appropriate words or phrases found behind this exercise.

1. 2D computer graphics are mainly used in applications that were originally developed upon traditional printing and drawing technologies such as_____.
2. A large form of digital art being pixel art is created through the use of _____software, where images are edited on the pixel level.
3. A sprite is a two-dimensional image or animation that is_____ into a larger scene.
4. Vector graphics consists in encoding information about shapes and colors that comprise the image, which can allow for more flexibility in_____.
5. 3D rendering is how a _____can be displayed.
6. Vector graphics formats are _____ to raster graphics.
7. To create the illusion of_____, an image is displayed on the computer screen then quickly replaced by a new image that is similar to the previous image, but shifted slightly.
8. Two-dimensional models give more direct control of the image than _____.
9. _____ graphics software will change with each keyframe.
10. In computer animation _____ allow an evaluated result returned from a user-defined logical expression.

    a. model
    b. integrated
    c. typograph
    d. expression-based techniques
    e. raster graphics
    f. movement
    g. complementary
    h. The 2D/3D
    i. rendering
    j. 3D computer graphics

10.3 GRAPHICS SOFTWARE (1)

1. Desktop Publishing

    Desktop publishing (DTP) grew naturally out of word processing though for a long time it was a separate activity[1]. Recently the two have shown all the signs of growing back together again. The difference between basic word processing (WP) and DTP can be seen by considering the traditional function of the author of a printed document, as compared with the function of the printer[2]. Before the advent of the desktop computer the author was responsible for producing a typescript—the process of assembling words in the right order. The printer then took those words and (perhaps with the aid of a designer or typographer) laid out the text in a particular manner, with or without appropriate illustrations, and printed them. The modern author does exactly what his predecessor did, but using a word processor, so that the words do not have to be retyped by the printer[3]. What DTP does is to automate most of the functions of the printer, using a desktop computer[4]. Four developments made this possible: the desktop computer with a GUI, DTP software, the laser printer and the page-description language-the PDL.

The importance of the laser printer was that a high-quality final product could be produced without the need for traditional typesetting processes (particularly the use of moveable type). The early laser printers, operating at 300~400dots per inch (dpi) could not rival traditional printing processes but could produce results that were acceptable for many everyday purposes, and at a much lower price[5]. The development of higher resolution laser printers and the digital type-setter (now called an image-setter) have since made it possible for work of almost any quality to be produced this way, though the term DTP, in some people's minds, is still associated with a poorer quality, amateur product[6].

The importance of the GUI lays in the fact that the function of DTP software is to lay out pre-prepared text and graphics and a GUI allows the user to see immediately an accurate representation of the final product[7]. This "what you see is what you get" (WYSIWYG) feature is vital to DTP though the slogan should be taken with a pinch of salt[8]. "what you see" on a 72 dpi screen can never be a wholly accurate representation of "what you get" on a 300~1200dpi (or better) printer. The slogan should, perhaps, be changed to WYSIANATTCMTWYG— "what you see is as near as the technology can manage to what you get". A necessary feature of a DTP package, therefore, is a zoom facility which displays a portion of the document at a larger size than normal so as to display it at something much closer to the resolution of the final printed product[9]. Unfortunately only a small part of the document can be seen at any one time in this magnified mode.

This idea that the software enables the user to lay out, electronically, the various items of text and graphics was an important aspect of the first DTP program—Aldus PageMaker (now Adobe PageMaker)[10]. The screen is used to represent a 'paste-board', items can then be dragged onto the paper and rearranged until a pleasing result is obtained. Each of the text and graphic items is contained in its own "frame"—a rectangular boundary—as if it were on its own separate piece of paper, and these are then moved and pasted into position to form the final design.

2. Electronic Publishing (CD-ROMs and the Internet)

In recent years more and more material has been published electronically rather than on paper. The two most important new media have been CD ROM and the Internet. As a result modern versions of many DTP packages and word processors provide the facility to output files in the formats which have been developed specifically for these two media—particularly HTML (Hyper-Text Markup Language) and PDF (Portable Document Format).

The hyper-text principle has also been extended to links to photographs, drawings, sounds, video, animations, tables of figures, maps etc, and the benefit to the user is that vast amounts of information can be made available—in such a way that the user can decide how much to access and in what order[11]. Print publishing on the other hand is still, essentially, a linear process. The reader reads the material in the order dictated by the writer, starting at the beginning and going on to the end. Hyperlinked material has no unique beginning and no unique end.

All this means that electronic publishing, though having many obvious overlaps with paper publishing, is very different medium needing different skills and different software[12].

New types of graphic-oriented software have been designed to produce electronic publications: web-publishing software; web-graphic design software; multimedia "authoring" software; PDF publishing software. One of the potential problems in establishing new software products of this kind is that of the proliferation of standards. Two standards, already referred to, are the PDF format—established by Adobe through its "Acrobat" suite of software. The other is the HTML format, now in the process of being extended to XML (Extensible ML), DHTML (Dynamic HTML) and VRML (Virtual Reality ML).

The PDF file format is used mainly for producing CD-ROM editions of existing printed products like computer manuals and magazines. To help establish the PDF format, Adobe created viewer software (called Acrobat "Viewer"), versions of which are available for all the major operating systems free of charge. PDF is, in fact, an extension of PostScript—Adobe's Page Description Language (PDL). For simple documents, all that is needed to produce a PDF document is a special "printer driver" that doesn't print—or at least, doesn't print on paper as one might expect but "prints" to a file[13]. The beauty of this approach is that, in principle, any software designed to produce text output—like a word processor—can be used to produce a PDF document. Once the special printer driver software (Acrobat "PDF Writer") has been installed as part of the operating system the user selects that driver at the output stage and "prints" to a PDF file any document created in that software[14].

The HTML format was designed as a way of describing the appearance of a Web page. Web viewers like "Netscape" and Microsoft's "Internet Explorer" are widely available either free or at a small charge and for exactly the same reason that PDF viewers are free—an attempt by software houses to establish their software as the dominant standard. HTML can also be thought of as a PDL (Page Description Language) which enables the user to specify what a Web page should look like and how it should link to other pages.

NOTES

[1] 主从复合句，Desktop…processing 是主句，though 引导一个让步状语从句，for a long time 是从句中的时间状语。

[2] The difference… and DTP 是主语，as…引导的是分词短语作状语。

[3] The modern…did 是主句，其中 what…是宾语从句，so that 引导一个目的状语从句。

[4] What 引导的是主语从句。

[5] The early…results 是主句，that…为定语从句，主句中 operating…是分词短语作定语。

[6] The development of…this way 为主句，though…product 为让步状语从句，in some people's minds 是从句中的插入语；since 为副词，意为从那时起；it 为先行代词，代替 to be produced…作宾语。

[7] that the function…and a GUI…product 是并列的同位语从句。

[8] with a pinch of salt 意为有些保留，though 引导的是让步状语从句。

[9] which 引导一个定语从句，so as to 引导一个目的状语。

[10] 主句为 The idea…was an…，that 引导的是同位语从句；Aldus…为同位语。

[11] 由 and 连接的两个并列句，后一句中 that 引导的是表语从句；破折号后面是同位语，其中 that 引导的是状语从句。

[12] 长句，that 引导的是宾语从句，中间插入 though 引导的让步状语从句，needing…分词短语作定语，overlap 重叠，重复。

[13] 两个 that 引导的都是定语从句，第一个从句修饰 all，第二个从句修饰 "printer driver"。

[14] once 引导一个条件状语从句。

KEYWORDS

GUI (Graphics User Interface)	图形用户界面
DTP (Desktop Publishing)	桌面出版
resolution	分辨率
image-setter	激光照排机
paste-board	粘贴板
HTML (Hypertext Markup Language)	超文本标记语言
PDF (Portable Document Format)	可移植文档格式
PDL (Page Description Language)	页面描述语言
graphics software	图形软件
WP(Word Processing)	（文）字处理
typescript	打印文稿
word processor	字处理器（软件）
laser printer	激光打印机

dpi (dots per inch)	每英寸点数
lay out	排版
Electronic Publishing	电子出版
video	视频
animation	动画
hyperlink	超链接
suite of software	软件套件

EXERCISES

1. Single choice (according to the text).

(1) DTP is_____.
    a. Digital transmitting processor   b. Desktop publishing
    c. Data transferring pipe

(2) GUI is_____.
    a. General User Interface   b. Graphics User Input
    c. Graphics User Interface

(3) WYSIWYG means_____.
    a. What you see is what you get   b. Where you see is where you get
    c. When you see is when you get

(4) PDF is_____.
    a. Portable Document Format   b. Programming Data Format
    c. Prolog Document Format

(5) HTML is_____.
    a. Hypertext Markup Logic   b. Hypertext Markup Language
    c. Hypertext Markup Link

(6) PDL is_____.
    a. Push Down List   b. Page Description Language
    c. Programmable Digital Logic

(7) Recently _____ have shown all the signs of growing back together again
    a. WP and GUI   b. WP and CD-ROM
    c. WP and DTP

(8) We can use _____ to establish the PDF format.
    a. Acrobat "Viewer"   b. Adobe's PDL
    c. Aldus PageMaker

(9) _____ is still a linear process.
    a. Hyperlinked material   b. Print publishing
    c. Web-oriented reading

(10) A frame in Desktop publishing is _____.

  a. a rectangular boundary   b. a ring boundary
  c. a software

2. True/False.

 (1) _____ The importance of the laser printer was that a high-quality final product could be produced without the need for traditional typesetting processes.

 (2) _____ The two most important new media are CD-ROM and Internet.

 (3) _____ Acrobat is the product of Microsoft.

 (4) _____ HTML can also be thought of as a PDL which enables the user to specify what a Web page should look like and how it should link to other pages.

 (5) _____ PDF format was designed to describe the appearance of a Web page.

 (6) _____ Electronic publishing has many obvious overlaps with paper publishing.

 (7) _____ PDF is an extension of Adobe's PDL.

 (8) _____ The early laser printers can operate at 300-1200 dpi.

 (9) _____ Using a word processor the author should ask the printer to retype.

 (10) _____ A special printer driver is used to produce a PDF document.

10.4 GRAPHICS SOFTWARE (2)

1. Computer-aided design (CAD)

  CAD is a classic example of a graphic application which has grown up on large powerful computers and at one time it would have been considered quite impractical to do any serious work of this type on a desktop computer[1]. It is true that much important CAD work is still done on graphic workstations but increasingly powerful applications are now also being run on ordinary desktop computers. One of the simplest types of computer-aided design is the preparation of 2-D plans—the sort of work done traditionally by draughtsman on a drawing board with the aid of tee-squares and compasses. This type of work can be done very rapidly with software which provides an electronic tee-square, rubber-banding, standardized parts and geometrical shapes, "instancing", grids, layers and a library of standard shapes[2].

  The more powerful CAD packages can be used to design and represent 3-D objects. These are sometimes called solid-modelers because they create a complete and unambiguous description of a 3-D object in the computer and can display images with hidden lines removed and exterior surfaces shaded. Because of the wide-spread interest in 3-D modeling, which has now spread well beyond the boundaries of CAD.

2. Illustration (drawing) software

  Though CAD software is associated with engineering design and other forms of technical drawing the facilities offered by such packages have proved useful in a wide variety of other applications. Illustration (or simply Drawing) software can be thought of as CAD software with a less specialized purpose. Like CAD software the shapes that are drawn are created and stored as vectors and so this type of software should be distinguished from "Painting software" which is used to create and modify bit-mapped images[3]. The differences between CAD and Illustration software lie in emphasis and direction rather than principle. In both types of software the user can assemble and edit a range of line-based geometric shapes; objects can be separately selected, moved, distorted and re-sized; bounded shapes can be filled with colours and patterns; objects can be grouped and treated as single objects and a variety of positioning aids like grids and snapping are available. The differences lie in a number of enhancements required, and a number of specialist features not required, by the typical user who might be using the software to produce semi-technical illustrations but might equally be producing nontechnical drawings for advertisements, leaflets and magazines[4]. In general such applications require a larger range of fonts, more colour and colour-effects, a completely different library of ready-made drawings and a more sophisticated range of curves. This images that are to be produced by this type of drawing package are less severely functional and more aesthetically pleasing—more like "art".

3. Business presentation software

This is another type of vector software with a very focused application—the rapid production of "presentations" (informative talks or short articles) which summarize as compactly as possible a set of ideas and/or data. The main features are simple drawing facilities (like illustration software), clip-art, attractively presented, small quantities of text and graphs. The target user, unlike the user of Illustration software or CAD, is not a designer and therefore it is important to be able to automate the production of an aesthetically-pleasing result. This is done by the provision of a wide range of "templates". A template, in the ordinary use of the term, is a pattern or guide that is used to help in the creation of some object—like a plastic sheet with the shape of letters of the alphabet cut in it which can be used to rapidly trace letters onto paper[5]. In graphics software a template is a design which the user can quickly adapt to their own purposes. For example, it could be a screen of "dummy" text which users adapt by substituting their own text for the one provided[6]. The point of the exercise is that a professional-graphic designer is employed in creating the templates—in choosing suitable fonts and colors, perhaps with a patterned background—and so the user does not need any design skills to produce a professional-looking result[7]. The process is also a lot quicker for the users than creating the screen from scratch.

NOTES

[1] 长句，which 引导的是定语从句；it 为先行代词，逻辑主语为 to do…; at one time 意思是曾经……。

[2] 长句，which 引导的是定语从句。instancing 动名词，意为"示例"。

[3] 本句句型较复杂，that 引出修饰 shapes 的定语从句，…and so…引出结果从句，which 引出的是修饰"Painting software"的定语从句。

[4] 本句中使用了两个过去分词 required 和 not required 作定语，分别修饰 enhancements 和 features，由于过去分词作定语时有被动的意味，由 by 引出行为主体；who 引导的定语从句修饰 user。

[5] 长句，主句是 A template…is，中间插入介词短语构成的状语；that 引导的是定语从句，破折号后面是同位语，cut in it 为过去分词短语作定语，修饰 letters，it 代表 plastic sheet；which 引导的定语从句，修饰 plastic sheet；trace 为描绘，描摹。

[6] it 代表 template，dummy 为虚的，伪的，one 代表 "dummy" text。

[7] that 引出表语从句；in creating… in choosing 是同位关系，so 引导的是目的状语从句。

KEYWORDS

bit-mapped image	位图图像
pattern	模式
vector	矢量，向量
template	模板
dummy	虚化
CAD (Computer-Aided Design)	计算机辅助设计
graphic workstation	图形工作站
2D (Dimension)-plan	二维平面
3D (Dimension) object	三维物体
illustration software	绘图软件
library of drawings	图形库
business presentation software	商业演示软件
pattern	图案，模式

EXERCISES

1. Fill in the blanks with appropriate words or phrases found behind this exercise.

(1) One of the simplest types of CAD is the preparation of_____ plans.

(2) Though CAD software is associated with engineering design and other forms of technical drawing, the facilities offered by such packages have proved useful in a wide variety of other_____.

(3) The differences between CAD and_____ lie in emphasis and direction rather than principle.

(4) In graphics software a_____ is a design which the user can quickly adapt to their own purpose.

(5) _____ is a vector software with a very focused application.
(6) _____ has now spread well beyond the boundaries of CAD.
   a. applications   b. 2D
   c. template   d. illustration software
   e. Business presentation software   f. 3-D modeling

2. True/False.

(1) _____ CAD work is only done on graphic workstation.
(2) _____ The more powerful CAD packages can be used to design and represent 3D objects.
(3) _____ Illustration (or simply Drawing) software can be thought of as CAD software with a less specialized purpose.
(4) _____ A template, in the ordinary use of the term, is a pattern or guide that is used to help in the creation of some object.
(5) _____ Only can Illustration software be used to assemble and edit a range of line-based geometric shapes.
(6) _____ A presentation should include informative talks or short articles.

10.5　DIGITAL IMAGE FILE FORMATS

Why do we need so many different types of image file formats? The short answer is that there are many different types of images and applications with varying requirements[1]. A more complete answer (which we will not go into here) also considers market share, proprietary information, and a lack of coordination within the imaging industry. However, some standard file formats have been developed, and the ones presented here are widely available. Many other image types can be readily converted to one of the types presented here by easily available image conversion software.

A field related to computer imaging is that of computer graphics. Computer graphics is a specialized field within the computer science realm that refers to the reproduction of visual data through the use of the computer. This includes the creation of computer images for display or print, and the process of generating and manipulating any images (real or artificial) for output to a monitor, printer, camera, or any other device that will provide us with an image. Computer graphics can be considered a part of computer imaging, insofar as many of the same tools the graphics artist uses may be used by the computer imaging specialist[2].

In computer graphics, types of image data are divided into two primary categories: bitmap and vector. Bitmap images (also called raster images) can be represented by our image model $I(r,c)$, where we have pixel data and the corresponding brightness values stored in some file format. Vector images refer to methods of representing lines, curves, and shapes by storing only the key points. These key points are sufficient to define the shapes, and the process of turning these into an image is called rendering. After the image has been rendered, it can be thought of as being in bitmap format, where each pixel has specific values associated with it.

Most of the types of file formats discussed fall into the category of bitmap images, although some are compressed, so that the I(r, c) values are not directly available until the file is decompressed. In general, these types of images contain both header information and the raw pixel data. The header must contain information regarding: (1) the number of rows (height), (2) the number of columns(width), (3) the number of bands, (4) the number of bits per pixel, and (5) the file type[3]. Additionally, with some of the more complex file formats, the header may contain information about the type of compression used and any other necessary parameters to create the image.

The simplest file formats are the BIN and the PPM file formats. The BIN format is simply the raw image data I(r, c). This file contains no header information, so the user must know the necessary parameters—size, number of bands, and bits per pixel—to use the file as an image. The PPM formats are widely used, and a set of conversion utilities are freely available (pbmplus). They basically contain raw image data with the simplest header possible. The PPM format includes PBM (binary), PGM (gray-scale), PPM (color), and PNM (handles any of the previous types). The headers for these image file formats contain a "magic number" that identifies the file type, the image width and height, the number of bands, and the maximum brightness value (which determines the required number of bits per pixel for each band)[4].

Two image file formats commonly used on many different computer platforms, as well as on the World Wide Web, are the TIFF (Tagged Image File Format) and GIF (Graphics Interchange Format) file formats. GIF files are limited to a maximum of 8 bits/pixel and allow for a type of compression called LZW. The 8 bits/pixel limitation does not mean that it does not support color images, it simply means that no more than 256 (2^8) colors are allowed in an image.

This is typically implemented by means of a lookup table (LUT), where the 256 colors are stored in one table, and a byte (8 bits) is used as an index (address) into that table for each pixel[5]. The GIF image header is 13 bytes long and contains the basic information required.

The TIFF file format is more sophisticated than GIF and has many more options and capabilities. TIFF files allow a maximum of 24 bits/pixel and support five types of compression, including RLE (run-length encoding), LZW, and JPEG (Joint Photographic Experts Group). The TIFF header is of variable size and is arranged in a hierarchical manner. TIFF is one of the most comprehensive formats available and is designed to allow the user to customize it for specific applications.

NOTES

[1] that 引导一表语从句，说明 The short answer 是什么。

[2] insofar as 在……范围内只要……，通常可写成 in so far as; the graphics artist uses

为省略 that 的定语从句，修饰 tools。

[3] the number of band 可译为频带宽度，但较为通用的是 bandwidth。

[4] that 引导的定语从句直到本句末尾，magic number 幻数。

[5] where 引导的是非限定性定语从句。

KEYWORDS

bitmap	位图
raster image	光栅图像
decompress	解压
gray-scale	灰度
magic number	幻数
LUT(look up table)	查找表
file format	文件格式
proprietary	所有人的，所有的，专利的
TIFF (Tagged Image File Format)	标记图像文件格式
GIF (Graphics Interchange Format)	图形交换格式
index	索引
RLE (Run-Length Encoding)	行程编程
JPEG (Joint Photographic Experts Group)	联合（静态）图像专家组

EXERCISES

Match the following terms to appropriate definition.

1. _____ graphics.
2. _____ vector.
3. _____ pixel.
4. _____ GIF.
5. _____ multimedia.
6. _____ header.
7. _____ compress.
8. _____ LUT.
9. _____ JPEG.
10. _____ TIFF.

   a. In computer graphics, a directed line segment.

   b. System-defined control information that precedes user data.

   c. A blending of media types: text, audio, visual and computer data in one convenient delivery medium.

   d. The branch of science and technology concerned with methods and techniques to present data in the form of drawing, picture and diagrams etc.

e. Graphics Interchange Format
f. Lookup table
g. To save storage space by eliminating gaps, empty fields redundancy, or unnecessary data to shorten the length of records or files.
h. Picture element
i. Tagged Image File Format
j. Joint Photographic Experts Group

CHAPTER 11
MODERN INDUSTRIAL AUTOMATION

11.1 OVERVIEW

Today's industries cannot survive worldwide competition unless they introduce new products with better quality (quality, Q), at lower cost (cost, C), and with shorter lead time (delivery, D). Accordingly, they have tried to use the computer's huge memory capacity, fast processing speed, and user-friendly interactive graphics capabilities to automate and tie together otherwise cumbersome and separate engineering or production tasks, thus reducing the time and cost of product development and production[1]. Computer-aided design (CAD), computer-aided manufacturing (CAM), and computer-aided engineering (CAE) are the technologies used for this purpose during the product cycle. Thus, to understand the role of CAD, CAM, and CAE, we need to examine the various activities and functions that must be accomplished in the design and manufacture of a product[2]. These activities and functions are referred to as the product cycle. The product cycle described by Zeid (1991) is presented here with minor modifications, as shown in Figure 11-1.

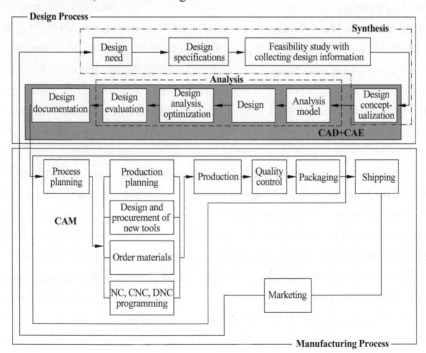

Figure 11-1 Product cycle

As indicated by the boxes bounded by solid lines in Fig.11-1, the product cycle is composed of two main processes: the design process and the manufacturing process. The design process starts from customers' demands that are identified by marketing personnel and ends with a complete description of the product, usually in the form of a drawing. The manufacturing process starts from the design specifications and ends with shipping of the actual products.

The activities involved in the design process can be classified largely as two types: synthesis and analysis. As illustrated in Figure 11-1, the initial design activities (such as identification of the design need, formulation of design specifications, feasibility study with collecting relevant design information, and design conceptualization) are part of the synthesis subprocess. That is, the result of the synthesis subprocess is a conceptual design of the prospective product in the form of a sketch or a layout drawing that shows the relationships among the various product components[3]. The major financial commitments needed to realize the product idea are made and the functionality of the product is determined during this phase of the cycle. Most of the information generated and handled in the synthesis subprocess is qualitative and consequently is hard to capture in a computer system.

Once the conceptual design has been developed, the analysis subprocess begins with analysis and optimization of the design. An analysis model is derived first because the analysis subprocess is applied to the model rather than the design itself. Despite the rapid growth in the power and availability of computers in engineering, the abstraction of analysis model will still be with us for the foreseeable future[4]. The analysis model is obtained by removing from the design unnecessary details, reducing dimensions, and recognizing and employing symmetry. Dimensional reduction, for example, implies that a thin sheet of material is represented by an equivalent surface with a thickness attribute or that a long slender region is represented by a line having cross-sectional properties[5]. Bodies with symmetries in their geometry and loading are usually analyzed by considering a portion of the model. In fact, you have already practiced this abstraction process naturally when you analyzed a structure in an elementary mechanics class.

Once a design has been completed, after optimization or some tradeoff decisions, the design evaluation phase begins. Prototypes may be built for this purpose. The new technology called rapid prototyping is becoming popular for constructing prototypes. This technology enables the construction of a prototype by deposition layers from the bottom to the top[6]. Thus it enables the construction of the prototype directly from its design because it requires basically the cross-sectional data of the product. If the design evaluation on the prototype indicates that the design is unsatisfactory, the process described is repeated with a new design.

When the outcome of the design evaluation is satisfactory, the design documentation is prepared. This includes the preparation of drawings, reports, and bills of materials. Conventionally, blueprints are made from the drawings and passed on to manufacturing.

As illustrated in Fig.11-1, the manufacturing process begins with process planning, using the drawings from the design process, and it ends with the actual products. The outcome of process planning is a production plan, a materials order, and machine programming. Other special requirements, such as design of jigs and fixtures, are also handled at this stage. The relationship of process planning to the manufacturing process is analogous to that of synthesis to the design process: It involves considerable human experience and qualitative decisions. This description implies that it would be difficult to computerize process planning. Once process planning has been completed, the actual product is produced and inspected against quality requirements. Parts that pass the quality control inspection are assembled, functionally tested, packaged, labeled, and shipped to customers.

NOTES

[1] 本句主要结构是 they have tried to use…to automate and tie…，其中 to use 有 3 个并列宾语；separate 此处为形容词，意思是"分开的""独立的"等。

[2] to understand…目的状语，主句为 we need…，其中 that 引导的是定语从句。

[3] in the form of a sketch or a layout drawing, 以草图或布局图的形式; that shows… 定语从句，修饰 a sketch or a layout drawing。

[4] Despite 引导的是让步状语从句，with us 指由我们去完成而不是用计算机完成。

[5] 长句，implies 后面由两个 that 引导的并列宾语从句构成。

[6] deposition, 沉积，此处表示一层层地构建原型。

KEYWORDS

CAD (computer-aided design)	计算机辅助设计
CAM (computer-aided manufacturing)	计算机辅助制造
CAE (computer-aided engineering)	计算机辅助工程
product cycle	产品周期
prototype	原型
feasibility	可行性
qualitative	定性的
symmetry	对称, 匀称
cross-sectional property	截面性质
design specification	设计规范（规格）
synthesis	综合
sketch	草图
layout	布局，规划，设计，草图
analysis model	分析模型
attribute	属性
optimization	优化

blueprint   蓝图

EXERCISES

1. Single choice (according to this text).

(1) CAD is an abbreviation of _____.
   a. Computer-Access Device   b. Computer-Aided Design
   c. Computer-Aided Diagnosis   d. Computer-Aided Detection

(2) CAM is an abbreviation of _____.
   a. Central Address Memory   b. Civil Air Management
   c. Computer-Aided Manufacturing   d. Computer-Aided Mathematical Analysis

(3) CAE is an abbreviation of _____.
   a. Computer-Aided Engineering   b. Computer-Aided Education
   c. Canadian Aviation Electronics d. Carbon Alcohol Extractions

(4) The product cycle is composed of _____ main processes.
   a. five   b. four   c. three   d. two

(5) The major financial commitments needed to realize a product idea are made in the phase of _____ process.
   a. analysis   b. manufacturing   c. synthesis   d. selling

(6) In analysis subprocess we should begin construct _____ first.
   a. a prototype   b. an analysis model
   c. an evolution model   d. a design model

2. True/False.

(1) _____ The manufacturing process starts from the design specifications and ends with shipping of the actual products.

(2) _____ Most of the information generated and handled in the synthesis subprocess is quantitative.

(3) _____ Once a design has been completed, after optimization or some tradeoff decision, the design evaluation phase begins.

(4) _____ The outcome of process planning is a production plan, a materials order, and machine programming.

(5) _____ When the result of the design is unsatisfactory the design documentation is prepared.

(6) _____ Once process planning has been completed, the actual product is produced.

11.2   USE OF CAD, CAM, AND CAE

We have described a typical product cycle. Now we will review it to show how the computer, or CAD, CAM, and CAE technologies, are employed in the cycle. As indicated earlier, the computer is not widely used in the synthesis phase of the design process because the computer does not handle qualitative information well. However, in the synthesis subprocess, for example, a designer might well collect the relevant design information for the feasibility study by using a commercial database and collect catalog information in the same way.

Nor is it easy to imagine how a computer might be used in the design conceptualization phase because the computer is not yet a powerful tool for the intellectual creative process[1]. The computer may contribute in this phase by physically generating various conceptual designs efficiently[2]. The parametric modeling or macroprogramming capability of computer-aided drafting or geometric modeling may be useful for this task. These packages are typical examples of CAD software. You may imagine a geometric modeling system to be a three-dimensional equivalent of a drafting system; that is, it is a software package by which a three-dimensional shape instead of a two-dimensional picture is manipulated[3].

The analysis subprocess of the design process is the area where the computer reveals its value. In fact, there are many available software packages for stress analysis, interference checking, and kinematic analysis. These software packages are classified as CAE. One problem with using them is the provision of the analysis model. It would not be a problem at all if the analysis model were derived automatically from the conceptual design. However, as explained previously, the analysis model is not the same as the conceptual design but is derived by eliminating unnecessary details from the design or by reducing its dimensions. The proper level of abstraction differs, depending on the type of analysis and the desired accuracy of the solution. Thus it is difficult to automate this abstraction process; accordingly the analysis model is often created separately. It is a common practice to create the abstract shape of the design redundantly by using a computer-aided drafting system or a geometric modeling system or sometimes by using the built-in capability of the analysis packages[4].

The analysis subprocess can be embedded in the optimization iteration to yield the optimal design. Various algorithms for finding the optimal solution have been developed, and many optimization procedures are commercially available. Optimization procedures could be thought of as a component of CAD software, but it is more natural to treat optimization procedures separately.

The design evaluation phase also can be facilitated by use of the computer. If we need a design prototype for the design evaluation, we can construct a prototype of the given design by using software packages that automatically generate the program that drives the rapid prototyping machine[5]. These packages are classified as CAM software. Of course, the shape of the prototype to be made should exist in advance in a type of data. The data corresponding to the shape are created by geometric modeling. Even though the prototype can be constructed conveniently with rapid prototyping, it would be even better if we could use a virtual prototype, often called digital mock-up, which provides the same valuable information[6].

As the analysis tools used to evaluate the digital mock-up become powerful enough to give an analysis result as accurate as that from the equivalent experiment on a real prototype, digital mock-ups will tend to replace real prototypes[7]. This tendency will increase as virtual reality technology enables us to get the same feeling from the digital mock-up as we get from the real prototype[8]. The activity of building digital mock-ups is called virtual prototyping. The virtual prototype can also be generated by a kind of geometric modeling that is specialized for that purpose.

The final phase of the design process is design documentation. In this phase, computer-aided drafting is a powerful tool. The file-handling capability of computer drafting systems also allows the systematic storage and retrieval of documents.

Computer technologies are also used in the manufacturing process. The manufacturing process includes the activities of production planning, design and procurement of new tools, ordering materials, NC programming, quality control, and packaging, as illustrated in Fig.11-1, so all the computer technologies for these activities can be classified as CAM. For example, computer-aided process planning (CAPP) software to aid the process planning activity is one type of CAM software. As mentioned previously, process planning is difficult to automate, and thus 100 percent automatic CAPP software is not available currently. However, there are many good software packages that generate the numerically controlled (NC) programs that drive NC machines. This type of machine creates a given shape when the shape exists in the computer in the form of data. This is similar to driving the rapid prototyping machine. In addition, also belonging to CAM are the software packages to program robot motion to assemble components or deliver them to the various manufacturing activities, or to program a coordinate measuring machine (CMM) to inspect the product[9].

NOTES

[1] 将否定词 nor 提至句首强调否定意义时，要求主谓倒装；because 引导一个原因状语从句；how 引导一个宾语从句。

[2] in this phrase，在这一阶段；conceptual designs，概念设计。

[3] that is 为插入语，it 代表 geometric modeling system；by which 引导的是定语从句，修饰 software package。

[4] It 为先行代词，实际的主语是 to create…；abstract shape 可译为抽象模型；by using…一直到句末为两个分词短语作状语。

[5] If 引导一个条件状语从句，we can… 为主句。by using… 为方式状语，第一个 that 引导定语从句修饰 packages，第二个 that 引导定语从句修饰 program。

[6] Even though 引导让步状语从句，it would be…为主句，if…为条件状语从句。often called…为过去分词短语作定语修饰 prototype，which 引导定语从句亦修饰 prototype；mock-up 为模型。

[7] As the analysis…是一个状语从句，其中又包含由 as…as…引导的同等比较状语

从句。

[8] 第一个 as 引导的是时间状语从句；第二个 as 引导的是方式状语从句。

[9] 倒装句，既强调 also belonging to CAM，又使长主语便于展开。正常语句应该是 the software package⋯are belonging to CAM。

KEYWORDS

interference checking	干扰检验
kinematic analysis	运动分析
iteration	迭代
virtual prototype	虚拟原型
NC (Numerical Control)	数控
CAPP (computer-aided process planning)	计算机辅助工艺计划
CMM (coordinate measuring machine)	坐标测量机
commercial database	商业数据库
macroprogramming	宏编程，宏程序设计
imbed(embed)	嵌入
optimal design	优化设计
type of data	数据类型
retrieval	检索
robot	机器人

EXERCISES

Fill in the blanks with appropriate words or abbreviations found behind this exercise.

1. The computer is not widely used in the synthesis phase of the design process because the computer does not handle_____ information well.
2. Nor is it easy to imagine how a computer might be used in the design conceptualization phase because the computer is not yet a powerful tool for the_____ creative process.
3. It would not be a problem at all if the analysis_____ were derived automatically from the conceptual design.
4. The analysis subprocess can be embedded in the optimization iteration to yield the_____.
5. If we need a design prototype for the design evaluation, we can construct a_____ of the given design by using software packages that automatically generate the program that drives the rapid prototyping machine.
6. The final phase of the design process is design_____.
7. Computer-aided process planning software to aid the process planning activity is one type of _____ software.
8. Process planning is difficult to automate, and thus 100 percent automatic_____

software is not available currently.
9. Depending on the type of analysis and the desired accuracy of the solution, the proper level of _____ differs.
10. A software package that programs _____ to inspect a product is belonging to CAM.

    a. CAPP   b. abstraction
    c. documentation   d. qualitative
    e. CMM   f. CAM
    g. prototype   h. model
    i. intellectual   j. optimal design

11.3 PRODUCT LIFECYCLE MANAGEMENT (PLM)

1. History

In industry, product lifecycle management (PLM) is the process of managing the entire lifecycle of a product from inception, through engineering design and manufacture, to service and disposal of manufactured products. PLM integrates people, data, processes and business systems and provides a product information backbone for companies and their extended enterprise.

2. Form

PLM systems help organizations in coping with the increasing complexity and engineering challenges of developing new products for the global competitive markets.

Product lifecycle management (PLM) should be distinguished from 'product life-cycle management (marketing)' (PLCM)[1]. PLM describes the engineering aspect of a product, from managing descriptions and properties of a product through its development and useful life; whereas, PLCM refers to the commercial management of life of a product in the business market with respect to costs and sales measures[2].

Product lifecycle management can be considered one of the four cornerstones of a manufacturing corporation's information technology structure. All companies need to manage communications and information with their customers (CRM-customer relationship management), their suppliers and fulfillment (SCM-supply chain management), their resources within the enterprise (ERP-enterprise resource planning) and their product planning and development (PLM)[3].

One form of PLM is called people-centric PLM. While traditional PLM tools have been deployed only on release or during the release phase, people-centric PLM targets the design phase.

As of 2009, ICT development (EU-funded PROMISE project 2004–2008) has allowed

PLM to extend beyond traditional PLM and integrate sensor data and real time 'lifecycle event data' into PLM, as well as allowing this information to be made available to different players in the total lifecycle of an individual product (closing the information loop)[4]. This has resulted in the extension of PLM into closed-loop lifecycle management (CL2M).

3. Benefits

Documented benefits of product lifecycle management include:
- Reduced time to market
- Increase full price sales
- Improved product quality and reliability
- Reduced prototyping costs
- More accurate and timely request for quote generation
- Ability to quickly identify potential sales opportunities and revenue contributions
- Savings through the re-use of original data
- A framework for product optimization
- Reduced waste
- Savings through the complete integration of engineering workflows
- Ability to provide contract manufacturers with access to a centralized product record
- Seasonal fluctuation management
- Improved forecasting to reduce material costs
- Maximize supply chain collaboration

4. Areas of PLM

Within PLM there are five primary areas:
- Systems engineering (SE)
- Product and portfolio m² (PPM)[5]
- Product design (CAx)[6]
- Manufacturing process management (MPM)
- Product data management (PDM)

5. Introduction to development process

The core of PLM is in the creation and central management of all product data and the technology used to access this information and knowledge. PLM as a discipline emerged from tools such as CAD, CAM and PDM, but can be viewed as the integration of these tools with methods, people and the processes through all stages of a product's life. It is not just about software technology but is also a business strategy.

Figure 11-2 shows a product lifecycle management.

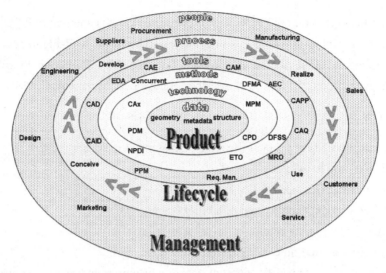

Figure 11-2 Product lifecycle management

For simplicity the stages described are shown in a traditional sequential engineering workflow. The exact order of event and tasks will vary according to the product and industry in question but the main processes are:

- Conceive

  -Specification   -Concept design

- Design

  -Detailed design   -Validation and analysis (simulation)   -Tool design

- Realise

  -Plan manufacturing   -Manufacture

  -Build/Assemble   -Test (quality check)

- Service

  -Sell and deliver   -Use

  -Maintain and support   -Dispose

The major key point events are:

- Order • Idea • Kickoff • Design freeze • Launch

The reality is however more complex, people and departments cannot perform their tasks in isolation and one activity cannot simply finish and the next activity starts. Design is an iterative process, often designs need to be modified due to manufacturing constraints or conflicting requirements. Where a customer order fits into the time line depends on the industry type and whether the products are for example, built to order, engineered to order, or assembled to order[7].

6. Pyramid of Production Systems

There are five long-term objectives that should be considered in production systems, as shows as Figure 11-3.

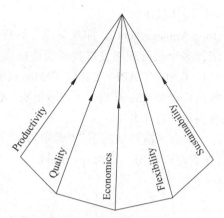

Figure 11-3　Pyramid of Production Systems

- Cost which can be measured in terms of monetary units and usually consists of fixed and variable cost.
- Productivity which can be measured in terms of the number of products produced during a period of time.
- Quality which can be measured, for example, in terms of customers' satisfaction.
- Flexibility, for example, ability of the system to produce variety of products.
- Ecological Soundness which can be measured in terms of biological and environmental impacts of the production system.

The relation between these five objects can be presented as pyramid in which its tip is associated with the highest productivity, the highest quality, the most economical, the most flexibility, and the most sustainability[8]. The points inside of this pyramid are associated with different combinations of five criteria. The tip of the pyramid is the ideal point but it is infeasible and the base of pyramid consists of the worst points.

NOTES

[1] PLM 和 PLCM 中文术语相同，都叫产品生命周期管理，只不过在 PLCM 中可注明"营销"。

[2] 长句，由连接词 whereas 连接的两个句子。

[3] CRM(customer relationship management)，客户关系管理，是一个不断加强与顾客交流，不断了解顾客需求，并不断对产品及服务进行改进和提高以满足顾客需求的连续的过程。

SCM(Supply Chain Management)，供应链管理，是一种集成的管理思想和方法，它执行供应链中从供应商到最终用户的物流的计划和控制等职能。

ERP 系统是企业资源计划（Enterprise Resource Planning）的简称，是指建立在信息技术基础上，以系统化的管理思想，为企业决策层及员工提供决策运行手段的管理平台。

[4] 长句,主句结构为 ICT development…has allowed PLM to extend…and integrate…, as well as allowing 现在分词短语作状语。

[5] PPM（Project Portfolio Management），即项目组合管理，是指在可利用的资源和企业战略计划的指导下，进行多个项目或项目群投资的选择和支持。

[6] CAx 是 CAD、CAM、CAE、CAPP、CIM、CIMS、CAS、CAT、CAI 等各项技术之综合叫法，因为所有缩写都是以 CA 开头，x 表示所有。CAx 实际上是把多元化的计算机辅助技术集成起来协调地进行工作。

[7] Where 此处为副词，意为"在该情况下""其中"。

[8] 长句。in which 引导的定语从句，修饰 pyramid。

KEYWORDS

Lifecycle	生命周期，生存周期
Inception	开始，开端
disposal	处理，清除
PLM (Product Lifecycle Management)	产品生命周期管理
PLCM (Product Life-Cycle Management)	产品生命周期管理
backbone	主干线，中心，中枢，供给中心
CRM(Customer Relationship Management)	客户关系管理
complexity	复杂性
cornerstone	基石
CL2M (Closed-Loop Lifecycle Management)	闭环生命周期管理
fulfillment	履行，实践，实现，达成
framework	框架，架构，体制组织
optimization	优化
revenue	收益，税收
fluctuation	波动，起伏
discipline	规程，规定，规范，要求，科目
portfolio	公文包，组合
strategy	战略，策略
conceive	构思，设想
validation	确认，生效
iterative	迭代的，重复的
pyramid	金字塔
flexibility	灵活性
ecological	生态的
soundness	健康，稳固
sustainability	可持续性
criteria	准则，标准

CHAPTER 11　MODERN INDUSTRIAL AUTOMATION

EXERCISES

Fill in the blanks with appropriate words or phrases found behind this exercise.

1. PLM integrates_____, data, processes and business systems and provides a product information backbone for companies and their extended enterprise.
2. PLM systems help organizations in coping with the increasing _____ and engineering challenges of developing new products for the global competitive markets.
3. Product lifecycle management can be considered one of the four_____ of a manufacturing corporation's information technology structure.
4. One form of PLM is called people-centric PLM. While traditional PLM tools have been deployed only on release or during the release phase, people-centric PLM targets the _____ phase.
5. The core of PLM (product lifecycle management) is in the _____ and central management of all product data and the technology used to access this information and knowledge.
6. PLM as a discipline emerged from tools such as CAD, CAM and PDM, but can be viewed as the _____ of these tools with methods, people and the processes through all stages of a product's life.
7. Design is an _____ process, often designs need to be modified due to manufacturing constraints or conflicting requirements.
8. There are _____ long-term objectives that should be considered in production systems.
9. In Pyramid of Production Systems the Cost means _____.
10. The main processes of a traditional engineering workflow are _____.

 a. cornerstones
 b. creation
 c. economics
 d. integration
 e. design
 f. five
 g. complexity
 h. Conceive, Design, Realize, Service
 i. iterative
 j. people

11.4　INDUSTRY 4.0

1. Name

The term was first used in 2011 at the Hanover Fair. In October 2012 the Working Group on Industry 4.0 chaired by Siegfried Dais (Robert Bosch GmbH) and Kagermann (acatech) presented a set of Industry 4.0 implementation recommendations to the German federal government. On 8 April 2013 at the Hanover Fair, the final report of the Working Group Industry 4.0 was presented.

2. Industry 4.0

Industry 4.0 is a collective term for technologies and concepts of value chain organization based on the technological concepts of cyber-physical systems, the Internet of Things and the Internet of Services, it facilitates the vision of the Smart Factory[1]. Within the modular structured Smart Factories of Industry 4.0, cyber-physical systems monitor physical processes, create a virtual copy of the physical world and make decentralized decisions. Over the Internet of Things, Cyber-physical systems communicate and cooperate with each other and humans in real time[2]. Via the Internet of Services, both internal and cross-organizational services are offered and utilized by participants of the value chain.

Meanwhile, in the United States, an initiative known as the Smart Manufacturing Leadership Coalition is also working on the future of manufacturing[3]. Smart Manufacturing Leadership Coalition (SMLC) is a non-profit organization of manufacturing practitioners, suppliers, and technology companies; manufacturing consortia; universities; government agencies and laboratories. The aim of this coalition is to enable stakeholders in the manufacturing industry to form collaborative R & D, implementation groups for development of the approaches, standards, platforms and shared infrastructure that facilitate the broad adoption of manufacturing intelligence[4].

Similarly, GE has been working on an initiative called 'The Industrial Internet'[5]. The Industrial Internet aims to bring together the advances of two transformative revolutions: the myriad machines, facilities, fleets and networks that arose from the Industrial Revolution, and the more recent powerful advances in computing, information and communication systems brought to the fore by the Internet Revolution[6]. According to GE, together these developments bring together three elements, which embody the essence of the Industrial Internet: Intelligent machines, advanced analytics and people at work.

3. Meaning

Characteristic for industrial production in an Industry 4.0 environment are the strong customization of products under the conditions of high flexibilized (mass-) production. The required automation technology is improved by the introduction of methods of self-optimization, self-configuration, self-diagnosis, cognition and intelligent support of workers in their increasingly complex work[7]. The largest project in Industry 4.0 at the present time is the BMBF leading-edge cluster "Intelligent Technical Systems OstWestfalenLippe (it's OWL)"[8]. Another major project is the BMBF project RES-COM, as well as the Cluster of

Excellence "Integrative Production Technology for High-Wage Countries"[9].

4. Difference

There are differences between a today's factory and an Industry 4.0 factory. In current industry environment, providing high-end quality service or product with the least cost is the key to success and industrial factories are trying to achieve as much performance as possible to increase their profit as well as their reputation. In this way, various data sources are available to provide worthwhile information about different aspects of the factory. In this stage, the utilization of data for understanding the current condition and detecting faults and failures is an important topic to research, e. g. in production, there are various commercial tools available to provide OEE (Overall Equipment Effectiveness) information to factory management in order to highlight root cause of problems and possible faults in the system[10]. In contrast, in an Industry 4.0 factory, in addition to condition monitoring and fault diagnosis, components and systems are able to gain self-awareness and self-predictiveness, which will provide management with more insight on the status of the factory. Furthermore, peer-to-peer comparison and fusion of health information from various components provides a precise health prediction in component and system levels and enforces factory management to trigger required maintenance at the best possible time to reach just-in time maintenance and gain near zero downtime.

5. What are the challenges

- Lack of adequate skill-sets to expedite the march towards fourth industrial revolution
- Threat of redundancy of the corporate IT department
- General reluctance to change by stakeholders

6. Role of Big Data and Analytics

Modern information and communication technologies like Cyber-Physical Systems, Big Data or Cloud Computing will help predict the possibility to increase productivity, quality and flexibility within the manufacturing industry and thus to understand advantages within the competition[11].

Big Data Analytics consists of 6Cs in the integrated Industry 4.0 and Cyber Physical Systems environment. 6C system that is consist of Connection (sensor and networks), Cloud (computing and data on demand), Cyber (model & memory), Content/context (meaning and correlation), Community (sharing & collaboration), and Customization (personalization and value). In this scenario and in order to provide useful insight to the factory management and gain correct content, data has to be processed with advanced tools (analytics and algorithms) to generate meaningful information. Considering the presence of visible and invisible issues in an industrial factory, the information generation algorithm has to capable of detecting and addressing invisible issues such as machine degradation, component wear, etc. in the factory floor.

7. Impact of the Industry 4.0

There are many areas that are foreseen to have an impact with the advent of the fourth industrial revolution. Of which four key impact areas emerge:
- Machine Safety
- Industry value chain
- Workers
- Socio-economic

NOTES

[1] cyber-physical systems，the Internet of Things，the Internet of Services，分别是指信息物理系统，物联网和因特网服务。其中cyber-physical systems（CPS）：即信息物理系统，又称信息物理融合系统，已成为国内外学术界研究的重要方向，被认为是继计算机、互联网之后，世界信息技术的第三次浪潮。

[2] 由and 连接的并列句，只不过humans 后面省略了谓语communicate and cooperate。

[3] initiative 为名词，在句中作主语，known as…过去分词短语作定语，修饰 initiative。

[4] 长句。句中 to enable…为表语，to form…为目的状语，最后的that引导的是定语从句。

[5] Industrial Internet，工业互联网，简称工业网络，是美国通用电器（GE）提出的概念。工业化创造了无数的机器、设施和系统网络。工业网络是指让这些机器和先进的传感器、控制和应用软件相连接，以提高生产效率、减少资源消耗。

[6] 长句。句子中冒号后面的句子是对 two transformative revolution 的解释，中间用 and 连接起来的两个成分中分别有that 引导的定语从句和过去分词 brought 短语。To the fore，为在前面，在显著地位。

[7] 长句，by…直到句末均为方式状语。

[8] Intelligent Technical Systems OstWestfalenLippe，是一种智能技术系统，该系统是2012 年德国联邦政府开展的称为"前沿集群竞赛项目（Leading-Edge Cluster Competition）"的五个获奖项目之一。

[9] RES-COM 是 BMBF（德国联邦教育与研究部）2011 年资助的课题。Cluster of Excellence，卓越集群。

[10] 是由两个句子构成的长句。

[11] BigData 大数据，见5.4 节。Cloud Computing 云计算，见5.3 节。

KEYWORDS

recommendation	推荐，劝告
collective	共同的，集合的

value chain organization	价值链组织
cyber-physical systems	信息物理系统
Internet of Things	物联网
Smart Factory	智慧工厂，智能工厂
decentralized	分散的
coalition	联盟
initiative	初步，主动，创始力
R&D (Research and Development)	研发
consortia	联盟，联合体
stakeholders	利益相关者
advocacy group	倡导团体，利益团体，宣传组
myriad	无数，巨大数量的
customization	定制，定型，用户化
cognition	认知
intelligent	智能的
OEE (Overall Equipment Effectiveness)	设备综合效率
peer-to-peer	对等
trigger	引起，触发
downtime	停工期，故障停机时间
Big Data	大数据
correlation	关联
scenario	方案，情境
degradation	退化，老化，降级
socio-economic	社会经济

EXERCISES

Fill in the blanks with appropriate words or phrases found behind this exercise.

1. On 8 April 2013 at the Hanover Fair, the final report of the _____ was presented.
2. _____ is a collective term for technologies and concepts of value chain organization Based on the technological concepts of cyber-physical systems, the Internet of Things and the Internet of Services.
3. Within the modular structured _____ of Industry 4.0, cyber-physical systems monitor physical processes, create a virtual copy of the physical world and make decentralized decisions.
4. Over the Internet of Things, _____ communicate and cooperate with each other and humans in real time.
5. Via the _____, both internal and cross-organizational services are offered and utilized by participants of the value chain.

6. Similarly, GE has been working on an initiative called _____.
7. The essence of the Industrial Internet are _____, advanced analytics and people at work.
8. The required automation technology is improved by the introduction of methods of self-optimization, self-configuration, _____, cognition and intelligent support of workers in their increasingly complex work.
9. Big Data Analytics consists of _____ in the integrated Industry 4.0 and Cyber Physical Systems environment.
10. Considering the presence of visible and _____ issues in an industrial factory, the information generation algorithm has to capable of detecting and addressing invisible issues such as machine degradation, component wear, etc. in the factory floor.

    a. Smart Factories
    b. The Industrial Internet
    c. Working Group Industry 4.0
    d. Industry 4.0
    e. Internet of Services
    f. Cyber-physical systems
    g. self-diagnosis
    h. Intelligent machines
    i. invisible
    j. 6Cs

11.5　3D PRINTING

3D printing (or additive manufacturing, AM) is any of various processes used to make a three-dimensional object. In 3D printing, additive processes are used, in which successive layers of material are laid down under computer control[1]. These objects can be of almost any shape or geometry, and are produced from a 3D model or other electronic data source. A 3D printer is a type of industrial robot. Figure 11-4 is an ORDbot Quantum 3D printer[2].

Figure 11-4　An ORDbot Quantum 3D printer

3D printing in the term's original sense refers to processes that sequentially deposit material onto a powder bed with inkjet printer heads[3]. More recently the meaning of the term has expanded to encompass a wider variety of techniques such as extrusion and sintering based processes. Technical standards generally use the term additive manufacturing for this broader sense.

1. General principles

(1) Modeling

3D printable models may be created with a computer aided design (CAD) package or via a 3D scanner or via a plain digital camera and photogrammetry software.

The manual modeling process of preparing geometric data for 3D computer graphics is similar to plastic arts such as sculpting. 3D scanning is a process of analyzing and collecting digital data on the shape and appearance of a real object. Based on this data, three-dimensional models of the scanned object can then be produced.

Regardless of the 3D modeling software used, the 3D model (often in .skp, .3ds or some other format) then needs to be converted to either a .STL or a .OBJ format, to allow the printing (a.k.a. "CAM") software to be able to read it[4].

Figure 11-5 is a 3D model slicing.

Figure 11-5   3D model slicing

(2) Printing

Before printing a 3D model from an STL file, it must first be examined for "manifold errors," this step being called the "fixup"[5]. Especially STLs that have been produced from a model obtained through 3D scanning often have many manifold errors in them that need to be fixed. Examples of manifold errors are surfaces that do not connect or gaps in the models[6]. Examples of software that can be used to fix these errors are netfabb and Meshmixer, or even Cura, or Slic3r[7].

Once that's done, the .STL file needs to be processed by a piece of software called a "slicer" which converts the model into a series of thin layers and produces a G-code file containing instructions tailored to a specific type of 3D printer (FDM printers) [8]. This G-code file can then be printed with 3D printing client software (which loads the G-code, and uses it to instruct the 3D printer during the 3D printing process). It should be noted here that in practice the client software and the slicer are often combined into one software program[9]. Several open source slicer programs exist, including Skeinforge, Slic3r, and Cura-engine as well as closed source programs including Simplify3D and KISSlicer[10]. Examples of 3D printing clients include Repetier-Host, ReplicatorG, Printrun and Cura[11].

(3) Finishing

Though the printer-produced resolution is sufficient for many applications, printing a slightly oversized version of the desired object in standard resolution and then removing material with a higher-resolution subtractive process can achieve greater precision[12].

Some printable polymers allow the surface finish to be smoothed and improved using chemical vapour processes.

Some additive manufacturing techniques are capable of using multiple materials in the course of constructing parts. These techniques are able to print in multiple colors and color combinations simultaneously, and would not necessarily require painting.

All of the commercialized metal 3-D printers involve cutting the metal component off of the metal substrate after deposition. A new process for the GMAW 3-D printing allows for substrate surface modifications to remove aluminum components manually with a hammer.[13]

2. Applications

AM technologies found applications starting in the 1980s in product development, data visualization, rapid prototyping, and specialized manufacturing. Their expansion into production (job production, mass production, and distributed manufacturing) has been under development in the decades since. Industrial production roles within the metalworking industries achieved significant scale for the first time in the early 2010s. Since the start of the 21st century there has been a large growth in the sales of AM machines, and their price has dropped substantially. According to Wohlers Associates, a consultancy, the market for 3D printers and services was worth $2.2billion worldwide in 2012, up 29% from 2011. There are many applications for AM technologies, including architecture, construction, industrial design, automotive, aerospace, military, engineering, dental and medical industries, biotech (human tissue replacement), fashion, footwear, jewelry, eyewear, education, geographic information systems, food, and many other fields[14].

Figure 11-6 Model of a turbine showing the benefits of 3D printing in industry.

In 2005, a rapidly expanding hobbyist and home-use market were established with the inauguration of the open-source RepRap [15]. Virtually all home-use 3D printers released to-date have their technical roots in the ongoing RepRap Project and associated open-source

Figure 11-6　Model of a turbine showing the benefits of 3d printing in industry

software initiatives[16]. In distributed manufacturing, one study has found that 3D printing could become a mass market product enabling consumers to save money associated with purchasing common household objects[17]. For example, instead of going to a store to buy an object made in a factory by injection molding (such as a measuring cup or a funnel), a person might instead print it at home from a downloaded 3D model[18].

NOTES

[1] laid down，为建造，铺设之意，laid 为 lay 的过去分词，which 代表 additive processes。

[2] ORDbot Quantum 3D printer，ORDbot 量子 3D 打印机。

[3] that 引导的是定语从句，修饰 processes。

[4] 长句，本段就是一句话，句中 to allow……是目的状语，to be able……是动词不定式作复合宾语。a.k.a. 即 also known as。.skp 通常指的是 SketchUp 这个软件，中文名为"草图大师"，这个软件在建筑设计中越来越多地被使用，它和传统的 3d 软件相比有可视性强、易操作等特点，在设计初期对建筑形体关系进行推敲的时候经常用到。

STL 文件是在计算机图形应用系统中用于表示三角形网格的一种文件格式。它的文件格式非常简单，应用很广泛。OBJ 是一种 3D 模型文件格式，由 Alias|Wavefront 公司为 3D 建模和动画软件 Advanced Visualizer 开发的一种标准，适合用于 3D 软件模型之间的互导，也可以通过 Maya 读写。3ds 即 3D Studio Max，常简称为 3ds Max 或 MAX，是 Discreet 公司开发的（后被 Autodesk 公司合并）基于 PC 系统的 3D 动画渲染和制作软件。其前身是基于 DOS 操作系统的 3D Studio 系列软件。CAM（Computer Aided Manufacturing）即计算机辅助制造，是利用计算机来进行生产设备管理控制和操作的过程。

[5] it 代表 3D model；fixup，准备，安排，选定，整顿。

[6] connect 此处是指 3D 打印连接，gaps 是指 3D 打印空白，此处系指补缺。

[7] netfabb 是一款常用的模型编辑软件，可以测量、修复以及检查模型文件。我们熟悉的 Shapeways，也是使用 netfabb 提供的服务，对用户上传的模型文件进行检查。

Cura 提供 3D 打印模型使用专门的硬件设备的简单方法。Slic3r 是一款用于将 STL 文件转换成 GCode 的开源软件，它具有更加快速生成，可配置参数更加灵活等特点。Meshmixer 是一款 Autodesk 开放的 3D 模型设计软件，主要用于修改、拼接模型，还可以对模型进行结构分析，添加支撑，以备打印。

　　[8] 长句，在 which 引导的定语从句中，有 converts 和 produces 两个谓语，第二个谓语后面的宾语 a G-code file 又先后有现在分词 containing 和过去分词 tailored 短语作定语，分别修饰其前面的名词。

　　slicer 此处是指 3D Slicer，是一个全面的、有效的开源软件包。G-code（易录宝）是一种自动录影制式。FDM (Fused Deposition Modeling)即熔融沉积成型工艺，熔融沉积制造工艺。

　　[9] that 引导的是主语从句。

　　[10] closed source program，封闭源程序。

　　Skeinforge 由 Enrique Perez 独立编写，是完整的工具链系统，能把 3D 模型转换为命令打印机的机器语言 G-code。Cura 是 Ultimaker 公司设计的 3D 打印软件，使用 Python 开发，并集成 C++开发的。Cura-engine 作为切片引擎，由于其切片速度快，切片稳定，对 3D 模型结构包容性强，设置参数少等诸多优点，拥有越来越多的用户群。

　　Simplify3D 是 2014 年 7 月 9 日德国 3D 打印公司 German RepRap 推出的一款全功能（All-in-One）3D 打印软件。Simplify3D 可取代 Repetier-Host 和 Slic3r 软件，支持导入不同类型的文件，可缩放 3D 模型，修复模型代码，创建 G 代码并管理 3D 打印过程。

　　KISSlicer 的全名是 Keep It Simple Slicer，意指"保持简单"的切片软件。

　　[11] ReplicatorG 的主要功能是把你的模型转换为打印机能识别的格式，但也可以用它对模型做一些基本的修改，使打印效果更好。Printrun 这款软件不仅有机器控制功能，还能和切片软件整合为一体（例如 slic3r），因此它可以独立完成从切片到打印的整个过程。它支持 Mac、Linux 和 PC 操作平台，几乎所有的开源 3D 打印机都可以使用这款软件。Repetier Host 是一款操作简单，将生成 Gcode 以及打印机操作界面集成到一起的软件，另外可以调用外部生成 Gcode 的配置文件，很适合初学者使用，尤其是手动控制的操作界面，可以很方便地实时控制打印机。

　　[12] Though 引导的是让步状语从句，主句是 and 连接的并列句。subtractive process，为 3D 打印中的减法处理。

　　[13] GMAW 3-D printing，熔化极气体保护焊 3D 打印。

　　[14] 长句，including…到末尾，列举了许多 AM 技术的应用。

　　[15] RepRap，是一个 3D 打印机原型机（或 3D 打印机），它具有一定程度的自我复制能力，能够打印出大部分自身的（塑料）组件。RepRap 是（replicating rapid prototyper）的缩写。

　　[16] technical roots，技术根基，技术基础。

　　[17] that 引导的是宾语从句，enabling…分词短语作定语。

　　[18] injection molding，注入成型。句中 instead of 作介词用，意为代替……，后面一个 instead 为副词，意为更换成……。

KEYWORDS

3D (3 Dimension)	三维
AM(Additive Manufacturing)	增材制造
process	过程，进程，处理
object	目标，对象，客体
successive layers	相邻层，连续层
shape	形状，整形
robot	机器人
deposit	沉积，沉淀
inkjet	喷墨
extrusion	挤压，伸长
modeling	建模，模型化，造型，模拟
CAD(Computer Aided Design)	计算机辅助设计
scanner	扫描仪
plain digital camera	普通数码相机
photogrammetry	摄影测量
plastic art	造型美术
sculpt	雕刻，雕像
manifold error	复印错误，印刷错误
clicer	限幅器，切片机
client	客户
closed source	封闭源
resolution	分辨率，清晰度，分解
precision	精度，精确，精密
prototyping	原型法
visualization	可视化，目视，显像
architecture	体系结构，结构，层次结构，总体结构
construction	构造

EXERCISES

Fill in the blanks with appropriate words or phrases found behind this exercise.

1. 3D printing uses a _____.
2. In 3D printing _____ of material are laid down under computer control.
3. In 3D printing objects are produced from a _____.
4. A 3D printer is a type of _____.
5. 3D printable models may be created with _____.
6. In order to prepare geometric data for 3D computer graphics, we need the _____.
7. _____ is a process of analyzing and collecting digital data on the shape and appearance of a real object.

8. CAM stands for _____.
9. The 3D model is often in _____ format.
10. Examining manifold errors is called as the _____.
11. Using _____ format can cause many manifold errors.
12. A _____ is a software that converts a model into a series of thin layers.
13. _____ file contains instructions tailored to a specific type of 3D printer.
14. In practice _____ and the slicer are often combined into one software program.
15. Cura-engine is a _____ slicer program.
16. The printer-produced resolution is sufficient for _____.
17. _____ can be used to smooth the surface of some printable polymers.
18. Some additive manufacturing techniques are able to print in _____ simultaneously.
19. All of the _____ involve cutting the metal components off the metal substrate after deposition.
20. Since the start of the 21st century the sales of _____ have been a large growth.
21. Virtually all home-use 3D printers have their _____ in the ongoing RepRap Project.
22. We can make a measuring cup at home by using a 3D model _____.

    a. manual modeling

    b. G-code

    c. AM machines

    d. fixup

    e. many applications

    f. downloaded from Internet

    g. additive processes

    h. .skp or .3ds

    i. 3D printing client software

    j. Chemical vapour processes

    k. 3D scanning

    l. 3D model

    m. technical roots

    n. Computer Aided Manufacturing

    o. CAD

    p. slicer

    q. commercialized metal 3D printers

    r. open source

    s. STL

    t. industrial robot

    u. multiple colors and color combinations

    v. successive layers

ANSWERS TO THE EXERCISES

1.1	1. f	2. h	3. o	4. n	5. i	
	6. m	7. d	8. p	9. g	10. a	
	11. l	12. e	13. k	14. b	15. c	16. j
1.2	1. a, b, c, d	2. a, b, d	3. b, c, d	4. a, c, d	5. a, b, c, d	
	6. b, c, d	7. b, d	8. a, b, c, d	9. b, c, d	10. b, d	
	11. a, c, d	12. a, b, c, d				
1.3	1. f	2. m	3. i	4. a	5. h	
	6. l	7. r	8. b	9. g	10. o	
	11. p	12. e	13. k	14. d	15. q	
	16. j	17. n	18. c			
1.4	1. (1) a, d	(2) a, b, c, d	(3) b, c, d	(4) a, c, d	(5) b, c, d	
	(6) a, b, c	(7) a, b, d	(8) b, c, d	(9) a, c, d	(10) a, b, c, d	
	(11) b, c, d	(12) a, c, d	(13) a, b, c, d	(14) a, b, c	(15) b, c, d	
	(16) a, c, d	(17) a, b, c, d				
	2. (1) f	(2) t	(3) t	(4) f	(5) t	
	(6) t	(7) f	(8) t			
1.5	1. d	2. m	3. j	4. z	5. l	6. x
	7. b	8. s	9. q	10. v	11. ee	12. o
	13. cc	14. g	15. p	16. r	17. f	18. w
	19. t	20. ff	21. y	22. h	23. c	24. aa
	25. i	26. u	27. a	28. dd	29. k	30. bb
	31. e	32. n				
2.1	1. t	2. f	3. t	4. f	5. f	
	6. t	7. t	8. t	9. f	10. t	
	11. f	12. t	13. f	14. f	15. t	16. t
2.2	1. (1) f	(2) t	(3) t	(4) f	(5) t	
	(6) t	(7) f	(8) f	(9) t	(10) t	
	2. (1) d	(2) g	(3) a	(4) i	(5) c	
	(6) e	(7) h	(8) j	(9) b	(10) f	
2.3	1. b, d	2. a, b, c, d	3. b, c, d	4. a, c	5. a, b, d	6. a, b, c, d
	7. a, b, c	8. a, b, c, d	9. a, b, d	10. b, d	11. c, d	12. a, b, c, d
2.4	1. d	2. l	3. g	4. k	5. e	6. b
	7. a	8. j	9. h	10. c	11. i	12. f
2.5	1. f	2. t	3. t	4. f	5. t	
	6. t	7. t	8. f	9. f	10. t	

	11. t	12. f	13. t	14. t	15. t	
	16. f	17. t	18. t	19. t	20. f	
	21. t	22. t	23. t	24. f		
2.6	1. m	2. n	3. k	4. j	5. i	
	6. b	7. a	8. e	9. h	10. d	
	11. l	12. g	13. c	14. f		
3.1.1	1. g	2. c	3. l	4. d	5. k	6. f
	7. j	8. a	9. h	10. e	11. i	12. b
3.1.2	1. a, b, c, d	2. a, b	3. b, c, d	4. a, b, c, d	5. b, c	
	6. a, d	7. a, b, c, d	8. a, b, c	9. b, c, d	10. a, b, c	
	11. a, c, d	12. a, b, c, d				
3.1.3	1. t	2. t	3. f	4. f	5. t	
	6. f	7. t	8. t	9. f	10. t	
3.2.1	1. a, b, d	2. a, b, d	3. b, d	4. a, b	5. b, c	6. a, c, d
	7. a, b	8. a, c, d	9. a, b, c, d	10. b, c, d	11. a, b, d	12. b, d
3.2.2	1. f	2. t	3. t	4. f	5. t	
	6. f	7. t	8. t	9. f	10. t	
	11. t	12. f	13. t	14. t		
3.3	1. a, b, c	2. a, b, d	3. b, c	4. a, c	5. b, d	
	6. b, c, d	7. a, c, d	8. b, c	9. a, b, c, d	10. b, c, d	
	11. a, b, c, d	12. a, c, d	13. a, b, c, d	14. b, c, d		
4.1.1	1. a, c, d	2. b, c, d	3. a, b, d	4. a, b, c	5. a, b, c, d	
	6. b, c, d	7. a, b, c, d	8. a, c, d	9. b, c, d	10. a, c	
4.1.2	1. f	2. f	3. t	4. f	5. f	
	6. t	7. t	8. f	9. t	10. f	
	11. t	12. t	13. f	14. t	15. t	16. t
4.2	1. c	2. i	3. f	4. m	5. b	
	6. j	7. d	8. n	9. g	10. a	
	11. l	12. h	13. k	14. e		
4.3.1	1. a, b, c, d	2. b, c, d	3. a, c, d	4. a, b, c, d	5. a, b, c, d	
	6. a, b, d	7. a, b, c, d	8. a, c, d	9. a, b, c, d	10. a, b, d	
	11. b, c, d	12. a, b, c, d	13. a, b, c, d	14. a, b, d	15. b, c, d	16. a, b, c, d
4.3.2	1. t	2. f	3. t	4. t	5. f	
	6. t	7. f	8. t	9. f	10. t	
	11. f	12. t	13. f	14. t	15. t	16. t
4.4.1	1. a, b, c, d	2. b, c	3. b, c	4. a, c, d	5. a, b, c, d	
	6. a, c, d	7. a, b, d	8. a, b, d	9. b, c, d	10. b, c, d	
	11. a, b, d	12. a, b, d	13. a, c, d	14. a, b, d	15. a, c, d	16. a, b, c, d

ANSWERS TO THE EXERCISES

4.4.2	1. j	2. d	3. a	4. f	5. h	6. g	
	7. l	8. e	9. b	10. k	11. c	12. i	
5.1.1	1. t	2. t	3. f	4. t	5. t		
	6. f	7. f	8. t	9. t	10. f		
	11. t	12. t	13. t	14. f	15. f	16. t	
5.1.2	1. b, d	2. a, b, c, d	3. a, b, c, d	4. c, d	5. a, d		
	6. a, b, c	7. c, d	8. a, c, d	9. a, b, c, d	10. a, b, c, d		
5.1.3	1. t	2. t	3. f	4. t	5. f		
	6. t	7. t	8. f	9. f	10. t		
	11. t	12. f	13. t	14. t			
5.1.4	1. c	2. g	3. e	4. i	5. h		
	6. d	7. j	8. b	9. a	10. f		
5.1.5	1. a, b, c	2. a, b, d	3. b, c, d	4. a, b, d	5. a, b, d		
	6. a, b, c, d	7. b, c, d	8. a, c, d	9. a, b, c, d	10. b, c, d		
5.1.6	1. c	2. g	3. l	4. j	5. b	6. i	
	7. f	8. a	9. e	10. h	11. k	12. d	
5.2.1	1. d	2. a	3. h	4. g	5. e	6. c	7. f
	8. b	9. j	10. n	11. l	12. k	13. m	14. i
5.2.2	1. t	2. t	3. t	4. f	5. t		
	6. f	7. t	8. t	9. f	10. t		
	11. f	12. t	13. t	14. f	15. t		
	16. t	17. f	18. t				
5.2.3	1. a, b, c, d	2. a, b, d	3. b, c, d	4. a, c, d	5. a, b, c, d		
	6. a, b, c	7. a, b, d	8. b, c, d	9. a, b, c, d	10. a, b, c, d		
	11. a, b, c, d	12. b, c, d					
5.3	1. a, b, d	2. a, b, c, d	3. b, c	4. a, b, c, d	5. b, c, d		
	6. a, b, c, d	7. a, c, d	8. b, c	9. b, c, d	10. a, b, c, d		
	11. a, c, d	12. a, c	13. b, c, d	14. a, b, d	15. a, b, c, d		
	16. a, c, d	17. a, b	18. a, b, c, d				
5.4	1. t	2. f	3. t	4. t	5. f		
	6. t	7. t	8. f	9. t	10. f		
	11. f	12. t	13. t	14. t	15. f		
	16. t	17. t	18. t	19. f	20. f		
5.5	1. a, b, d	2. a, b, c, d	3. b, c, d	4. a, b, c	5. a, b, c, d		
	6. a, c, d	7. b, c, d	8. b, d	9. a, b, c, d	10. a, c, d		
	11. a, b, c, d	12. a, b, c, d	13. a, b, c, d	14. a, b, c, d			
6.1	1. t	2. t	3. t	4. f	5. t		
	6. t	7. f	8. t	9. f	10. t		

6.2	1. a, b, c, d	2. a, b, c, d	3. b, c, d	4. a, b, c	5. a, b, d	
	6. a, b, c, d	7. a, c, d	8. b, c	9. a,c,d	10. a, b, c, d	
6.3	1. c	2. f	3. n	4. k	5. a	6. p
	7. j	8. o	9. g	10. l	11. b	12. r
	13. i	14. q	15. d	16. h	17. m	18. e
6.4	1. f	2. t	3. f	4. t	5. f	
	6. t	7. t	8. t	9. t	10. f	
	11. f	12. t	13. f	14. t	15. t	
	16. t	17. f	18. f	19. t	20. t	
7.1	1. a, b, c, d	2. a, b, d	3. a, b, c, d	4. b, c, d	5. a, b, c, d	
	6. a, b, c, d	7. a, b, d	8. a, b, c, d	9. b, c, d	10. a, b, d	
	11. a, b, c, d	12. b, c, d				
7.2	1. t	2. t	3. t	4. t	5. f	
	6. t	7. f	8. f	9. t	10. t	
7.3	1. d	2. l	3. n	4. h	5. a	
	6. i	7. p	8. c	9. m	10. f	
	11. k	12. o	13. b	14. j	15. g	16. e
8.1	1. a, b, c, d	2. a, b, d	3. a, b, c, d	4. b, c, d	5. a, c, d	6. a, b, d
	7. b, c	8. a, b, d	9. a, b, c, d	10. a, c, d	11. a, b, c, d	12. a, b, c, d
8.2	1. d	2. j	3. g	4. o	5. a	
	6. m	7. i	8. e	9. n	10. c	
	11. l	12. h	13. p	14. b	15. k	16. f
8.3	1. t	2. f	3. t	4. f	5. t	6. t
	7. t	8. f	9. f	10. f	11. t	12. t
	13. f	14. f	15. f	16. t	17. t	18. t
9.1	1. b, d	2. a, b, c, d	3. a, b, d	4. b, c, d	5. a, b, c, d	
	6. b, c, d	7. a, c, d	8. a, b, c, d	9. c, d	10. a, b, c, d	
9.2	1. f	2. c	3. i	4. e	5. a	6. l
	7. b	8. d	9. g	10. k	11. j	12. h
9.3	1. f	2. g	3. d	4. a	5. j	
	6. i	7. c	8. e	9. h	10. b	
10.1	1. t	2. f	3. t	4. t	5. f	
	6. t	7. f	8. t	9. t	10. f	
10.2	1. c	2. e	3. b	4. i	5. a	
	6. g	7. f	8. j	9. h	10. d	
10.3	1. (1) b	(2) c	(3) a	(4) a	(5) b	
	(6) b	(7) c	(8) a	(9) b	(10) a	
	2. (1) t	(2) t	(3) f	(4) t	(5) f	

ANSWERS TO THE EXERCISES

		(6) t	(7) t	(8) f	(9) f	(10) t	
10.4	1. (1) b	(2) a	(3) d	(4) c	(5) e	(6) f	
	2. (1) f	(2) t	(3) t	(4) t	(5) f	(6) t	
10.5	1. d	2. a	3. h	4. e	5. c		
	6. b	7. g	8. f	9. j	10. i		
11.1	1. (1) b	(2) c	(3) a	(4) d	(5) c	(6) b	
	2. (1) f	(2) t	(3) t	(4) t	(5) f	(6) t	
11.2	1. d	2. i	3. h	4. j	5. g		
	6. c	7. f	8. a	9. b	10. e		
11.3	1. j	2. g	3. a	4. e	5. b		
	6. d	7. i	8. f	9. c	10. h		
11.4	1. c	2. d	3. a	4. f	5. e		
	6. b	7. h	8. g	9. j	10. i		
11.5	1. g	2. v	3. l	4. t	5. o		
	6. a	7. k	8. n	9. h	10. d		
	11. s	12. p	13. b	14. i	15. r		
	16. e	17. j	18. u	19. q	20. c		
	21. m	22. f					

BIBLIOGRAPHY

1. Timothy J. O'Leary, Linda I. O'Leary. Computing Essentials 2014. 计算机专业英语[M]. 北京：机械工业出版社，2015.
2. 刘艺，王春生. 计算机英语[M]. 4版. 北京：机械工业出版社，2013.
3. 司爱侠，张强华. 计算机英语教程[M]. 6版. 北京：电子工业出版社，2014.
4. https://en.wikipedia.org/wiki/Windows_10.
5. https://en.wikipedia.org/wiki/Microsoft_Office_Mobile.
6. https://en.wikipedia.org/wiki/MySQL.
7. https://en.wikipedia.org/wiki/Wi-Fi.
8. https://en.wikipedia.org/wiki/Bluetooth.
9. https://en.wikipedia.org/wiki/Massive_open_online_course.
10. https://en.wikipedia.org/wiki/Flipped_classroom.
11. https://en.wikipedia.org/wiki/Big_data.
12. https://en.wikipedia.org/wiki/Industry_4.0.
13. https://en.wikipedia.org/wiki/Internet_of_Things.
14. https://en.wikipedia.org/wiki/WeChat.
15. https://en.wikipedia.org/wiki/Home_network.
16. https://en.wikipedia.org/wiki/Storage_area_network.
17. https://en.wikipedia.org/wiki/Internet_area_network.
18. https://en.wikipedia.org/wiki/Twitter.
19. https://en.wikipedia.org/wiki/Cloud_computing.
20. https://en.wikipedia.org/wiki/Computer_graphics.
21. https://en.wikipedia.org/wiki/Product_lifecycle.
22. https://en.wikipedia.org/wiki/3D_printing.
23. https://en.wikipedia.org/wiki/Multimedia.

参考译文

第一部分 计算机体系结构和计算机网络

第1章 计算机组成和部件

1.1 计算机组成

1. 计算机的组成

计算机是一种可编程的电子设备,它接收输入、完成对数据的运算或处理,并输出和存储结果。因为它是可编程的,这些被称为程序的指令告诉计算机去做什么。计算机的这4种主要操作(输入、处理、输出和存储)之间的关系如图1-1所示。

完成这些任务的相应设备是输入设备、处理器、输出设备和存储器。

(1)输入设备

输入设备是任一种给计算机提供资料的设备。最常用的输入设备是键盘和鼠标,如图1-2所示。其他的有图像和条形码扫描器、操作杆、触摸屏、数码相机、电子笔、指纹阅读器和麦克风等。立体系统的输入设备是CD播放器和天线。

(2)处理器

任何计算机系统的核心都是中央处理器(CPU),放在计算机主机箱或系统单元中。

处理器由两个功能部件(控制部件和算术逻辑部件)与一组称为寄存器的特殊工作区组成。

图1-3描述了处理器的结构,其中,CPU的内部互连机构提供了控制部件、算术逻辑部件和寄存器之间的通信。

控制部件是负责监控整个计算机系统操作的功能部件。

控制部件从存储器中取出指令,并确定这些指令的类型或对其进行译码。然后将每条指令都分解成一系列简单的小步骤或动作。这样,它就控制了整个计算机系统的步进操作。

算术逻辑部件(ALU)是为计算机提供逻辑及计算功能的部件。控制部件将数据送入算术逻辑部件,然后由算术逻辑部件完成指令所要求的某些算术或逻辑运算。

寄存器是处理器内的存储单元。控制部件中的寄存器用来跟踪正在运行的程序的总体状态。控制器寄存器存储诸如当前指令、下一条将要执行的指令地址以及该指令的操作数等信息。在算术逻辑部件中,寄存器存放要进行加、减、乘、除以及要比较的数据项。而其他寄存器则存放算术和逻辑运算的结果。

（3）输出设备

与输入设备类似，输出设备也是人与各类计算机系统之间进行解释和通信的设备。输出设备从 CPU 中取出机器代码形式的输出结果，然后将其转换成人们可读的形式（例如打印或显示报告）或另一处理周期的机器输入代码。

在个人计算机系统中，常用的输出设备是显示器和台式打印机。比较大型的计算机系统通常要配备更大、更快的打印机、多台在线工作站和磁带机等。

（4）存储设备

存储设备是计算机的一部分，主要是用于存储诸如指令、程序和数据信息的。

存储器有两种类型，一是内存储器（有时称作主存储器），另一种是二级存储器。主存储器置于系统单元内，那里还有 CPU 和其他部件。二级存储器包括存储介质和驱动器，我们将在本书 1.4 节中叙述。

任意一台台式计算机（不一定必须是 PC）的框图如图 1-4 所示。它有一个很大的主存储器，用于保存操作系统、应用程序和数据，以及一个用于连接海量存储设备（磁盘、数字视盘/光盘只读存储器）的接口。它有各式各样的输入输出设备，用户用来实现输入（键盘、鼠标和音频设备）、输出（显示接口和音响设备）以及连接功能（连网和外围设备）。快速处理器需要一个系统管理程序去监视它的内核温度和供电电压，并可进行系统复位。

2. 计算机的种类

计算机有 4 种类型：超级计算机、大型计算机、中型计算机和微型计算机。

（1）超级计算机是功能最强的计算机，这些机器的性能特别高，通常由大单位使用。IBM Blue Gene 被许多人视为是世界上最快的计算机。

（2）大型计算机安装在专门布线和有空调的房间内，尽管它的功能不如超级计算机，但它有很高的处理速度和很大的数据存储量。例如，保险公司用大型计算机去处理上百万投保人的信息。

（3）中型计算机，也称为服务器，是功能比大型计算机弱，但比微型计算机强的计算机。最初，这种计算机由中型公司或大型公司的部门使用，用于其业务需要。现今，中型计算机更多地服务于终端用户，用于从数据库检索数据，或访问应用软件等特殊需求。

（4）微型计算机，是功能最弱而使用最广和增长最快的计算机。有 4 种微型计算机类型：台式、笔记本、平板和手持计算机（见图 1-5）。台式计算机很小，可以放在桌面上或桌边，但搬动起来比较困难。笔记本计算机，又称膝上电脑，是很轻的、便携式的，可以放在公文包里。平板电脑，又称平板计算机，是最新型的计算机，它比笔记本计算机更小、更轻，耗电也少。与笔记本计算机一样，平板电脑也有一个平面显示屏，但一般没有标准键盘，它使用一个在显示屏上出现的虚拟键盘，并且是触摸式的。最出名的平板电脑是苹果公司的 iPad 。手持计算机是最小的，可以放在手掌上。这些系统都是一个完整的计算机系统，包括电子部件、二级存储器和输入输出设备。个人数字助理（PDAs）和智能电话是使用最多的手持电脑。智能电话是蜂窝式电话，采用无线方法连入因特网。过去几年间，这些设备数量剧增。

1.2 微处理器和系统板

1. 微处理器

在微型计算机系统中，中央处理器（CPU）或称处理器，包含在一个芯片上，又称为微处理器。微处理器既可以安装在插入到系统板上的承载插件上，也可以安装在插入到系统板专用插槽上的盒上，如图 1-6 所示。微处理器是计算机系统的"大脑"。它有两个基本部件：控制器和算术-逻辑部件。这两个部件的功能已在本书的 1.1 节中介绍过。

2. 微处理器芯片

芯片容量通常用字长表示。字是 CPU 一次可以访问的位数（如 16、32 或 64），一个字的位数越多，计算机的功能就越强，速度就越快。正如前面所提到的，8 位组成一个字节。32 位的计算机一次可以访问 4 个字节。64 位的计算机一次可访问 8 个字节，因而能处理 64 位字的计算机是更快的。

早期的微机每秒钟能处理几百万个数据和指令，或称微秒级。比较新的微机处理数据和指令更快，为每秒几十亿个或称毫微秒级。相比之下，超级计算机运算速度以微微秒度量，速度比微型机快千倍以上，如图 1-7 所示。

微处理器两个最新最重要的发展是 64 比特处理器和双核芯片。直到最近，64 比特处理器只用于大型和超级计算机中。而当 64 比特处理器在当今功能较强的微机中都在使用时，就表明一切都在变化之中。

这种新型双核芯片有两个分开且独立的 CPU。这些芯片允许一台计算机在同一时间运行两道程序。例如，在端用户用 PowerPoint 去制作多媒体演示文稿时，另一程序可同时搜索一个大型数据库。而更重要的功能是以前只能在大型和超级计算机上运行的非常复杂的程序也可在微机上运行。这要求把程序专门设计成可以在每个 CPU 上单独处理的两部分。这种方法称为并行处理。

3. 多核处理器

多核处理器是有两个或多个独立的中央处理器（称为"核"）的单个计算机部件。图 1-8 为通用双核处理器框图。

制造厂商通常把几个核集成在一个集成电路模片上（称为单片多处理器-CMP），或集成在多个模片上，再封装成一个芯片盒。

多核处理器可以有 2 核（双核 CPUs，例如 AMD Phenom Ⅱ X2 and Intel Core Duo）、4 核（4 核 CPUs，例如 Intel's i7 processors）、6 核、8 核或更多的核。

多核处理器广泛用于多个领域，包括通用领域，嵌入式，网络，数字信号处理以及图形学领域。

商业上，Adapteva 公司的 Epiphany 架构，是一个多核处理器体系，可以在一个芯片上有多达 4096 个处理器，尽管推向市场的只是 16 核的版本。

4. 系统板

系统板也叫母板（主板）。系统板是整个计算机系统的通信媒体。系统部件中的每个部件都连向系统板。它的作用就像数据通路一样，允许各个部件之间相互通信。外部设备，如键盘、鼠标和监视器没有系统板也不能与系统部件通信。

台式计算机的系统板是在系统部件底部或在一侧。它是一块大的电路板，上面布满各种电子部件，包括插座、插槽和总线，如图1-9所示。

（1）插座是称为芯片的小型专用电子部件的连接点。芯片是由蚀刻在像砂子一样称为硅材料的方形薄片上的电路板组成的。这些电路板可能比你的指尖还小。芯片也可称为硅片、半导体器件或集成电路。芯片安装在承载插件内（见图1-10）。这些插件可以直接插入系统板的插座上或放在卡上，然后再将这些卡插到系统板的插槽上。插座用于将系统板与不同类型的芯片，包括微处理器和存储器芯片连接起来。

（2）插槽是专用卡或电路板的连接点。这些卡为计算机系统提供扩展功能。例如调制解调器卡插入系统板的槽中，可以与因特网连接。

（3）连接线，又称为总线，是连接各个电子部件之间的通信通路，这些电子部件安装在系统板上或附着在系统板上。

笔记本计算机、平板电脑和手持PC的系统板比台式机的小，但它们实现的功能和台式机系统板一样。

1.3 存储器

1. 存储系统的需求

存储系统有以下3项需求。

（1）容量：无限大，对程序和数据集大小没有任何约束。

（2）速度：无限快，等待时间在现有存储器技术下是最短的。

（3）价格：每位的价格在可使用的技术中应最低。

很明显，由于这3个需求相互制约，所以很难全部满足。不过随着半导体和磁存储技术的发展，这些需求几近得到满足。

2. 存储器

存储器是存放数据、指令和信息的地方。与微处理器一样，存储器是安装在与系统板连接的芯片上的。大家熟知的存储器芯片有3种类型：随机存储器（RAM）、只读存储器（ROM）和互补金属—氧化物半导体（CMOS）。

（1）随机存储器（RAM）

随机存储器芯片保存CPU正在处理的程序（指令序列）和数据。由于数据在处理之前，或程序在运行之前，它们必须先保存在RAM中，因而有时也称RAM为主存储器。RAM又称暂时或易失性存储器，这是由于大多数这类存储器当微机关闭时，其所存储的各种信息都会丢失。如果电源出故障或其他因素中断了对微机的供电，信息也会丢失。

随机存储器分为动态随机存储器（DRAM）和静态随机存储器（SRAM）两种类型。

① 动态随机存储器（DRAM）

最近生产的DRAM芯片有以下三种类型。

- 同步DRAM（SDRAM），比常规RAM芯片要快、要贵。SDRAM芯片采用流线型操作方式，方法是对该芯片和系统部件中的其他部件之间的数据和指令的传送进行协调同步。
- 双数据速率 SDRAM（DDR SDRAM），又称为 SDRAMII，它比 SDRAM 更快、更可靠、更贵。DDR SDRAM 芯片与 SDRAM 相比，在相同时间内传输数据的速率要快一倍。
- 直接 DRAM，是最快、最贵的一种芯片。

当今大多数微机都采用这几种 DRAM 芯片的组合方式。

② 静态随机存储器（SRAM）

静态 RAM 与 DRAM 一样，要求稳定的供电。与 DRAM 比较，SRAM 不需要太高的电源功率，但是更快更可靠。SRAM 也很贵，通常是专用的，这些应用之一是高速缓存或 RAM 缓存。

③ 高速缓冲存储器（缓存）

高速缓冲存储器改善了存储性能，其方式是将其作为存储器和 CPU 之间的一个暂时高速存储区。在有缓存的计算机中（并非所有计算机都有），计算机检测 RAM 中哪些信息是最常用的，并将其复制到缓存中，在需要时 CPU 能很快将这些信息从缓存中取出。

有三种不同类型或级别的缓存。
- 1 级（L_1），也称为主缓存和内部缓存，它放在微机处理器芯片上。
- 2 级（L_2），也称外部缓存，它比 1 级速率慢，但容量更大。在旧一点的微处理器构成的计算机中，2 级缓存是放在插入系统板的芯片上的。由较新的微处理器构成的新型计算机有内置在微处理器上的 2 级缓存。这种安排有时是指先进的传输缓存，它比安装在系统板上的缓存反应速度更快。
- 3 级（L_3），为最新的一种缓存，它与专用微处理器 L_2 缓存共同使用。L_3 缓存并未置入到微处理器中，而是使用系统板上的同步动态 SDRAM。

当今大多数微机都有两种或三种缓存，功能最强的有全部三种缓存。

④ 闪存

闪存芯片可以在电源断开后仍然能保存数据。这种随机存储器是最贵的，主要用在专用设备中，如数字移动电话、数码相机和便携式计算机中。

（2）只读存储器（ROM）

只读存储器（ROM）芯片上有在工厂制造时写入的程序。与 RAM 芯片不同，ROM 芯片不是易失性的，用户不能更改它上面的内容。"只读"是指 CPU 只能读或检索写到 ROM 芯片上的数据和程序。但计算机不能写（即编码或改变）ROM 中的信息或指令。

通常 ROM 芯片上有用于计算机具体操作的专用指令。例如，ROM 中的指令可以启动计算机，为键盘上的键安排专用的控制功能，并将字符显示在屏幕上。

（3）互补金属氧化物半导体（CMOS）

互补金属氧化物半导体（CMOS）芯片能使计算机系统更灵活，更具可扩充性。它含有当计算机系统启动时所需的基本信息。该芯片提供当前日期和时间、RAM 容量、键盘、鼠标、监视器和磁盘驱动器的类型等信息。与 RAM 不同，CMOS 是由电池供电的，当电源关断以后，其内容不会丢失。与 ROM 也不同，它上面的内容可以更改，以反映计算机系统有了变化，如增加了 RAM 和新的硬件设备等情况。

1.4 二级存储器

由于 RAM 是一种暂时或易失性存储器，因而更需要永久或非易失性存储器来存储数据和程序。又由于用户需要比通常的主存或 RAM 更大的存储容量，因此也需要外部存储器。

二级存储器是永久和非易失性存储器。使用像硬磁盘驱动器那样的二级存储器，数据和程序在计算机关闭之后仍得以保存。二级存储器的操作是通过向这些存储设备写入文件或从这些设备读出文件来实现的。

1．软磁盘

软磁盘是可移动存储媒体。这种软盘通常用来存储和传送相对小的字处理、电子表格和其他类型的文件。

2．硬磁盘

与软磁盘相比，硬磁盘存储和检索信息更快，容量更大。

有三种类型的硬盘：内部硬盘、盒式硬盘和硬盘组。

（1）内部硬盘

内部硬盘放在系统部件内部。对大多数微机系统，将一个内部硬盘驱动器设计为 C 驱动器，用于存储程序和大的数据文件。例如，几乎每台微机都用其内部硬盘存放操作系统和 Word 与 Excel 等主要应用程序。

（2）盒式硬盘

虽然内部硬盘访问速度快，但它们的存储容量是固定的，并且不能轻易地从机箱中取出。盒式硬盘很容易取出，就像从录像机中取出录像带一样。

盒式硬盘的主要用途是作为内部硬盘的补充。因为盒式硬盘易于取出，所以对保护敏感信息特别有用。

用于台式计算机的盒式硬盘的典型容量是 20～100GB。称为 PC 卡式硬盘的信用卡大小的盒式硬盘，用于笔记本计算机，通常容量可达 5GB。

（3）硬盘组

硬盘组也是可移动存储设备，用于存储海量信息，其容量远远超过其他类型的硬盘。虽然你可能从未见过这种硬盘，但你肯定用过它们。可以访问因特网的微机、小型计算机或大型计算机一般通过通信线路访问外部硬盘组。银行和信用卡公司用这种硬盘组记录金融信息。

有三种方法能改善硬盘性能，它们是磁盘高速缓存、独立磁盘冗余阵列（RAID）和文件压缩／解压缩。

3．光盘

现今的光盘可以存储大于 50 GB 的数据，相当于将几百万页打印稿或一个中等规模的图书馆都存在一张光盘上。

（1）高密度磁盘

高密度磁盘更应该称为光盘，是应用最广的光学形式之一。光盘驱动器在很多微机系统中是标准的。一般光盘存储容量（单面）为 650 MB 到 1 GB。

光盘有三种基本类型：只读、写一次和可重写型。

① 只读 CD-ROM，是光盘只读存储器，类似于商用音乐光盘。只读表明用户不能去写或擦除。因此，作为用户，你只能读由出版商刻上去的数据。CD-ROM 用于存储大型数据库和参考文献，也用于存储大型应用软件包。

② 写一次 CD-R，是一种可录制却只能写一次的光盘。此后可读多次而不损伤光盘，但不能再写和擦除。CD-R 常用于从因特网上下载数据和音乐的保存。

③ 可重写 CD-RW，代表可重写光盘，也称为可擦除光盘。CD-RW 与 CD-R 非常类似，但是这种盘表面在数据记录上以后是可以改变的。因为内容可以改变，CD-RW 常用于制作和编辑多媒体演示文稿。

（2）数字通用光盘（DVD）

数字通用光盘 DVD 又称为数字视频光盘。作为标准光盘这是一种可以取代 CD 的较新的形式。DVD 与 CD 非常类似，只是在相同的空间中可以容纳更多的数据。在一张 DVD 盘上可以存储 4.7GB 到 17GB 数据，容量比 CD 大 17 倍。与 CD 一样，DVD 也有三种类型：只读、写一次和重写 DVD。

（3）高清晰光盘

虽然 CD 和 DVD 是当今应用最广的光盘，但今后的要求是容量更大。虽然 DVD 盘满足了记录标准分辨率电影和音乐的容量要求，但不能满足记录高分辨率视频的要求，它要求的容量比 DVD 盘大约高 4 倍。高清晰光盘是下一代光盘，又称为 HiDef 光盘，容量远大于 DVD。与 CD 和 DVD 相同，HiDef 光盘也有三种基本类型：只读、写一次和可重写。

4. 其他类型的二级存储器

对于一般的微机用户，这三种基本存储选项——软盘、硬盘和光盘是相互补充各有所长的。现今几乎所有微机至少都配备了一个软驱、一个硬驱和一个光驱。对很多用户，这些二级存储设备会进一步配备更专用的存储器，如固态存储器、因特网硬驱和磁带机。

（1）固态存储器

迄今为止所讨论的每一种二级存储器都有活动部分。例如，硬盘旋转、读/写头移进移出。与这些设备不同，固态存储器没有活动部分。数据和信息以电的形式直接存入这些设备或从这些设备检索出来，很像访问传统计算机存储器一样。尽管这种类型的存储器比其他存储器更贵，但更可靠，功耗低。由于这些原因，这一技术正广泛用于专用二级存储器中。

① 闪存卡是信用卡大小的固态存储器，广泛用于笔记本计算机。闪存也用于各种专用输入设备中，用来捕获并把数据传送到台式机中。例如，闪存用于存储从数码相机捕获的图像，然后把这些图像传送到台式和其他计算机中。闪存也用于像 iPod 那样的数字媒体播放器中，用于存储和播放音乐和视频文件。

② USB 驱动器又称 USB 闪存驱动器，这类驱动器很小巧，可以放在钥匙链或项链上携带。这些驱动器可以很方便地直接连入计算机 USB 端口，传送文件，其容量一般为 2GB。2013 年 1 月，已经有 512GB 的驱动器可用。而 2013 年 1TB 的驱动器，在国际电子学和技术商展上首次公开，并在这一年后期提供使用。2TB 存储量的驱动器也在计划

之中，并在每单位容量的芯片面积和价格上有稳定的改善。而某些驱动器允许十万个读/擦除周期，这取决于所用的存储芯片的具体类型，并有十年的备用存储器时间。

（2）因特网硬盘驱动器

在万维网专用服务网站上向用户提供存储空间，这种存储器称为因特网硬盘驱动器如图1-11所示。

比起其他类型的二级存储器，因特网硬盘驱动器的优点是费用低，以及能从任何使用因特网的站点访问信息的灵活性。由于所有信息必须跨因特网传输，因此访问速度较低。另一个要考虑的问题，是用户要依赖服务站点提供的服务能力以及安全问题。由于这些因素的限制，因特网硬盘驱动器一般用作专用二级存储器，不用于存储高私密性和敏感信息。

（3）磁带机

为了在一盘录音带上找一首歌曲，你可能要播放几英尺带子。相比之下，在一个高密度音盘上找一首歌曲就很快。你选择某一磁道，磁盘播放器就直接移到该道上。简言之，这代表了两种不同的访问外部存储器的方法。这两种方法称为顺序访问和直接访问。

磁带机是慢速顺序访问设备。虽然访问磁带上的信息速度低，但它仍是备份数据的有效和常用工具。

5. 海量存储器

海量存储器是指大型机构要求的，能存储特别大量数据的二级存储器。海量存储器就是能满足这种单位数据需求的，专用大容量二级存储器。

很多大型机构都有一种称为企业存储系统的策略，以鼓励在他们的机构内有效和安全地使用跨网数据（见图1-12）。支持这种策略的一些海量存储设备是：

（1）文件服务器；

（2）RAID 系统；

（3）磁带库；

（4）DVD-ROM 和 CD-ROM 光盘机；

（5）单位因特网存储器。

1.5 输入与输出设备

1. 概述

输入和输出设备能使人与计算机通信。输入设备把人们能理解的数据转换为计算机能处理的格式。这种设备把人们经常读写的字母、数字和其他自然语言符号翻译成计算机能处理的一串0和1二进制数。输入设备也能输入其他类型的数据，如照片、语音和视频。

另一方面，输出设备把处理过的一串0和1转换成人们可以理解的格式。这些设备通常在屏幕上或纸上输出。输出设备以硬拷贝或软拷贝形式产生结果。术语硬拷贝一般是指输出在容易携带的永久性媒体上，如纸上。术语软拷贝一般是指输出成具有有限移植性的、暂时的形式，如计算机屏幕上。

有些设备，如调制解调器既是输入又是输出设备。

2．输入设备

（1）键盘输入

输入数据最常用的方法之一是使用键盘。键盘把人们能理解的数字、字母和专用符号转换成电信号。这些信号送往系统部件并由系统部件进行处理。大多数键盘采用名为 QWERTY 的方式对各个键做出安排。这一名字实际上是键盘从下往上数第 4 行前 6 个字母的布局。

键盘有多种不同的类型，从全尺寸的到小型的，从刚性的到柔性的，最常用的有：

- 传统键盘（见图 1-13）
- 柔性键盘（见图 1-14）
- 人体工程学键盘（见图 1-15）
- 无线键盘
- 个人数字助理（PDA）键盘

（2）定位设备

定位设备提供了最舒适的与系统部件的接口，其方法是接受定位光笔指令，然后将其转换成机器可读的输入信息。有多种不同的定位设备，包括鼠标、操作杆、触摸屏、光笔和记录笔等。

① 鼠标

鼠标控制在监视器上显示的指针。鼠标指针通常以箭头形式出现。当然可以根据需要经常改变指针的形状。鼠标可以有一个、两个或多个按钮，用于选择命令选项和控制屏幕上的鼠标指针。某些鼠标有一个轮式按钮，可以使屏幕上显示的信息滚动。虽然有多种不同的鼠标类型，但基本类型只有三种：机械鼠标、光学鼠标、无线鼠标。

② 操纵杆

操纵杆是计算机游戏中最常用的输入设备。通过改变操纵杆的压力、速度和方向来玩游戏。其他的控制，如按钮和触发器则用于指定命令或开始一些规定的动作，如图 1-16 所示。

③ 触摸屏

触摸屏是一种带有清晰塑料外层的特殊的监视器，在此层后面是不可见的红外线十字形光柱。这种结构可使人们能用手指接触屏幕去选择菜单项或命令。在人们需要很快获得信息时，触摸屏很有用。触摸屏通常用在饭店、自动取款机和信息中心等地方，如图 1-17 所示。

④ 光笔

光笔是由光敏元件做成的像笔一样的设备。使用时要将光笔对准监视器。这样它就闭合了光电子电路，因而确定输入或修改数据的一个点。于是，光笔就可用于编辑数字图像和图形了。

⑤ 记录笔

记录笔类似一种笔，常与平板电脑和个人数字助理一起用，如图 1-18 所示。记录笔用压力在屏幕上画图。通过手写识别软件，记录笔与计算机交互。该软件把手写注记翻译成系统能处理的格式。

（3）扫描设备

扫描仪在文本和图像上面移动。扫描设备把扫描过的文本和图像转换成系统部件能处理的格式。有四种扫描设备：光学扫描仪、读卡器、条形码阅读器和字符与标记识别设备。

① 光学扫描仪

光学扫描仪简称扫描仪，它接受由文本和图像组成的文档，并将其转换为机器可读的格式。这些设备不能识别单个字母和图像。相反，它们能识别亮、暗和构成单个字母或图像的着色区域。通常，文档扫描后存储为能进一步处理、显示、打印的文件，或存储起来供以后使用。有两种基本的光学扫描仪：平板和便携式。

② 读卡机

几乎每个人都使用信用卡、借贷卡、进出卡（停车或进入楼宇）以及其他类型的识别卡。这些卡上通常都有用户名、某种类型的识别码和嵌入在卡内的签名。另外，也常把编码信息存储在卡上。读卡机解释编码信息。有两种类型的读卡机：

- 磁卡阅读机

  磁卡阅读机，是迄今为止最常用的。编码信息存储在卡片背面的薄磁条上。当卡通过磁卡阅读机时，信息被读出。

- 射频读卡机

射频读卡机虽然不太常用，但更为方便，因为不要求卡与读卡机实际接触。这种卡有一个小型射频识别（RFID）微芯片，它包含用户编码信息。无论何时，只要卡距读卡机几英寸之内通过时，用户信息就被读出。

③ 条形码阅读器和 QR 码

你可能很熟悉日杂店中的条形码阅读器或扫描仪，如图 1-19 所示。这些设备或是手持棒阅读器或是平面扫描仪。这些设备含有光电元件，用于扫描或读条形码，或扫描印在产品包装盒上的垂直斑马条标识。

QR 码（快速响应码），是日本首次为汽车工业设计的矩阵条形码（或称二维条形码）。条形码是一种可读的光学标记，它含有相关信息。QR 码采用四种编码模式（数字、字母数字、字节/二进制和日本汉字）以有效地存储数据。QR 码也具有扩展功能。

由于 QR 码系统比起标准的 UPC 条形码，读出得更快，存储能力更强，所以很快在汽车工业以外普及。其应用包括产品跟踪，项目确认，时间跟踪，文档管理和市场经营。

QR 码是在白底上，由安排成正方形栅格的黑模块（方形点）组成，它可用图像设备（如照相机）去读，然后用 Reed-Solomon 错误纠正软件去处理，直到该图像能被解释为止。然后从图像的水平和垂直元素所呈现的图案中提取所需要的数据，如图 1-20 所示。

④ 字符和标记识别器

字符和标记识别器是用于识别专用字符和标记的扫描仪。这些设备是某些应用场合下的专用工具。有三种类型：

- 磁墨水字符识别（MICR）；
- 光符识别（OCR）；
- 光标识别（OMR）。

（4）图像捕获设备

图像捕获设备像传统复印机那样，是一种可对原件复印的光学扫描仪。例如，光学扫描仪可将照片做数字拷贝。另一方面图像捕获设备可生成或捕获原始图像。这类设备包括数码相机和数字视频相机。

（5）音频输入设备

音频输入设备能把声音转换成系统部件可以处理的格式。至今，应用最广的音频输入设备是麦克风。音频输入可以有多种形式，包括语音和音乐。

语音识别系统使用麦克风、声卡和专用软件。这些系统能使用户操作计算机，用语音命令制作文档。医生、律师和其他一些人都用便携式语音识别系统去录制口述语音（见图1-21）。这些设备可以录制几个小时的语音，然后再连入计算机系统去编辑、存储和打印口述信息。有些系统甚至可把口述语音从一种口述语言翻译成另一种语言，如从英语翻译成日语。

3．输出设备

（1）监视器

最常用的输出设备是监视器，又称为显示屏。这种监视器展现的是文本和图形画面。这种输出常称作软拷贝。监视器因尺寸、形状和价格不同而有很多种。几乎所有监视器都有几个明显的特性。

监视器最重要的特性是清晰度。清晰度是指所显示图像的质量和清晰程度。它是监视器几个特性参数的函数，包括分辨率、点距、刷新速率和尺寸。

- 分辨率是最重要的特性之一。在监视器上，图像是由一系列点或像素（图形元素）组成的。分辨率用这些点或像素组成的矩阵表示。例如，当今很多监视器的分辨率是1600行和1200列像素，总像素为1 920 000个。监视器的分辨率越高（更多的像素），图像越清晰。图1-22为最普通的监视器分辨率。
- 点距是每个像素之间的距离，比较新的监视器的点距为0.31mm或更小。点距越小，图像越清晰。
- 刷新速率，表示监视器上的图像多长时间更新或重新刷新一次。多数监视器的刷新速率是75 Hz，即监视器每秒钟重新刷新75次。当刷新速率小于75 Hz时，监视器上的图像会出现闪烁，因而会引起眼睛疲劳。刷新速率越快（图像重新刷新得更快），所显示的图像质量越好。
- 尺寸是指监视器观测区的对角线长度，通用的尺寸有15、17、19和21英寸，监视器的尺寸越小，所显示的图像质量越高。

监视器有几种类型，其中最常用的是阴极射线管（CRT）和平面监视器。当今大多数平面监视器是液晶显示器。

还有其他一些监视器，其中有：

- 电子图书是显示文本和图形的手提的、书本大小的设备；

- 数据投影仪是类似于幻灯投影仪的专用设备；
- 高清电视。

（2）打印机

你大概经常会用打印机去打印布置的家庭作业、照片和网页。打印机把系统部件处理过的信息翻译好，然后在纸上打印出来。打印机输出通常称为硬拷贝。

打印机有几个基本特性，它们是：

- 分辨率，与监视器类似，分辨率是表明所产生图像的清晰度的。打印机的分辨率是用每英寸点数（即 dpi）来表示的（见图 1-23）。大多数个人使用的打印机分辨率一般为 1200 dpi。dpi 越高，所打印的图像质量越高。
- 大多数打印机现在都有色彩功能。用户通常都可以选择是打印黑色，还是打印彩色。
- 速度是用每分钟打多少页来表示的。通常个人用的打印机打单色（黑色）的速度是每分钟 15～19 页；彩色是 13～15 页。
- 打印机内的存储器，是用来存储打印指令和等待打印的文档的。打印机内的存储器越大，其打印大型文档的速度越快。

有多种类型的打印机，它们是：

- 喷墨打印机；
- 激光打印机；
- 热敏打印机。
- 还有其他类型的打印机，包括点阵式打印机、绘图仪、照片打印机和便携式打印机。

（3）音频输出设备

音频输出设备把从计算机输出的音频信息翻译成人能听懂的声音。应用最广的音频输出设备是扬声器和耳机。这些设备与系统部件的声卡相连。声卡用于捕获和播放所录制的声音。音频输出设备用于演奏音乐，口头上将一种语言传译成另一种语言，以及从计算机系统把信息传给用户。

产生语音输出不像识别和解释语音输入那样困难。实际上，语音输出很普通，在清凉饮料饮品机、电话和汽车上常用。它也用于学习，如用作帮助学生学习外语的强化工具。

4. 输入和输出组合设备

很多设备把输入和输出功能组合起来了。这样做有时是为了节省空间，有时是用在专用设备上。常用的组合设备包括传真机、多功能设备、因特网电话和终端。

第 2 章　系统软件

2.1　计算机软件

正如前面已经提到的那样，术语"计算机软件"是指告诉计算机硬件去做什么的程序或指令。软件通常是以一种软件包形式购买的。这种软件包应该包括 CD 或 DVD 盘上的程序、指导和帮助材料、印刷的操作指南和用户手册，以及使用该软件的许可证——这些都是封装在压缩包装盒中或塑料袋中。你可以在商店里，通过邮购或通过因特网购买软件。如果你从因特网上下载这种软件，你就不会收到实际的软件包，而是直接以电子格式把这些内容（程序许可证、用户手册等）下载到你的计算机上。

计算机有两种基本软件类型：应用软件和系统软件。应用软件帮助你执行一种专门的任务。而系统软件则是指操作系统，以及所有在低层次管理计算机资源的公用程序。形象些说，就是应用软件在系统软件的"顶层"。计算机只有在拥有操作系统和系统公用程序时，才能执行应用软件。

1. 应用软件

应用软件通常被称为生产力软件。应用软件包括为最终用户设计的程序。三种应用软件是通用、专用和移动应用软件。

通用应用软件几乎为所有行业使用。这些应用软件是与计算机的处理能力密切相关的各类程序。其中一个应用是因特网导航、搜索和查找信息的浏览器软件。三个使用最多的浏览器是 Mozilla 的 Firefox、Microsoft 的 Internet Explorer 和 Google 的 Chrome。

专用应用软件有几千种，多集中用于学科和行业中。两类最著名的专用应用软件是图形学软件和万维网编著程序。

移动应用软件是为移动设备，如手机、平板电脑和其他移动设备设计的小程序。这种程序约有 50 余万种。最常用的移动应用软件是文本消息、因特网浏览和社交网络方面的。

2. 系统软件

系统软件是一组程序，它们互相合作并控制计算机系统的资源和操作。系统软件能使计算机系统的很多部件相互通信。系统软件有三种类型：操作系统、实用程序和语言翻译程序。

（1）操作系统

操作系统为用户或应用程序和计算机硬件之间提供一个接口。操作系统软件有许多品牌和版本。每一种操作系统软件都是为配合一个或多个具体的处理器工作而设计的。例如，像 Windows 这样的操作系统就是为配合英特尔公司生产的处理器设计的。许多 IBM PC 兼容计算机就含有这种品牌的处理器。大多数 Macintosh 计算机含有摩托罗拉公司生产的处理器。Windows 操作系统则不能和这种摩托罗拉处理器搭配工作。

（2）实用程序

实用程序帮您处理日常事务。使用这些程序可以完成一些专门任务——管理计算机资源，管理文件等。有一些实用程序是操作系统的一部分，而另外一些则是自含程序。下面列举一些实用程序的功能：

格式化磁盘——磁盘格式化实用程序提供给计算机如何格式化的指令。

从硬盘复制文件到软盘——文件管理实用程序则提供相应的指令给计算机。

做一份硬盘的文件备份——使用备份实用程序。

（3）语言翻译器

语言翻译器将类似英语的软件程序转换成计算机能理解的机器语言。Vista

Multimedia 聘请了一位程序设计师，编写一个软件来为商店所有库存编制目录。该程序师用一种称为 VB 的编程语言编写。程序语句指导计算机执行规定的动作。

但是，计算机不能识别 VB 程序语句，因为这些语句是用我们能读懂的语言编写的，这就是翻译器发挥作用的地方。翻译器将每一个 VB 程序语句转换成机器语言。在高级语言中的一条语句能代表几条机器语言指令。现在，这些语句能被执行，Vista Multimedia 库存目录也能被处理。

2.2 操作系统概述

在过去 30 年间操作系统朝着两个主要目标发展：第一，为程序的开发和执行提供一个方便的环境；第二，操作系统试图对计算任务进行调度以确保计算系统性能良好。

如前所述，操作系统对于其生产厂家及其运行的硬件环境通常是特有的。一般说来，安装一台新计算机系统的同时也购买了与该硬件相应的操作系统。用户需要能有效地支持其处理工作，且可靠的软件。

尽管各厂家的操作系统软件各不相同，但特性都是相似的。现代的硬件，由于其复杂性，要求其操作系统满足某些特定的标准。例如，考虑到该领域的现状，操作系统必须支持某种形式的联机处理。通常，与操作系统软件相关的功能有：

（1）作业管理；
（2）资源管理；
（3）I/O 操作控制；
（4）差错排除；
（5）存储管理。

1．作业管理

任何一个操作软件的一个非常重要的职责是对计算机系统要处理的作业进行调度。这是作业管理功能的主要任务之一。操作系统建立了程序处理的顺序，并确定了具体作业执行的序列。术语"作业队列"常用于描述等待执行的作业序列。操作系统在排列作业队列时权衡各方面因素，包括当前正在处理的作业，正在使用的系统资源，即将处理的程序所需的资源，与其他任务相比该作业的优先级，以及系统必须响应的任何特殊处理要求等。

操作软件必须评估这些因素并控制作业处理的顺序。

批处理系统通过一个驻留监控程序使作业自动排序，这大大提高了计算机的总体利用率。计算机不再需要等待人工操作。但是，由于 I/O 设备的速度相对于 CPU 仍然很低，因此 CPU 的利用率一直很低，于是就尝试了低速设备的脱机操作。缓冲技术是另一种改善系统性能的方法，它是通过将单个作业的输入、输出和计算的重叠操作来实现的。最终，假脱机（spooling）技术使得 CPU 将一个作业的输入同其他作业的计算和输出重叠起来操作。

假脱机技术还提供了一个作业池，其中的作业已读出并等待运行，这个作业池支持多道程序设计的概念。在多道程序下，几个作业可同时保存在内存中，CPU 在它们之间来回切换服务，以便提高 CPU 的利用率，减少执行一个作业实际所需的总时间。

为提高系统性能而开发的多道程序设计技术同样也允许分时操作。分时操作系统允许多个用户（从一个到几百个）同时交错使用一个计算机系统。其他类型的操作系统还有实时系统和多处理器系统。

2. 资源管理

计算机系统中的资源管理是操作系统的另一个主要方面。显然，程序对不能使用的设备是无能为力的。正如我们所看到的，操作系统软件监督着所有程序的执行，它还监视正在运行的设备。为达到此目的，它建立了一张表，将程序与其正在使用或将要使用的设备相匹配，操作系统通过检查该表允许或拒绝使用某一设备。

3. 控制 I/O 操作

系统资源的分配与控制 I/O 操作的软件密切相关。由于 I/O 操作开始之前往往需要对特定设备进行访问，因此操作系统必须协调 I/O 操作和所使用设备之间的关系。实际上操作系统建立了一个执行程序和完成 I/O 操作必须使用的设备的目录。使用控制语句，作业可以调用指定设备，这使得用户可以在指定站点上读取数据或在选定的办公室内打印信息。利用此性能的优点，读自某一设备的数据可以分布于整个计算机化的系统中。

为便于 I/O 操作的执行，大多数操作系统都有一套标准的控制指令集来处理所有输入输出指令。这些标准指令称为输入输出控制系统（IOCS），是大多数操作系统不可分割的部分，它们简化了所有被处理的程序所承担的 I/O 操作。

实际上是正在执行的程序通知操作系统需要一个 I/O 操作，去使用某一 I/O 设备。控制软件调用 IOCS 软件去实际完成 I/O 操作。由于在大多数程序中还要考虑到 I/O 操作的级别，因此 IOCS 指令是至关重要的。

2.3 WINDOWS 10

Windows 10 是微软作为 Windows NT 操作系统系列的一个成员开发并即将推出的操作系统。Windows 10，2014 年 4 月在 Build 会议上首次展现，并安排在 2015 年中期公布，而现在由 Windows Insider 程序进行公开的 beta 测试。在其提供使用的第一年内，对于合法的 Windows 7 和 Windows 8.1 用户，可以免费升级到 Windows 10。

Windows 10 的目标是围绕通用内核统一 Windows PC、Windows Phone、Windows Embedded 和 Xbox One 产品系列。这些产品都共享一个通用的、"万能"的应用架构和 Windows Store 生态系统，该生态系统扩展到由 Windows 8 引出的 Windows Runtime 平台。Windows 10 还将推出一种新的、绑定的万维网浏览器 Microsoft Edge，用它取代 IE 浏览器。

图 2-1 为 Windows 10 屏幕快照。

1. 用户界面与桌面

Windows 10 改进了 Windows 8 的用户界面，这一界面按照使用的设备类型和所用的输入方法改变用户的习性。当一个键盘接入时，就会问用户是否希望切换到鼠标和键盘优化的用户界面模式，或停在触摸优化模式上。启用了一个新的重复开始菜单的方法，其方法是，在左侧有一个应用程序列表和"全部应用程序"按钮，在右侧有一个实时图形显示块。该菜单可以缩放，可以扩展成全屏幕显示，在触摸环境下，它是默认项。

一个新的称为任务查看的虚拟桌面系统加入进来。单击任务栏上的 Task View 按钮，或从屏幕左侧敲击，都会显示所有打开的窗口，并允许用户在这些窗口之间切换，或在多个工作空间之间切换。以前 Windows Store Apps 只能用在全屏幕方式，现在可用于桌面视窗方式或全屏幕方式。通过拖曳，可把程序视窗移到屏幕角上，只占屏幕的 1/4。当一个视窗被快速移动到屏幕的一侧时，用户就可以选择第二个视窗去填充未用的另一侧，这称为抓取助手。

2．特性

Windows 10 的主要特点是，在不同级别的设备之间，在 Windows 8 首次推出的 Windows 用户界面中出现的寻址不足问题上，集中协调用户的经验。

Windows Store app 生态系统被修订进 "Windows apps" 中。它们可以跨多个平台和设备级运行，包括智能电话、平板电脑、Xbox One 和其他与 Windows 10 兼容的设备。Windows apps 跨平台共享代码，有满足于设备要求和用于输入的响应式设计，能使 Windows 10 各设备之间的数据同步（包括通知，资历证明，并允许跨平台多人玩游戏），并通过统一的 Windows Store 进行分配。开发者也能"交叉购买"，这时购买的 app 许可证，可用于所有的用户兼容设备，而不是只能用于所购买的那个设备（即一个用户购买一个 PC 上的 app，也能在智能手机上使用，而无须额外付费）。

Windows 10 也允许 Web apps 和桌面软件（使用 Win32 或 .NET Framework）包装在一起，在 Windows Store 上销售。通过 Windows Store 销售的桌面软件将使用 App-V 系统包装，以允许沙箱工作。

3．版本

和以前一样，我们将提供适合于各种设备使用的不同 Windows 版本。这些版本满足不同客户的专门需求，从消费者到小型企业，到最大的企业。

- Windows 10 Home　是客户专用的桌面版，它为 PC、平板电脑和二合一电脑提供熟悉的个人使用环境。
- Windows 10 Mobile　用在像智能手机、平板电脑那样的小型、移动、触摸式设备上。
- Windows 10 Pro　是 PC、平板电脑和二合一的桌面版。它既有 Windows 10 Home 版相似的特性，也有改革之处。它有很多附加功能，能满足小型企业的各种需求。
- Windows 10 Enterprise　是构建在 Windows 10 Pro 上的，增加了先进功能，以满足中大型组织的需求。
- Windows 10 Education　是构建在 Windows 10 Enterprise 上的，以满足学校员工、管理人员、教师和学生的需求。
- Windows 10 Mobile Enterprise　是为企业客户的智能手机和小型平板电脑提供最佳手段的版本。

2.4 UNIX 和 LINUX

1. UNIX

UNIX 操作系统最早是由美国电话电报公司（AT&T）贝尔实验室的丹尼斯·里奇和肯·汤普森开发的操作系统，允许计算机同时接纳多个用户并处理多道程序。从 20 世纪 70 年代 UNIX 开发以来，许多个人，特别是加利福尼亚大学伯克利分校的计算机科学家对它进行了改进（通常称为伯克利软件版本 UNIX 或 BSD UNIX）。这种操作系统在各类计算机系统，从个人计算机到大型机上广泛使用，并可以用其他相关形式使用。AIX 是运行在 IBM 工作站上的，A/UX 是在 Macintosh 计算机上运行的图形版本；Solaris 在英特尔微处理器上运行。

特性

（1）UNIX 系统能支持多用户和多任务

（2）UNIX 系统内核

内核是 UNIX 操作系统的心脏，负责控制计算机的资源和调度用户作业，以便每一个用户都能合理地共享资源。程序与内核的交互作用是通过带有熟知名字的专用函数实现的，这称为系统调用。

（3）外壳

外壳是一种命令解释程序，其作用如同用户与操作系统之间的接口。当你在终端上输入一个命令时，外壳解释该命令并调用你所要求的程序。

（4）设备无关的输入和输出

对于 UNIX 程序，设备（如打印机或终端）和磁盘文件都是作为文件出现的。当你给 UNIX 操作系统一个命令时，你可以指挥操作系统把输出送到任何设备或文件上去。这种变更称为输出重定向。

2. Linux

Linux 是一种类似 UNIX 的、基本上兼容 POSIX 的计算机操作系统，是由免费和开源软件开发模型和 Linux 发行版汇集而成的。Linux 定义的部件是 Linux 核，是 1991 年 10 月 5 日由 Linux Torvalds 首先公布的操作系统的核。自由软件基金会（FSF）使用 GNU/Linux 这个名字去描述该操作系统，但引起了争论。

Linux 最初是为 Intel x86 型个人电脑开发的免费操作系统，但此后被移植到更多的硬件平台上，其数量超过任何其他操作系统。在服务器和其他大型机以及超级计算机中，Linux 是个领先的操作系统，但是用在台式机上仅占约 1.5%。 Linux 也运行在嵌入式系统中，在这些系统中，操作系统是典型的固件，并与整个系统是高度融合的。安卓，这种在平板电脑和智能手机中应用最广的操作系统，就构建在 Linux 核之上。

（1）硬件支持

Linux 核是被广泛移植的操作系统核，它运行在多种计算机架构上，包括基于 ARM 的手持 iPAQ 和 IBM 大型系统 z9 或 z10 上——覆盖范围从手机到超级计算机，如图 2-2 所示。

（2）使用

除了在台式机和服务器上使用的通用 Linux 发行版外，还可以为不同的应用制作专用的发行版，这包括支持某种计算机体系结构，嵌入式系统，稳定性，安全，为某一地区或某种语言的本地化，目标指向一个专门用户组，支持实时应用或委托给一个给定桌面环境等。到 2015 年，已实际开发了 400 多种 Linux 发行版，其中有大约 12 种是最普通的应用。

（3）桌面

Linux 在标准台式计算机和膝上电脑的普及程度与日俱增。目前最多的应用是带有两个最普通环境的图形用户环境：GNOME（它可以使用另外的 shell，如默认的 GNOME shell 和 Ubuntu Unity），和 KDE Plasma Desktop。

并没有单个的正式的 Linux 桌面存在：需要从免费和开源软件池中根据桌面环境和 Linux 发行版去选择各种模块，用这些模块去构建一个具有或多或少精准设计指南功能的图形用户接口。例如，GNOME，作为一个设计指南，它有自己的人机接口指导原则，它给人机接口一个重要指导，而不只是做这个图形设计，它要考虑人的有限能力，甚至考虑到安全问题。

图 2-3 为 Linux 桌面栈可视化软件模块。

2.5 MAC 操作系统

Mac 操作系统是由苹果公司（以前是苹果计算机公司）开发的，基于图形用户接口的操作系统系列，用于他们的 Macintosh 计算机系统系列。Macintosh 用户体验到了该机的大众化图形用户接口。

Mac 操作系统的早期版本只与基于 Motorola 68000 的 Macintosh 机器兼容。当苹果公司推出 Power PC 硬件时，该操作系统也被移植过来，支持这一体系结构。Mac OS 8.1 是运行在 68000 类处理器（68040）上的最新版本。Mac OS X，又称为"虎"，已经取代了"经典"Mac OS，通过 10.5 版（豹）与 Power PC 和英特尔处理器兼容。10.6 版（雪豹）只支持英特尔处理器。

Mac OS 可以分为两个系列

- Mac OS 经典系列，是基于苹果公司自己的代码的。
- Mac OS X 操作系统，是由 Mac OS 经典系列发展而来，还有 NeXTSTEP，是基于 UNIX 的。

1."经典"Mac OS

"经典"Mac OS 的主要特点是一个完全的图形操作系统，没有一行命令行。Mac OS 直到 System 4 版，一次只能运行一个应用程序。即使这样，由于容易使用也很受关注。Mac OS 的 System 5 有了合作式多任务功能，运行在 Mac SE 和 Macintosh Ⅱ 上。

2. Mac OS X

Mac OS X 是苹果公司最新的 Mac OS 操作系统系列。尽管其正式设计的是简单的 Mac OS 10 版，但历史上它与早期的 Mac OS 各版本相比其独立性最强。

Mac OS 10 操作系统是 Mac OS 9 和"经典"Mac OS 的后继者。它是一种 UNIX 操作系统，是基于 NeXTSTEP 操作系统和 Mach Kernel 的，Mach Kernel 是苹果公司收购 NeXT 计算机公司之后形成的。Mac OS X 也使用 BSD 代码格式。有 6 个重要的客户版本。最新的是 Mac OS X 10.6 版，称为雪豹版。2010 年 10 月 20 日苹果公司的"返回 Mac"事件上，看到了 Mac OS X 10.7 狮子版，展示了包括 Mac App Store 在内的功能改进和增强。

与客户版一样，Mac OS X 也有 6 个重要的服务器版，称为 Mac OS X Server。服务器版在体系结构上与客户版相同，不同之处在于所包含的服务器管理工具，包括管理基于 Mac OS X 的工作组、邮件服务器和万维网服务器工具以及其他工具。Mac OS X 现在对 Xserve 服务器硬件是默认的操作系统，对于 Mac Mini 是选件，对大多数其他型号的 Mac 机也可以安装。不像客户版，Mac OS X Server 也可以运行在使用如 Parallels Desktop 仿真软件的虚拟机上。

Mac OS X 也是 iOS 的基础（以前是 iPhone OS），用于苹果的 iPhone、iPod Touch 和 iPad 上，同时也是用在苹果电视机上的操作系统的基础。

3. OS X Yosemite

OS X Yosemite 是 OS X 的第 11 版，是为苹果公司的苹果计算机设计的台式和服务器操作系统。

OS X Yosemite 是 2014 年 6 月 2 日在 WWDC2014 会议上向开发者公布的，在 2014 年 7 月 24 日又向公共 beta 测试者公布了。Yosemite 向客户公布是在 2014 年 10 月 16 日。

很多 Yosemite 的新特性集中在继承性上，在提高与其他苹果机和平台，如 iOS 和 iCloud 的集成度上做文章。其 Handoff 功能使得该操作系统能在蓝牙 LE 和 Wi-Fi 上与 iOS8 设备集成，用户可将他们的 iPhone，用作管道，去接听电话，发送和接受文本信息，激活个人热点，直接将正在移动应用上操作的项目（如邮件手稿或数字表格）加载到他们的桌面设备上。

2.6 安卓操作系统

安卓是一种用于智能电话和平板电脑这样的触摸屏移动设备的操作系统，它是由 Open Handset Alliance 开发的，该组织是一个由谷歌公司领导的，由硬件、软件和通信公司组成的联合体。

安卓由基于 Linux kernel 的核构成，并配有中间件、库和用 C 语言写的应用程序接口，以及运行在包括基于 Apache Harmony 的、与 Java 兼容的库的应用框架上。

自 2013 年 7 月以来，Google Play 商店已经公布了 100 多万个安卓应用软件，下载应用软件已超过 500 亿个。一个开发者对 2013 年 4 月至 5 月间的调查表明，约 71%的移动开发者使用安卓进行开发。Google I/O 2014 会议上，谷歌公司透露，每个月的安卓用户都超过 10 个亿，而截至 2013 年 6 月，用户数为 5.38 亿。到 2015 年，安卓会成为最大的，所有通用操作系统使用的基础。

尽管大多数安卓设备最终都是以开源和专用软件的组合形式售出的，这包括由谷歌开发并授权的专用软件,但安卓的源代码是在开源许可下由谷歌公布的。

　　安卓对于那些需要成熟的、低价位的、客户化的,用于高技术设备的操作系统的技术公司来说,是很普通的。安卓的开放特性,促使大量开发者和热衷者,将这种开源码作为共同驱动课题的基础,这又会为资深用户增加新功能,或将安卓引入已经正式公布的运行其他操作系统的设备中。该操作系统的成功,又会使其成为各技术公司之间进行所谓"智能电话大战"一部分的专利诉讼的目标。

　　在当前的手持设备中,安卓的功能和技术规范包括:

（1）平台

　　该平台能适应较大型 VGA、二维图形库、基于 Open GL ES 2.0 规范的三维图形库和传统智能电话。

（2）存储

　　用于数据存储的是 SQLite,是一个小型关系型数据库。

（3）连通性

　　安卓支持的连通技术可连接 GSM／EDGE、IDEN、CDMA、EV-DO、UMTS、蓝牙和 Wi-Fi 等。

（4）消息

　　消息的可用格式有 SMS 和 MMS,包括线程的文本消息。现在安卓的云到设备消息（C2DM）也是 Android Push Messaging 服务的一部分。

（5）Web 浏览器

　　安卓里的 Web 浏览器是基于开源 WebKit 引擎的,是与 Chrome's V8 Java Script 引擎相配合的。

（6）Java 支持

　　虽然大多数安卓应用程序是用 Java 编写的,但在此平台上不使用 Java 虚拟机,也不执行 Java 字节代码。

（7）媒体支持

　　安卓支持音频／视频／静态媒体格式,包括 MP3、MIDI、WAV、JPEG、PNG、GIF、BMP 等。

（8）流媒体支持

　　支持 RTP／RTSP Streaming,HTML 连续下载,Adobe Flash 流（RTMP）和 HTTP Dynamic Streaming 等。

（9）附加硬件支持

　　安卓可以使用电视（Android TV）、汽车（Android Auto）、腕表（Android Wear）、静态/数码相机、游戏操作台、触摸屏和全球定位系统等。

（10）视频呼叫

　　安卓不支持本地视频呼叫,但某些手持设备有定制的、支持视频呼叫的操作系统版本。

　　另外一些支持手持设备的功能和规范有,多种语言、多任务、多触摸、蓝牙、基于音频的功能和屏幕捕捉等。

第 3 章　计算机网络

3.1　局域网

3.1.1　以太网

局域网是一种计算机网络，它在有限范围内，如家庭、学校、计算机房或办公楼内，把计算机用网络媒体相互连接起来。与广域网相比较，局域网的特点是它的较小的地域范围，以及不使用租用通信线路。

ARCNET、令牌环和其他技术标准是以前使用过的，但在双绞线上的以太网和 Wi-Fi 是目前构建局域网的两个最常用的技术。

网络拓扑叙述的是在设备和网段之间相互连接的布局格式。在数据链路层和物理层，局域网有很多拓扑结构，包括环型、总线型、网型和星型。但现今最常用的局域网拓扑是交换式以太网。在高层，互联网协议（TCP/IP）已经成为标准，取代了 NetBEUI、IPX/SPX、AppleTalk 和其他协议。

简单的局域网一般由一或几个交换机组成，一台交换机可以连接一台路由器、电缆调制解调器或 ADSL 调制解调器，以接入因特网。复杂的局域网由于采用具有生成树协议的交换机，且使用冗余链路，从而防止了回路效应，也具有通过服务质量管理不同类型流量的能力，也能用虚拟局域网技术去隔离流量。一个局域网可以包括很多网络设备，如交换机、防火墙、路由器、负载平衡器以及传感器。

局域网通过租用线路、租用设备或使用虚拟专用网技术的因特网，保持与其他局域网的连接。

1. 网络交换机

网络交换机（又称为交换式集线器、桥式集线器，正式称为 MAC 桥）是在一个计算机网络内将设备连在一起的计算机网络设备（见图 3-1），是采用分组交换方式去接收、处理并向目的设备转发数据的。与落后的网络集线器不同的是，网络交换机只向一个或多个需要数据的设备转发数据，而不是向所有端口广播同一个数据。

2. 10 千兆比特以太网

10 千兆比特以太网（10GE、10GbE 或 10GigE）是以 10Gb/s 速率（10×10^9 或每秒 100 亿位）传送以太网帧的一组计算机网络技术。这是首次由 IEEE802.3ae—2002 标准定义的。与前面的以太网标准不同的是，10 千兆位以太网只规定使用由网络交换机连接的、全双工点对点链路；并未使用以前的共享媒体的 CSMA/CD 操作的以太网标准。半双工操作和集线器在 10GbE 中也不用了。

3. 太比特以太网

脸书网站和谷歌在众多公司中表示需要太比特以太网。有些人想，400 千兆比特比太比特（1000Gb/s）更实际。2011 年，研究人员预测，太比特以太网（1Tb/s）在 2015 年出现，而 100 太比特（100Tb/s）以太网会在 2020 年出现。

3.1.2 Wi-Fi 和蓝牙

1. Wi-Fi

Wi-Fi（通常用 WiFi，但不妥），是允许电子设备使用 2.4GHz UHF 和 5GHz SHF ISM 无线频带加入计算机网络的一种无线局域网技术。

Wi-Fi 联盟将 Wi-Fi 定义为任何"基于 IEEE 802.11 标准的无线局域网产品"。但是，术语"Wi-Fi"通常在英语中是"无线局域网"的同义语，这是因为大多数现在的无线局域网都是基于这些标准的。Wi-Fi 是 Wi-Fi 联盟的商标。这种"认证了的 Wi-Fi"商标只能由完全通过 Wi-Fi 联盟互操作性认证测试的 Wi-Fi 产品使用。

很多设备都可以使用 Wi-Fi，如个人电脑、视频游戏控制台、智能电话、数码相机、平板电脑和数字语音播放器。这些设备可以通过无线网络接入点接入像因特网那样的网络资源。这样的接入点（或称热点）室内的作用范围为 20 米（66 英尺），室外范围要大一些。热点覆盖形成的区域小可小到一间房间，其围墙阻挡了无线电波，大可大到几平方公里的范围，这时要使用重叠接入点技术。图 3-2 是笔记本计算机和打印机之间通过接入点进行通信的过程。

Wi-Fi 比起有线连接，如以太网，安全性要差一些，因为入侵无线网的人不需要连线。使用安全套接层（SSL）的网页是安全的，但未经加密的因特网访问，很容易被入侵者检测出来。鉴于此，Wi-Fi 采取了多种加密技术。早期的 WEP 加密很容易被攻破，后来加入了高质量协议（WPA，WPA2）。在 2007 年又加入了称为 Wi-Fi 保护设置（WPS）的选项，但它的严重瑕疵使得入侵者可以恢复路由器的密码。Wi-Fi 联盟后来修订了其测试计划和认证程序，以确保新的认证了的设备可以抵御攻击。

2. 蓝牙

蓝牙是一种用于在固定和移动设备之间短距离数据交换的无线技术标准（使用 ISM 波段中的短波 UHF 无线波段，范围为 2.4～2.485GHz），也用来构建个人区域网。这项技术是由电信公司爱立信在 1994 年研发的。最初的设想是用无线技术替代 RS-232 数据电缆的。它能连接几种设备，克服了一些同步问题。

蓝牙由蓝牙特殊兴趣组（SIG）管理，该组织的成员是电信、计算机、网络和消费电子学领域中的 25000 多家公司。IEEE 为蓝牙制订的标准是 IEEE 802.15.1，但它并未保持这个标准。蓝牙 SIG 关注着该规范的发展，管理认证过程，保护该商标。制造商所生产的设备必须遵守蓝牙 SIG 标准并标识其为蓝牙设备。

蓝牙是一个标准的取代连线的通信协议，主要用于低功耗、短距离环境下，在每台设备中，都有一个低价位的收发器微芯片。因为这些设备使用无线（广播）通信系统，它们相互之间不必具有可视视距，但准光学的无线通道必须是可见的。作用范围由功率决定，不过，有效范围还随实地情况而变，见图 3-3。

3. 蓝牙与 Wi-Fi 对比

蓝牙与 Wi-Fi 有一些相似的应用：组建网络，打印或传输文件。Wi-Fi 作为高速电缆的替代物，通常用于局域网接入。这种类型的应用，有时也称为无线局域网。蓝牙主要用于手持设备。这类应用可视为在无线个人区域网中的应用。蓝牙在各种个人手持应用中，全部取代了电缆，并能在固定位置上应用，如在家庭中的智能能源应用（自动调温器等）。

Wi-Fi 和蓝牙在它们的应用中，在某些方面是互补的。Wi-Fi 的接入通常采用以点为中心的方式，具有非对称客户-服务器连接，所有通信量都通过该接入点路由，而蓝牙则在两个蓝牙设备之间采用对称方式。蓝牙在简单的应用中，如两个设备需要用最小的设置进行连接，如按下按键、头戴耳机和远程控制这些情况工作得很好，而 Wi-Fi 更适合于客户配置有一定档次，需要高速传输的应用场合，特别是通过接入节点接入网络的情况。

3.1.3 家庭网

家庭网或称家庭区域网，是一种局域网，其目标是使家庭内部或邻近家庭的数字设备之间能够通信。加入到这种网络的设备，例如智能设备，如网络打印机和手持移动计算机，常常通过它们的交互能力而获得意想不到的功能。这些额外的功能可以在多方面提高家庭的生活质量，如重复工作的自动化，提高人们的效率，增加家庭的安全性以及容易参加娱乐活动等，见图 3-4。

为建立物理层、数据链路层和网络层的连接，无论是内部还是户外网，一个家庭网络通常都要有下列设备中的一种：

- 调制解调器，一般由因特网服务提供商提供，它有一个以太网接口，可通过提供商的电信设施连向广域网。在家庭网络中，这些设备通常是 DSL 调制解调器或电缆调制解调器。
- 路由器，管理广域网和家庭网之间的网络层连接。大多数家庭网都配备了一个小型的、无源冷却台式设备，并带有集成化的无线接入点和四端口以太网交换机。这些设备应尽可能地帮助用户自动地、友好地、即插即用地安装、配置和管理家庭网络。
- 网络交换机，用于家庭网络设备通过以太网相互通信。虽然大多数家庭网络需要 Wi-Fi 或路由器内置交换功能，但有时则要求特殊的交换机，例如：
  - ▲ 当路由器的交换能力不足时。因为大多数路由器仅配有 4～6 个以太网端口。
  - ▲ 当一些设备，如 IP 相机和 IP 电话，要求使用有源以太网。
  - ▲ 当远端房子里有大量相邻很近的连线设备时。
- 为把无线设备连接到网络中，需要一个无线接入点。大多数家庭网络使用"无线路由器"组合设备去实现这一功能。
- 相互连接两个网络接口的网桥，只用于连线设备，例如 Xbox，允许其接入无线网络媒体。

3.2 广域网

3.2.1 广域网概述

广域网是一种覆盖地域范围很广的网络。很多广域网把两个或多个地域上分散的局域网连在一起。按这种定义，因特网是一个非常大的广域网。广域网可以像因特网那样是公用的，也可以是私人拥有和使用的。

有很多对广域网分类的方法，其中之一是按传输设备上的信息流分类。使用这种方法可将广域网分成三种类型：电路交换网、专用网和分组交换网。

1. 分组交换网

现在大多数广域网以分片或分组形式发送数据，这些分组一个个送往广域网，在其到达目的站以后再重新进行装配。

分组的最大长度由网络进行设置。比较长的传输可以分成多个分组。每个分组不仅包含数据，也包含带有控制信息的首部（如优先级代码和源地址以及目的地址）。分组通过网络从一个节点到一个节点传送。在每个节点，分组被暂时存储，然后按其首部信息寻找路径。

有两种常用的分组交换技术：数据报和虚电路（见图 3-5）。

（1）数据报方式

在分组交换的数据报方式中，每一分组都独立于其他所有分组单独处理。即使是一个分组只代表多个分组传送的一部分，这种网络（和网络层功能）也把它作为单独存在部分处理。这种技术中的分组称为数据报。

这种方法可能会使这些数据报不能按顺序传送到目的地。在大多数协议中，在将数据报传送给目的端口之前，恢复它的顺序是运输层的任务。

（2）虚电路方式

在分组交换的虚电路方式中，属于一个报文或会话的所有分组之间的关系是规定好了的。在会话的开始阶段，在发送方和接收方之间就已经选择好了一条路径。当数据被送出去以后，所有传送的分组沿着这条路径一个接一个地传送出去。

那么电路交换与虚电路之间的区别是什么？虽然电路交换可以在端用户级使用多路复用技术，但在交换机上不进行多路复用。然而，在虚电路方法中，我们可以在交换机上使用多路复用技术。

目前，虚电路传输有两种实现方式：交换式虚电路（SVC）和永久虚电路（PVC）。

2. 专用网

这是一种通过租用线路连接的专用网。租用线路是由租用客户专用的一条通信线路，不必进行交换机间的交换。租用或私人线路是用户专用的。专线的优点是终端或计算机一直连在租用线路上。这种服务的访问时间极短。其他由因特网服务提供商构建的网络，可将一个单位的局域网接入因特网。包括 TCP/IP 协议在内的网络协议提供传输和寻址功能，而服务提供商常用的协议包括 Packet over SONET/SDH、MPLS、ATM 和帧中继，这些协议都提供广域网的链路。X.25 是早期广域网的重要协议，并常被视为帧中继祖父级的协议，也是很多其他协议的基础协议，X.25 的功能现在还一直为帧中继使用。

3. 路由器

路由器已经有几个名字了。回到过去称为 ARPANET（现在的因特网）的那个时代，路由器称为互联网报文处理机（IMP）。最近路由器又被称为网关；这一术语的痕迹仍可在边界网关协议（BGP）和内部网关路由协议（IGRP）中见到。在开放系统互联（OSI）领域，路由器被称为中间系统（IS）。

路由器这个名字，最能表达这些设备所实现的功能。路由器沿两个网络之间的路由——路径传送信息。这条路径可以经过一个路由器或很多路由器。进而，在有多条路径指向同一目的地的互联网中，现代路由器用一组程序去决定并使用最佳路由。如果这一路由变得不是最佳或完全不可用，该路由器则选择下一个最佳路径。由路由器使用的，用于确定和选择最佳路由并与其他路由器共享网络可达性和状态信息的程序，总的称为路由选择协议。

正如数据链路可以直接连接两个设备一样，路由器也在两个设备之间建立一个连接。不同的是，如图3-6所示，共享一条公共数据链路的两个设备之间的通信路径是一条物理路径，而由路由器提供的，在不同网络上的两个设备之间的通信路径是高层逻辑路径。

这一概念对于了解路由器的功能是非常重要的。注意，图3-6中两个设备之间的逻辑路径即路由，经过了几种类型的数据链路：一个1Tb以太网、两条串行链路和一个10Gb以太网。正如前面所提到的，要在数据链路的物理通路上传送，数据必须封装成一个帧，即一个数字封装。同样，要在通过路由的跨网络逻辑路径上传送，数据也必须封装；由路由器使用的数字封装是分组。

3.2.2 存储区域网和因特网区域网

1. 存储区域网

存储区域网是一种对联合存储的块级数据访问的专用网络。存储区域网主要用来提升存储设备，如磁盘阵列、磁带库和自动点唱机，对服务器访问的方便性，因而这些设备对操作系统来说都似乎是在本地一样。典型的存储区域网有其自己的存储设备网，而其他设备一般是不能通过这种局域网去访问的。在2000年初期，存储区域网的价格和复杂性都有所下降，因此能比较广泛地用于中小型企业环境。

存储区域网不提供文件提取业务，只能对块级数据操作。但是，构建在存储区域网顶部的文件系统提供文件级访问，并将其称为存储区域网文件系统或共享磁盘文件系统。

2. 因特网区域网

因特网区域网是一个通信网的概念，它在IP上的云环境内连接语音和数据端点，并取代现在的局域网、广域网和公共电话交换网。

从倡导者的眼光来看今后的联网模型，因特网区域网可以通过公共万维网安全地连接各个端点，因此，这些端点之间可以通信，交换信息和数据，而不需要限定在一个具体位置上。

与局域网和广域网的传输方式都不同，因特网区域网对整个网络没有地域的限制，因为它的应用和通信业务都是虚拟化的。端点需要的仅仅是跨因特网的宽带连接。

一个因特网区域网平台，托管在由服务提供商管理的云中，通过因特网连接，向用户提供可在任何地点、任何时间对信息的安全访问。用户也可以从任何所连接的端点拨打电话，传送语音邮件、电子邮件和传真。对于企业，托管的模式降低了IT和通信费用，防止了数据丢失和灾难性死机，同时，通过提高雇员的工作效率和降低通信费用实现了对投资的更高的回报。

3.3 因特网

因特网是世界上最大的和最著名的计算机网。在技术上，因特网是一个网络的网络，这是因为单个用户都可以连接到由因特网接入提供商或因特网服务提供商（ISP）所建立的网络上，而该网络又可以连接到更大的网络上，而这个网络又连向了比它更大的网络上。所有这些网络的网络在一起就称为因特网。由于因特网上的所有网络是互联的，因此具有因特网访问功能的任何计算机都能在因特网上相互通信，而不必关心所使用的ISP。

1．域名系统

建立域名系统是为了集中管理把网络名变换成地址的任务，并使这种翻译功能自动化。早期的因特网，由中央单位（在加利福尼亚曼佬公园 SRI 网络信息中心斯坦福研究院）负责维护 HOSTS 文件，该文件包含了每一个因特网上的主机名连同它的地址。管理员必须把改变的内容传送给 SRI NIC，而且这些改变会定时合并到该文件中。当然，这意味着还必须把文件分发到每一个主机，以使它能有更新的版本。

DNS 采用一种遍布因特网的，跨越很多计算机的层次型分布式体系结构。根服务器保持有关顶级域的信息（像.COM、.EDU 和.GOV），并且整个因特网上的每个域都有一个域名服务器，负责该域中的计算机名和地址的对应。当客户计算机需要获得主机的地址时，它们查询 DNS 服务器。如果本地 DNS 知道该地址，它把该地址返回给客户机。如果它不知道，它把该查询送往 DNS 服务器链，直到查找到一个能分析这个名字的 DNS 服务器并提供一个真正有效的名字为止。

在 DNS 层次中，最顶级的项称为根域，并且它前面有一个句号（.）。在这个根域下面是顶级目录，它分成两组：地理域和组织机构域。地理域常用于指明国家。例如，.au 是指澳大利亚，.cn 是指中国。在每一个地理域下面，你能找到组织机构域。

组织机构域你可能熟悉，它包括下列各项：

（1）com 用于商业组织。

（2）edu 用于教育机构。

（3）gov 用于美国政府实体。

（4）mil 用于美国军事机构。

（5）Int 国际性组织。

（6）net 用于网络组织，像因特网服务提供者。

（7）org 用于非营利的组织。

（8）arpa 用于逆向地址查找。

域名系统的结构和一棵倒长着的树很相似。在图 3-7 中，你可以看到顶端是根域，在它下面有 com 到 cn 的各个域。在 com 域下面是单独的商业机构，他们都有自己的域。在任一具体的域下面，都可能有子域。

2．连入因特网

（1）拨号连接

拨号连接一般是在普通电话线上连接。为了连向因特网，你的调制解调器（或其他接口设备）要拨号并与你的 ISP 计算机上的调制解调器相连。在你连接时，要对此次会话为你的 PC 机安排一个临时 IP 地址。在每次因特网会话结束时，要与你的 ISP 计算机断开，以便其他用户使用该地址进行连接。标准因特网拨号业务使用常规的拨号调制解调器，其最高数据传输率为 56kb/s。

（2）ADSL 连接

目前应用最广泛的另一种接入因特网的拨号连接，是非对称用户线 ADSL。有时 ADSL 又称宽带接入，因为它比普通调制解调器所提供的带宽更宽。

ADSL 是 20 世纪 80 年代电信行业为电缆行业提供视频点播的一种解决方案，首先开发出来的。然而到 90 年代中期就认识到，作为可行的技术，应该能访问像因特网这样的高速网络。ADSL 提供非对称传输速率，典型的顺流可达到 9Mbps（从中心局到用户设备——下行），逆流为 16Kbps～640Kbps（从用户设备到中心局——上行），如图 3-8 所示。像所有的铜线传输系统一样，速率越高，作用范围越短。ADSL 传输的限制是距离，它只能工作在距电话交换局 3 英里的范围之内，而距离越是接近 3 英里，其传输速率下降得越厉害。

（3）专线连接

拨号连接只是在你需要接入因特网时才连接到你的 ISP 计算机上，而专线连接一直使你连到因特网上。使用专线连接通常要为你的 PC 机安排一个静态（不变化）IP 地址，以便通过因特网来回传送数据。

专线因特网连接包括通过学校或办公室的局域网，以及 ADSL、电缆、卫星和固定无线连接等连接方式。

（4）无线连接

无线连接不使用具体的介质去连接接收和发送设备，而是使用空气本身。用于无线连接的主要技术有红外、广播无线电、微波、卫星和移动无线连接。

① 红外：红外使用红外光波进行短距离通信。由于这种光波只能直线传播，因此它有时也叫做视距通信。这就要求收、发设备能相互看清楚而中间无任何物体遮挡视线。一种最常用的技术是从笔记本计算机或个人数字助理这样的便携设备向台式计算机传输数据和信息。

② 广播无线电：广播无线电使用无线信号与无线设备通信。例如，蜂窝电话和很多可访问万维网的设备，通过广播无线电用电话呼叫与万维网连接。某些最终用户把他们的笔记本或手持电脑与移动电话连接，从远处访问万维网。这些可访问万维网的大多数设备都采用称为 Wi-Fi（无线保真）的标准。这一无线标准广泛用于计算机相互连接，并连向因特网。

③ 微波：微波通信采用高频无线电波。与红外一样，微波由于只能直线传播，因而也是一种视距通信。因为微波只能短距离传输，所以它是城市中或大型校园内楼宇之间传输数据的很好的媒体。

蓝牙是短距离无线通信标准，使用微波进行短距离数据传输，最长距离约 33 英尺。与传统微波不同的是，蓝牙不要求视距通信。它使用的无线电波可以透过墙壁和其他非金属壁垒。

④ 卫星：卫星用于收、发大量数据。术语上行链路是指向卫星发送数据，下行链路是指从卫星接收数据。卫星通信的主要缺点是恶劣天气有时会使数据流中断。

卫星通信最有趣的一种应用是全球定位。这是一个由国防部（美）拥有和管理的，由 24 颗卫星构成的网络，它连续地向地球发送位置信息。全球定位系统（GPS）设备使用该信息去确定本设备的唯一地理位置。

⑤ 移动无线连接

与卫星和固定无线连接使用电缆将调制解调器连向某种固定收发器不一样，移动无线连接允许设备从一处移向另一处。因此大多数手持 PC 机和其他移动设备（像具有 Web 功能的蜂窝式电话）都可以使用移动无线连接并通过类似于蜂窝电话和报文设备那样的无线网络访问因特网。

第二部分　　因特网应用

第 4 章　传统因特网应用

4.1　万维网概述

4.1.1　关于万维网

万维网是一种由多个因特网服务器组成的大型网络，它向运行客户应用程序（如浏览器软件）的终端，提供超文本和其他服务。

万维网允许用户从动态链接信息的全球联网服务器系列中搜索、访问和下载信息。Web 客户通常通过 Web 浏览器，向服务器传送用户需要信息的请求。服务器与客户通过传输协议，通常是超文本传输协议（HTTP）进行通信。然后，服务器用统一资源定位符（URL）访问网页。搜索引擎可用来简化访问，允许用户输入某一题目的搜索条件，从而使若干 URL 返回有关所需信息的网页。

1．提供商

访问因特网最常用的方法是通过因特网服务提供商（ISP）。这些提供商已经连入因特网并为个人访问因特网提供通路或连接。你们的学院或大学，很可能通过局域网或电话拨号连接，免费让你访问因特网。有一些公司也提供免费的因特网访问服务。

最广泛使用的商业因特网服务提供商是全国性的提供商（例如美国在线（AOL））和无线提供商。

2．浏览器

浏览器是访问万维网资源的程序。这种软件把你连向远端计算机、打开并传输文件、显示文本和图像，并在一种工具中提供访问因特网和万维网文档的简单界面。浏览器能使你很容易地从一个网站移动到另一个网站，对万维网进行探索或冲浪。三个著名的浏览器是 Mozilla Firefox、Netscape Communications 和微软公司的因特网浏览器（见图 4-1）。

浏览器是一种基于图形用户接口的超文本客户应用程序，用于通过万维网和因特网，对无数个远程服务器上的超文本文档和其他服务进行访问。

如图 4-2 所示，当单击超链接时，所发生的是一连串十分精彩的事件，不仅包括 Web 浏览器软件，而且包括某处的 Web 服务器，并且所包含的这些事务紧紧地依赖于 HTML 语言。

3. 个人网站

你想和全世界共享某件事情吗？你是否喜欢个人网站，但又不想学习 HTML 语言？在因特网上创建你自己的主页很容易，有很多服务能帮助你开始这项工作。

万维网上的服务网站提供制作个人网页的工具。在注册到该网站之后，你就可以使用所提供的工具制作你的网页了。一旦网页制作完成，该服务网站就作为你个人网站的宿主机使用了，而且其他网站可以从世界上任何地方免费访问你的个人网站。

4.1.2 搜索引擎

1. 概述

可以想象，当走进一个杂乱无章、堆满了书的图书馆时，要找到所需的资料几乎是不可能的。随着 WWW 的迅速发展，同样也需要对网页信息进行分门别类的管理，需要记住"是什么"和"在哪里"。在 Web 技术发展的初始阶段很难查找有用的信息。

在 WWW 的早期，美国斯坦福大学的两名研究生 Jerry Yang 和 David Filo 提出按照目录来组织超链接的方法，并发现该方法确实可行。1993 年后期，这种方法被称为"WWW Jerry Yang 向导"。这个名字不久被改为 Yahoo!，第一个搜索工具就这样诞生了。

当今有许多搜索工具能帮助我们在 Web 上迅速且方便地找到所需的信息。这些工具不断地演变，不断地优胜劣汰。与其逐一解释每个搜索引擎是如何工作的，不如只对其中的少部分进行介绍，并且给出一些网页演示，以回顾现行所有可用的搜索工具。

2. 搜索引擎是如何工作的

搜索引擎按下列顺序操作：
① 慢慢浏览万维网；
② 索引；
③ 搜索。

搜索引擎要存储很多网页信息，这些信息都是用 html 本身检索的。网页是用浏览器（有时也称为网络蜘蛛）检索的——一种能自动跟踪每次网站链接的万维网浏览器。然后分析每个网页的内容，以确定如何去索引（例如，从标题或称为元标签的专用字段中提取一些字）。有关网页的数据，存储在索引数据库中，便于以后查询时使用。查询可以按单个字进行。索引的目的是尽可能快地找到信息。有些搜索引擎，如谷歌，存储所有或部分源网页（称为高速缓存）和有关该网页的信息，而其他一些搜索引擎如 AltaVista 则存储它们能找到的每页中的每个字。由于高速缓存的页是实际索引的页，故它保存的总是实际搜索的文本，因此当当前网页的内容被更新以后并且搜索的条目不再包含在其中时，这是很有用的。

当用户往搜索引擎中输入一个查询时（通常用关键字），该引擎检查它的索引，并按其自己的标准列出最匹配的网页，通常还包含文档标题的简短摘要，有时是部分文本。

3. 有关 Google Chrome

Google Chrome 是谷歌公司开发的免费万维网浏览器，直到第 27 版，它都使用 WebKit 布局引擎。随着 iOS 各个版本出现异常，从第 28 版开始及以后的版本，都使用 WebKit fork Blink 引擎。

在 2008 年 9 月 2 日，首次为微软 Windows 公布了一个 beta 版，同年 12 月 11 日则成为正式公开版本。

2015 年 1 月，StatCounter 对 Google Chrome 进行评估，结果是该浏览器在世界范围内的共享率为 51%，这表明它是世界上使用最多的浏览器。

作为开源课题 Chromium，谷歌公开了 Chrome 的绝大部分源代码。而很重要的，内置在 Adobe Flash Player 中的代码是非开源的。

Chrome 的特点是极小的用户界面，而且以后还会植入到其他一些浏览器中。例如，将地址栏和搜索栏合并到 omnibox 中去。Chrome 由于其强势浏览器性能，口碑不错。

4.2 电子邮件

电子邮件是在因特网上传送电子报文的手段。从前，电子邮件只由文本报文组成。现在，电子邮件一般都包括图形、照片和很多不同类型的文件附件。世界上很多人都在相互发送电子邮件。你可以向你的家庭、同事发电子邮件。所有你需要发送和接受的电子邮件，都要使用电子邮件账号，都要访问因特网，都要使用电子邮件程序。两个使用最广的电子邮件程序是微软公司的 Outlook Express 和 Mozilla Thunderbird。

典型的电子邮件报文有三个基本部分：头部，报文和签名（图 4-3），头部最先出现，一般包括以下内容。

① 地址：发信和收信人的地址，选项是接受拷贝的人。电子邮件地址有两部分（图 4-4）。第一部分为用户名，第二部分为域名，包括顶级域名。在我们的例子中，dcaots 是 Dan 的用户名。为 Dan 提供电子邮件服务的服务器是 usc.edu。该顶级域表明，提供服务的是教育机构。

② 主题：一行字，用于描述报文的题目。主题行一般在收件人检查其邮箱时显示。

③ 附件：很多电子邮件程序允许附加如文档和表格一类的文件。如果报文有附件，该文件名出现在附件行上。

紧接着下面是信件或报文。信件一般较短而且是扼要的。最后签名行是有关发件人的信息，一般包括发件人的姓名、地址和电话号码。

在你个人或业务活动中，电子邮件是有价财产。当然，像其他很多有价技术一样，电子邮件也有一些不足之处。美国人每年都会收到几十亿个不希望和未经请求的电子邮件。这种不受欢迎的邮件称为垃圾邮件。垃圾邮件不仅是一种骚扰，而且还能造成危害。

例如计算机病毒或者有害的程序，经常附着在未经请求的电子邮件上。

在控制垃圾邮件方面，反垃圾邮件已经加入到法律体系中。例如，最近制定的 CAN-SPAM Act 要求每个与营销有关的电子邮件提供一个退出选项（opt-out option）。当这一选项被选中，则接收者的电子邮件地址就从今后的邮件地址表中删除。然而这种方法影响很小，因为超过 50%的垃圾邮件来自美国境外的服务器。另一种更为有效的方法是开发并使用垃圾邮件拦截器（见图 4-5），这些程序使用各种不同的方法去识别和删除垃圾邮件。

4.3 电子商务和物联网

4.3.1 电子商务

1. 什么是电子商务

电子商务是一个系统，它不仅包括可以直接产生收益的那些商品交易和服务，而且也包括那些支持产生收益的事务处理，例如为那些商品和服务所产生的需求，提供销售支持和客户服务（见图 4-6）或提供商业伙伴之间的通信等。

电子商务是在传统商业的优势和结构上，增加了计算机网络提供的种种灵活性而建立的。

电子商务产生了一些新的商业模式和一些新的运营方式。例如，亚马逊 Amazon.com 是一家在华盛顿州西雅图的图书经销商。这家公司没有实际的书店，而是通过因特网售书，由与其合作的出版商直接发送图书，因而，该公司不必拥有任何库存。而像 Kantara 和 software.net 这样的公司在此方面更进了一步。

因为他们的所有产品（商业软件包）都是电子的，都能存储在相同的计算机中，这些计算机也用于处理订单，并作为 Web 服务器使用，其库存则全部是数字化的。作为另外一个例子，AMP 公司正向其客户提供这样的机会：客户可以直接从它的 Web 服务器目录上选购电子接插件和相关配件，而无须基于电子数据交换 EDI 的订购和确认。

2. 电子商务业务模式

一个公司的政策、运作和技术决定了它的业务模式。本质上，一个公司的业务模式描述了该公司如何产生收益。有很多标准电子商务业务模式，包括企业对消费者（B2C）、企业对企业（B2B）、消费者对消费者（C2C）和企业对政府（B2G）等模式。下面讨论这些最常用的模式。

（1）企业对消费者（B2C）

企业采用 B2C 模式向单个消费者销售物品或提供服务。B2C 模式是最早定义并通过万维网实现的主要电子商务业务模式之一。B2C 企业的几个例子包括亚马逊（Amazon.com）、宾恩（L.L.Bean）、沃尔玛（Walmart.com）和 Polo.com（见图 4-7）。这些企业可能只是因特网商店，也可能是既在网上又有店铺的混合经营商店。

（2）企业对企业（B2B）

企业对企业的应用包括发生在两个企业之间的任何类型的电子商务交易。B2B 的收益一直在增长，并有望在今后几年内持续大幅度增长。

（3）消费者对消费者（C2C）

消费者对消费者——有时也称为个人对个人（P2P）——其业务模式大多数只是由消费者拍卖而形成，即这里的消费者向其他消费者销售产品。易趣（eBay）购物网是当今最大的 C2C 电子商务企业之一，每天有上百万件产品被拍卖。

（4）企业对政府（B2G）

美国政府每年开支超过 5000 亿美元，这种趋势并未见减弱，B2G 组织正在变得异常重要。这些组织（见图 4-8）向本地、州和联邦政府购买者销售产品和提供服务。一般来讲，政府部门接受在线购物的程度比私人要差。

与 B2G 有关的一些措施包括美国有些州允许市民网上付款，如付税、更新驾照等。这有时称为客户对政府（C2G）电子商务。图 4-8 所示的联邦政府新网站 Pay.gov 有望允许在联邦范围内进行网上交易。

3．电子商务的应用

（1）在线购物

多数人将电子商务与在线购物等同起来，这是由于通过将安全和密码技术嵌入到流行的浏览器中，使得在线购物成为可能。根据近期的调查，只有不到一半的 Web 网络用户使用过在线购物，但是这个比例每年都在增加。

在线购物的巨大优势之一，是建立商店的资产投资低。用不到 2000 美元的启动费用，就可以开设一个 Web 网络店面并且开始在线销售产品。举例来说，亚马逊（Amazon.com），这个"世界最大的书店"，如同它自己描述的一样，在西雅图一栋办公大楼中除了有几间办公室外什么也没有。

如果用一个词语来总结所有这些网站上共享的东西，那就是减少了中间环节。所有这些在线商店的成功事例都表明他们都跳过了中介。这些网站能使客户接触无比丰富的信息资源。他们能使客户自己进行选择，而没有限制在现有存货（或售货员的干预）上。

（2）在线银行

银行以不同的方式实现在线业务。一个方法是利用支票簿程序，如微软的 Money 或 Intuit 的 Quicken 程序，它们能使客户自动地结算他们的支票簿。然而，不利之处是，你必须在装有 Money 和 Quicken 所有数据的计算机上访问你的在线银行账目，而且这笔数据能够被任何访问你的计算机的人看到。其优点是 Money 和 Quicken 有强大的功能，能为你的开支做预算，分析你的消费习惯。

只需要一个程序（Web 浏览器）的、基于 Web 的新系统，是比较容易使用的。所有数据存储在银行而不是在你自己的计算机中，这意味着你可以在任何地点访问你的账户。基于 Web 的在线银行也是非常容易使用的。然而，基于 Web 的系统不提供像预算和消费分析这样的先进特性。那么对银行来说这样做意味着什么呢？的确，要设那么多银行分行和出纳员该花多少钱！

4．社交电子商务

社交电子商务是电子商务的一个子集，它包含支持社会交互活动的社会媒体和在线媒体，也包括在线购销产品和服务的广大用户。

更简单地说，社交电子商务就是电子商务交易中对社交网络的使用。

社交电子商务这一术语是 2005 年 11 月由雅虎提出来的，当时它描述了一套在线合作购物工具，如共享商品目录，用户信誉度和其他用户产生的内容——共享在线产品信息。

社交电子商务的概念由戴维·贝塞尔发展了，他说明了在电子商务网站上的用户产生的宣传报道的内容，而史蒂夫·鲁比尔所发展的则包括合作电子商务工具，它使购物者"从诚信的个人那里得到通知，找到货物和服务，然后购买"。业已证明，发放这些忠告的社会网络增加了顾客对某个零售商的信任度。

社交电子商务帮助公司在以下几方面受益。第一，社交电子商务帮助公司，按照顾客的消费习惯制作商标，以吸引顾客。第二，它刺激顾客返回他们的网站。第三，它为顾客提供一个平台，用于在他们的网站上讨论他们的产品品质。第四，提供给顾客所有要研究、比较的信息，并最终让顾客在竞争对手面前选择了你，并从你那里而不是别处购买。

当今，社交电子商务已经扩展到包括社交媒体工具和用于电子商务活动的内容，特别是在时尚行业。社交电子商务的例子有顾客信誉度和回顾，用户建议及委托，社会购物工具（共享在线购物活动），论坛和社区，社交媒体最优化，社会应用和社会广告。像强化现实这些技术已经集成到社交电子商务中了，它使得购物者通过社交媒体工具，看到自己身上和反馈的服饰内容。

某些学者已经在设想区分"社交电子商务"和"社会购物"，前者是指在线商店的合作网络，后者是指在线购物的合作行动。

4.3.2 物联网

1. 物联网概述

物联网是具体的物品或"东西"的网，这些物品嵌入了电子器件、软件、传感器和连接，通过与制造厂商、操作人员或其他连接设备交换数据，获得更大的价值和服务。每件物品都通过嵌入式计算机系统进行识别，而且在现在的因特网设施内都是可以互操作的。

典型的物联网是有望提供先进的设备、系统和服务的连通性，并超越机器对机器间的通信，覆盖各种协议，覆盖各个域和应用。这些嵌入设备（包括智能物品）的互联，有望在自动化领域的各个方面起引领作用，同时也使像智能网格这样的技术有实现的可能。

物联网中的东西，可以是各式各样的设备，如心脏监护移植设备、安在农场动物身上的生物芯片收发报机、海水中的电子蚌、带有内置传感器的汽车、帮助消防队员搜救的现场操作设备。这些设备在现有的各种技术的协助下，收集有用的数据，这些数据以后自动在其他设备之间自动流动。当前市场上的产品有智能调温系统，以及利用 Wi-Fi 的远程监视的洗衣机和烘干机。

除了因特网与自动化系统的连接已经扩展到许多新的应用领域之外，物联网有望从各个地方产生大量的数据，并且数据聚集的速度非常之快，因而需要制作索引，要存储和处理这些数据。

2. 物联网的体系结构

物联网系统理应有一个事件驱动的体系结构。图 4-9 是物联网发展的三层体系结构。

顶层是由驱动应用程序组成。这个应用空间非常大。底层代表不同类型的传感器：主要是射频识别标签、ZigBee 或其他类型的传感器，以及标识路线图的 GPS 导航仪。收集在这些传感器上的信号或信息，通过中间层上的云计算平台与应用程序相链接。

在中间层的移动网络上，在因特网主干网上和各种信息网上，构建了信号处理云。在物联网中，感知事件的含义是不遵循确定性或句法模型的。事实上，这里采用的是面向服务的体系结构模型。大量传感器和滤波器用来收集原始数据。各种计算和存储云以及网格用于处理数据，并将其转换为信息和知识格式。所感知的信息可共同为智能应用构建一个决策支撑系统。也可以把中间层视为语义万维网。有些行为体（服务、组件、计算机化身）是自引用的。

3. 应用

按照 Gartner 公司（技术研究和顾问公司）的说法，到 2020 年物联网上的设备将接近 260 亿台，ABI 的研究估计，到 2020 年将超过 300 亿台设备是用无线连入物联网的（万物互联）。PewResearch Internet 课题最近的观察和研究表明，绝大多数技术专家和有回应的因特网用户（约占 83%）同意这种观点，即到 2025 年，物联网和物联云，嵌入式和可穿戴计算设备（以及相应的动态系统）将有广泛的、重大的影响。因此，很清楚，物联网将由非常大量的连入因特网的设备组成。

具有有限 CPU、存储器和电源的网络嵌入设备的能力表明，物联网可在任何领域找到应用。这些系统应该负责收集所设置的信息，范围从自然生态系统到建筑物和工厂，并由此，在环境监测和城市规划领域中找到应用。另一方面，物联网系统也应负责完成不仅是感知物体的任务，例如，智能购物系统，还应该通过跟踪用户手机，监视某些用户在商店的购物习惯。然后专门向用户提供他们喜欢的产品，甚至报告给他们所需物品在什么地方。另外一些感知和执行的应用例子是与热、电和能量管理有关，还有旅游辅助交通系统等。

当然，物联网应用不仅限于这些领域，也还有其他物联网专用场合。在这里把其他一些最突出的应用领域概述一下。基于应用，物联网产品总体可分为五类：智能穿戴、智能家庭、智能城市、智能环境和智能企业。在这些市场中，每一类物联网产品和解决方案都有各自的特性。

4.4 计算机和网络安全

4.4.1 计算机罪犯和犯罪

1. 计算机罪犯

计算机犯罪是一种犯罪人使用计算机专业知识的非法行为。通常计算机罪犯可以是雇员、外界人员、黑客、破坏者、有组织的犯罪成员或恐怖分子。

什么是黑客和破坏者？有些人把这两组人视为同一种人，但二者是不同的。黑客是一种没有被授权而能访问计算机系统的人，其目的是对计算机开玩笑或进行挑战。破坏者做同样的事，但目的是恶意的。他们都试图窃取技术信息或将他们称之为炸弹的程序引入计算机系统，以破坏计算机程序。

2．计算机犯罪

美国联邦调查局估计每年由于计算机犯罪造成的商业损失超过 1.5 万亿美元。过去两年这些犯罪的数量增加了三倍。计算机犯罪有多种形式，包括制作恶意程序，拒绝服务攻击，因特网欺诈、窃取和数据操纵。

（1）恶意程序

黑客和破坏者在制作和传播恶意程序方面是臭名昭著的。这些程序被称为不良件，它是一种短的恶意软件。这些程序是专门破坏计算机系统的。三种常见的不良件是病毒、蠕虫和特洛伊木马。

1）病毒

① 什么是病毒？

病毒（更正式地应称为计算机病毒）是一种计算机程序，它在你不知道发生了什么的情况下，能够进行自我复制。它可以从硬盘的一部分自我复制到另一部分，或者是从一台机器自我复制到另一台机器上。

许多病毒不只限于复制自身。有些会造成实质性的损害，比如说，删除硬盘上的文件。而有些只是骚扰用户。当你在机器上工作时，这些病毒可能会在你的显示器上闪出一条信息或者出现一些莫名其妙的事情。所有病毒的危害在于它们恶意地自行其是而你并不知道发生了什么，并且通过无法控制的传播引发一些问题。

② 病毒的类型。

尽管有数千种病毒在遍布全世界的机器和网络上传播，但大部分病毒都具有共同的特点，可归为以下几类：

- 引导扇区病毒
- 文件感染病毒
- 宏病毒

③ 你是如何感染病毒的？

当今很多病毒是通过因特网来传播的。宏病毒是传播速度最快的病毒，因为它通常利用电子邮件的附件在机器间传递。然而，也可以通过软盘、文件传输及 Web 下载来传播病毒。另外，宏病毒可以感染某一应用程序中的所有打开的文件。

因此，如果机器里有一个微软字处理宏病毒，它将感染机器里已打开的所有字处理文档。如果将那些已感染的文件保存到一张软盘上，当你将软盘的内容拷到另一个硬盘时，或在另一台机器上使用相同应用程序打开它时，病毒都会伴随着文件一起进入该机器。

另外，使用 Web 下载文件是感染病毒的另一种常见途径。那些允许用户利用因特网共享信息的站点，如 Napster 及 Gnutella，是计算机病毒潜在的繁殖地，就像现实生活中的病毒容易在高流动人群的商场和学校滋生繁殖一样。

2）蠕虫

蠕虫是病毒的一种特殊类型，它并不把自己附着在程序和数据库上。

（2）拒绝服务（DoS）

像蠕虫一样，拒绝服务攻击，力图使计算机系统或网络降低运行速度或停机。与蠕虫的自复制不一样，拒绝服务攻击可使计算机或网络中的信息和数据泛滥成灾。这些攻击的目标通常是因特网服务提供商（ISP）和专用网站。一旦受到攻击，ISP上的服务器或网站就被这些服务请求信息所淹没，因而对合法用户也不能做出反应。结果IPS或网站实际上已关闭。

（3）因特网欺诈

欺诈是一种欺骗行为，用来欺骗个人，使他们浪费时间和钱财而很少或根本没有回报。因特网欺诈是使用因特网的一种简单欺诈。因特网欺诈正在成为一种严重的问题，已经给成千上万的人造成经济和法律问题。几乎所有欺诈都是向信任的人发送大量邮件开始的。

（4）窃取

窃取有多种形式——硬件、软件、数据、计算机时间的窃取。窃贼盗取设备和程序，当然也可以是白领犯罪。这些犯罪包括窃取以机密信息格式保存的数据，如受惠的客户清单。另一种常见的犯罪是使用（盗用）公司计算机时间，方式是某一雇员运行另外的商业程序。

（5）数据操纵

找到进入某人计算机网络的入口，并留下恶作剧者的信息，看起来像在开玩笑，为什么黑客要这样做。这一直是不合法的。更有甚者，如果操纵数据看起来是无害的，这会引起很大的忧虑，并且浪费网络用户的时间。

4.4.2 保护计算机安全

安全涉及保护信息、硬件和软件不被非法使用，没有受到干扰、破坏和自然灾害的损害。因为有很多原因造成计算机系统和数据会受到损害，因此我们明白了为什么安全问题越来越重要。保护计算机安全的一些基本方法是加密、限制访问、预防灾害、数据备份和防火墙。

1. 消息加密

任何时候通过网络发送信息，都可能遇到非授权访问。电文传输的距离越长，安全风险越大。例如，一封电子邮件电文在局域网上，只会被一些用户接收，而他们工作的办公室环境是可控的。而跨越国家在因特网上传输的电子邮件电文，电文被截获的机会更大。

商业上的电文加密或编码已经有很多年了，最广泛使用的个人加密程序之一是Pretty Good Privacy。

2. 限制访问

安全专家经常发明出保护计算机系统不被个人非法访问的方法。有时安全是一种方法，把防护放在公司的计算机房中，检查每个人身份的合法性。有时采用生物扫描设备，如指纹和虹彩（眼睛）扫描仪。

认真安排口令也是一种常用的方法，而且当人们离开公司时应该更改口令。口令是密码字或数字，为获得访问系统的权利，口令必须键入到计算机中。

3. 数据备份

设备总是要更换的。但是公司的数据是不被替换的。很多公司都有一些方法试图保持软件和数据在最初就不被篡改。这些方法包括保护口令、时时检查数据和程序。但是实质性的方法是经常备份数据并将其保存到遥远的地方。

4. 防火墙

网络防火墙的作用相当于在网络周围设置一层外壳，用于防止连入网络的系统受到各种威胁。

防火墙可以通过滤掉某些不安全的网络业务，而降低网络系统的风险。例如网络文件系统（NFS）可以通过封锁所有进出网络的 NFS 业务，而防止网络外部人员使用。这就保护了各个主机，使其仍可提供服务，这在局域网环境如内部网络中很有用。可以采用的办法是对访问网络进行过滤，同时仍允许用户访问"外部世界"。典型的网络防火墙如图 4-10 所示。

在这种配置中，用一个防火墙网关把内部网和外部网分开。网关一般用于实现两个网络之间的中继服务。防火墙网关还提供过滤业务，它可以限制进出内部网络主机的信息类型。有三种基本防火墙技术：包过滤、电路网关和应用网关。通常可采用上述的多种技术，以提供完整的防火墙业务。

第 5 章 因特网新应用

5.1 即时消息

5.1.1 即时消息概述

即时消息（IM）是电子邮件的扩展，它能使两人或多人通过实时通信直接相互联系。为了使用即时消息软件，你要有一个你朋友的通讯录，向即时消息服务器注册。无论何时连入因特网，一种专用软件会通知你的消息服务器，你已经在线了。作为回答，该服务器提示你是否有联系人在线，同时还通知你的朋友你已经在线了。此后，你们就可以相互来回发消息了。大多数即时消息程序都包含视频会议、文件共享和远程辅助功能。许多商业，日常都使用这些即时消息功能。

即时消息是聊天的大众化变种，在这种应用中，当你的好友通讯录上有人已经在线时，你会得到通知，好友通讯录是一个你认定的朋友和同事名字的目录；当他们也在线时，你可以向他们发送消息，这一消息会立刻出现在他们的屏幕上（见图 5-1）。于是你们就可以用实时敲键的方法进行会话了。

自 2010 年以来，社交网提供商就经常提供即时消息功能，脸书交谈就是一种即时消息，而推特可以认为是 Web 2.0 的即时消息系统。大多数约会网站，如 OKCupid 或 Plenty of Fish，都有类似于服务器端的交谈功能。广泛使用的智能电话和类似的设备在 2000 年后期与传统的即时消息产生日益激烈的竞争，其起因是文本消息服务到处都在使用。

很多即时消息服务提供视频呼叫，如 IP 上的语音和 Web 会议服务等功能。Web 会议服务可以和视频呼叫与即时消息功能集成在一起。某些即时消息公司也提供对话音和视频的桌面共享功能，如 IP 无线电和 IP 电视。

现在，每个即时消息服务一般都提供它自己的客户服务，可以是单独安装一个软件，也可以是基于浏览器的客户软件。这些软件通常只能与供应商公司的服务软件一起工作，尽管有些客户端可以使用其他公司的有限的服务功能。也有第三方客户应用软件，可以连接大多数主要即时消息服务。几种常用的第三方客户软件是 Adium、Empathy、Miranda IM、Pidgin、Qnext 和 Trillian。

对即时消息应用软件标准的补充，能够提供像文件传输、通讯录、同时进行几起会话等功能。这些可能都是小型公司所需的功能，而更大一些的组织则要求能在一块工作的、复杂的应用软件。解决这个应用能力的方案是企业版的即时消息应用软件。这些软件包括像 XMPP、Lotus Sametime、Microsoft Office Communicator 等，这些软件通常与工作流系统这样的企业级应用软件集成在一起。这些企业级应用软件，或称企业应用集成，是按一定条件构建的，主要用于按一般格式存储数据。

5.1.2 QQ

1. QQ 概述

腾讯 QQ 通常称为 QQ，是中国大陆上最普通的免费即时消息计算机程序。到 2010 年 9 月 30 日，QQ 即时消息实际用户数大约 6.37 亿，可能是世界上最大的在线团体。同时在线的 QQ 用户数超过 1 亿。2011 年 2 月，QQ.com 在 Alexa 因特网排行榜中名列第 10，仅排在第 9 位的 Twitter 之后。该程序由 Tencent Holding Limited（HKEX 0700）维护。Naspers 有部分版权。自从 QQ 进入中国家庭以来，很快就成为现代的一种文化现象，现在正成为一种大众文化。除了这个聊天程序之外，QQ 还开发了许多包括游戏、虚拟宠物、铃声下载、音乐、购物、博客、微博、成组语音聊天等在内的子功能。

QQ 当前版本是 QQ 2010 beta 2。腾讯周期性地公布 QQ 专用版，以配合如奥林匹克运动会或中国新年这些事件。

正式客户运行在微软的 Windows 上，而 beta 大众版是为 Mac OSX 10.4.9 版或更新版开发的。万维网版、WebQQ（整版）和 WebQQ Mini（小型版），是用 Ajax 编写的，这些 QQ 现在都可以使用。

截至 2015 年 1 月，腾讯已拥有 8.29 亿账户，最高峰同时在线 QQ 用户为 1.76 亿。

2. QQ 国际

（1）Windows

从 2009 年开始，QQ 通过专用英文门户网站，将其 Windows QQ 国际客户端扩展到国际服务。

QQ 国际是为非本土语言交谈者提供使用与中国本土使用者相同的所有功能的一种环境。通过交谈和视频呼叫与其他 QQ 用户接触，并向非本土用户提供访问腾讯社交网（Qzone）的界面。该客户端支持英语、法语、西班牙语、德语、韩语、日语和普通汉语。第三方应用软件这一财富与 QQ 国际绑在一起有助于更方便地与国内外用户跨文化交流。

QQ 国际的一个主要功能是有一个选项，它可以自动地对所有交谈进行机器翻译。

（2）Android

QQ 国际的安卓版是 2013 年 9 月公布的。该客户端界面可使用英语、法语、西班牙语、德语、韩文、日文和普通汉语。除文本消息外，用户间可以相互传送图像、视频和语音媒体消息。另外，用户通过此客户端的 Qzone 界面与所有接触的人共享多媒体内容。

现场翻译功能对所有进来的消息都可用，并支持 18 种语言。

（3）iOS/ iPhone

2013 年底，用于 iPhone 和 iOS 设备的 QQ 国际公布了，与安卓所宣布的完全一致。

3．Web QQ

2009 年 9 月 15 日，腾讯正式宣布它的基于 Web 的 QQ，最新版本是 3.0。与很少用的基于 Web 的即时消息不同，Web QQ 3.0 的功能更像它自己的操作系统，并在桌面上可以加入 Web 应用软件。

4．开源与跨平台客户端

使用逆向工程技术，从事开源任务的人们能够很容易地了解 QQ 协议，并着力去实现与更多的用户友好的客户端相兼容的客户核心库的功能。这些客户端大多数是跨平台的，因此它们能够使用官方客户端不支持的那些操作系统。然而，这些功能只是官方客户软件功能的子集，因而功能有限。

5.1.3 脸书

1．特性

脸书是一种社交网络服务，也是一个网站，是 2004 年 2 月公布和运作的，属于 Facebook 公司。到 2014 年 6 月，脸书的实际用户已超过 13 亿。用户可以创建带有照片、个人兴趣目录、联系信息和其他个人信息的个人配置文件，还可将其他用户作为朋友加进来，当他们更新他们的配置文件时，可以交换信息，包括自动公告。另外，用户可以加入共同感兴趣的用户组，这些是由车间、学校或院校组织的，也可以按其他特点去组织。

脸书有很多用户可以进行交互的特性：

（1）Wall，是每个用户配置文件页上的一个空间，允许他的朋友向该用户发送信息，供查看。

（2）Pokes，允许用户相互发送虚拟"戳"（一个公告，用于告诉用户他们已经存入了"戳"）。

（3）照片，用户可以上传照相册和照片。

（4）状态，用户可以告之他们的朋友，他在哪里，在干什么。

根据隐私设置，可以查看用户配置文件的任何人，都可以看那个用户的 Wall。

2010 年 2 月 23 日，脸书在它的 News Feed 的某些方面被授予专利，该专利覆盖可以提供链接的 News Feed，因此一个用户可以参加另一个用户的同一活动。这一专利可以促使脸书对违反专利的那些网站采取行动，这些网站很可能包括像推特那样的网站。

脸书最常用的应用之一是照片应用——用户可以上传相册和照片，脸书允许用户上传无限量的照片，而其他图像托管服务，如 Photobucket 和 Flickr 则限制了用户上传照片的数量。在前几年，脸书限制用户每本相册 60 幅照片。到 2009 年 5 月，每本相册可有

200 幅照片。

2. 技术

脸书是用 PHP 构建的，这里 PHP 是用 HipHop for PHP 这种"源代码翻译程序"编译的，而这个源代码翻译程序又是由脸书工程师编写的。他们又将 PHP 转换成 C++。据报告，使用了 HipHop 之后，脸书服务器的 CPU 占用时间减少了 50%。

脸书使用了基于 HBase 的组合平台，在分布式机器上存储数据。由于采用拖尾架构，新事件可以以日志文件形式存储，而且日志是拖尾的。该系统弹出这些事件并将他们写入存储器。此后，用户界面将这些数据提出并将其显示给用户。脸书习惯处理 AJAX 请求，这些请求用 Scribe（由脸书开发的）写入日志文件。

5.1.4 推特

推特是一种社交网在线服务，它能使用户发送和读取称为"推文"的 140 个字符的短消息。

注册了的用户可以读取和发送推文，但非注册用户只能读取消息。用户可以通过推特网站界面、短信服务（SMS）或移动设备 App 访问推特。推特公司位于美国旧金山，在全世界有超过 25 个办公室。

推特创建于 2006 年 3 月，其业务迅速普及到全世界，2012 年用户超过 1 亿，每天发送推文 3.4 亿个，每天处理 16 亿个搜索请求。2013 年，推特成为 10 个访问最多的网站之一，并作为"因特网的短信服务"来表述。到 2014 年 12 月，推特用户已超过 5 亿，其中超过 2.84 亿是活跃用户。

1. 特性

推特在默认情况下是公开可见的，但发送者可以限制消息发送给某些人。用户可以通过推特网站，与外部兼容的应用（如智能电话），或通过某些国家的短信服务使用推文。用户通过推特向前转发推文的过程叫做转推。无论是推文还是转推文，都可以被跟踪，以观察哪些用户是用得最多的。由于这种服务是免费的，通过短信服务访问它，可能会产生电话业务提供商的费用。

推特允许用户用手机去更新他们的配置文件，可以通过智能电话和台式机中的文本消息，或应用程序去更新。

作为一个社交网络，推特围绕着跟进者这个概念在发展。当你打算去跟进另一个推特用户时，那个用户所发布的推文就会按照逆向顺序出现在你的推特主页上。假设你跟进了 20 个人，那么你可能在滚动页面上看到这 20 个用户的混合在一起的推文，比如会有有关谷物早餐的消息、感兴趣的网址连接、音乐推荐，甚至对未来教育的假想。

2. 实现

推特在很大程度上依赖于开源软件。推特 Web 界面采用 Ruby on Rails 框架，安排在 Ruby 的增强型企业版功能上。

到 2011 年 4 月 6 日，推特工程师们决定将他们的 Ruby on Rails 搜索栈转换到他们称为 Blender 的 Java 服务器上。

该服务的应用程序界面允许其他 Web 服务和应用与推特集成在一起。

5.1.5 微软服务网 MSN

MSN（最初是微软网络）是因特网网站的集合，是由微软公司提供服务的。1995 年 8 月 24 日微软网络是作为在线服务和因特网服务提供者首次亮相的，与 Windows 95 版操作系统同时推出。

MSN 除了其最初的拨号服务以外，微软这些年来使用"MSN"商标名来标明各种产品和服务，主要是 Hotmail（现在是 Outlook.com）、Messenger（在因特网俚语中曾经是"MSN"的同义词，现被 Skype 取代）和它的搜索引擎（现在是 Bing），以及其他几个另外的商标和间断的服务。

现在由 MSN 提供的网站和应用软件系列是 2014 年由微软首先推出的，是重新设计和再次公布的全线产品的一部分。MSN 的基础在美国，它为全世界几十个国家提供门户网站的国际版。

2012 年，MSN 在它的博客上宣布，10 月 26 日将公布 MSN.com 主页的新版本，且只用于 Windows 8，并表明这个新版将是"清晰、简单、采用触摸方式的"。微软说，由于 IE10 的速度，该网页更像 App 程序。更新的功能包括"前向翻转"（"Flip Ahead"），该功能允许用户从一个题目切换到另一个题目。

2014 年 9 月 30 日，微软宣布一个完全重新编写和重新设计的 MSN 网站，使用公司的现代设计语言。这一新 MSN 门户网站有一个新版的徽标，其格式类似于当前其他微软产品。该网站不再提供原始内容，取而代之的是让编辑者根据普通和信赖的组织参与的人物，重新安排内容。大多数 MSN 上现有的内容被取消，因为该网站被简化成有一个新主页和目录，而且目录中都有相应的应用程序，它们是：

- 新闻
- 天气
- 娱乐
- 体育（运动）
- 金钱
- 生活状态
- 健康与肥胖
- 饮食
- 旅行
- 汽车
- 视频

该主页顶部提供对微软服务的访问，它们是 Bing、Outlook.com、Skype、Office Online、OneNote、OneDrive、BingMaps XboxMusic 以及通用社交媒体服务脸书和推特。用微软账号签名进入 MSN 就能出现个性化内容，还能使该网站的各种设备同步以及能使相应的应用软件同步。该网站的重新设计，使得 MSN 长期个性化主页服务 My MSN 被终止，因为新的网站不再支持用户指定的 RSS 内容，虽然这个服务曾经给客户提供了丰富站点摘要（RSS）功能。新的网站增加了按规格定制功能，使得主页上的每个目录都可重新排序或隐藏起来。

2014 年，微软还宣布 MSN 现在支持反应式设计，并去掉了单独的移动网站。MSN 的这种重新设计被证明是明智的，因为在两个月之后，每天的访问人数增加了 1000 万。

在 2014 年重新设计 MSN 万维网门户网站的同时，微软再次宣布了很多 Bing 应用软件，这些软件原先是作为 MSN 软件系列与 Windows 和 Windows 电话一起供货的。2014 年 12 月，这些新的应用软件可跨所有主要移动设备平台使用，也可以用在 iOS、Android 和 Fire OS 平台上。

5.1.6 微信

微信是由中国腾讯公司开发的一种移动文本和话音消息通信业务，2011 年 1 月首次推出。按每月的实际用户来看，它是最大的独立消息应用软件。

这项应用可以在安卓、iPhone、黑莓、Windows Phone 和 Symbian 电话中使用，并可以在基于 Web 和 OS X 的客户端使用，但这些都要求用户在对该应用有授权的手机上使用。截至 2014 年 8 月，微信用户已有 4.38 亿，其中 0.7 亿在境外。

用户注册微信可以用脸书账号或电话号码。目前用电话号码注册的国家超过 100 个。不能通过腾讯 QQ 直接注册。但通过电话号码注册之后，用户就可以用腾讯 QQ 账号连接他们微信的账号了。

使用微信，可以传送文本消息，进行"按住说"通话，广播（一对多）消息，共享照片和视频以及位置共享。微信通过蓝牙技术让人们相互联系，并在需要时随机提供联系的各种手段，还可以和社交网络服务集成，比如通过脸书和腾讯 QQ 运行的那些微信。照片还可以用滤波软件和文字说明去修饰，也可以使用机器翻译服务。

微信支持用户使用公共账号去注册，这使得用户将馈送送给其他用户，与他们互动，为他们提供服务。到 2014 年底，微信公共账户已达到 800 万。

在中国，微信公共账户已成为政府、新闻媒体和公司的普通业务，也是推广的平台。专用公共账户的用户使用该平台可进行医院预约、签证更新或信用卡等业务。

2014 年 9 月 30 日，公布了带有视力捕获和共享等新功能的 WeChat6.0。

按 GlobalWebIndex 统计，微信是世界上第五个用得最多的智能电话应用软件，2013 年 8 月，它紧随 Google Maps、Facebook、YouTube 和 Google+之后。微信还宣布，他有 1 亿个国际注册用户，这是仅用了 3 个月，就从 5000 万上升到 1 亿的，而中国用户为 3 亿。

按新华社的统计，2013 年 10 月，全世界微信用户为 6 亿，而且全部用户的 30%是在国外。

5.2 社交网络服务

5.2.1 社交网络服务概述

社交网络服务（又称社交网站）是在那些有共同兴趣、共同活动、共同背景或现实生活中有联系的人之间，构建社交网络或社会关系的一个平台。一个社交网络服务由下面的内容构成：每个用户的个人情况介绍（通常是一个配置文件）、他们的社会联系和各种附加服务。社交网站是基于万维网的，它允许个人创建公共配置文件，以在大家之间产生一个通讯录去共享他们的联系，在该系统内查看并展现这些联系。大多数社交网服务为用户提供多种在因特网上进行交互活动的方法，如电子邮件和即时消息。社交网站是多样的，他们接收新信息，使用移动连接，使用摄影、视频、共享和博客等通信工具。在线社区服务有时也可视为社交网络服务，尽管广义上讲社交网服务通常被认为是以个人为中心的服务，而在线社区服务是以群为中心的。社交网站允许用户在他们的网络内与人们共享思想、图片、邮件、活动、事件和感兴趣的事。

社交网服务的主要类型包含服务类场所（如以前上学的年代或同学），与朋友联系的方法（通常为自述页）和与信任相关的推荐系统。现在通常都将这些方法组合起来使用，在美国就有在世界上广泛使用的服务，如脸书、谷歌+、LinkedIn、Instagram、Reddit、Pinterest、Vine、Tumblr 和推特。

因特网发展最快的应用之一是社交网络技术，即个人的相互连接。社交网站有三种基本类型：再重聚网站、朋友之朋友网站和共同兴趣网站。

1. 再重聚网站

再重聚网站是为联络相互认识但又失去联系的人设计的；例如：几年未见面的高中老同学。你只要连入一个再重聚网站并且给出你的年龄、性别、高中学校的名字等简要信息，你就加入这一社交网络了。这些信息加入到该再重聚网站的成员数据库中。网站的成员可以搜索该数据库以找到这些同学。很多这种网站，当有新人加入，而他的某些信息与你的简要信息部分相吻合时（如高中班级），就会给你提示。最知名的两个再重聚网站是 Classmaters Online 和脸书。

2. 朋友之朋友网站

朋友之朋友网站是为把两个相互不认识，但又有一个共同的朋友的两个人联系在一起设计的。基本原理是，如果你们共同有一个朋友，那么很可能你们也会成为朋友。例如，你认识的一个人通过提供其自己的简要信息和朋友通讯录，应该能启动这种网站。你也应该能访问你相识人的网站，因而就能够联系到你相识人的朋友。你甚至可以加入到该网站所提供的朋友通讯录中。两个知名的朋友之朋友网站是 Friendster 和 MySpace（见图 5-2）。

3. 共同兴趣网站

共同兴趣网站把有共同兴趣和爱好的人联系在一起。你可以选中基于一个具体兴趣的网站。例如，如果你想共享图像，应该进入 Flickr 网站。如果你想与商业接触，你应该进入 LinkedIn 网站。如果你想找一个或建立一个专门兴趣组，你应该进入 Meetup 网站。

在向社交网站或个人提供任何信息之前，应仔细考虑你会暴露什么。不要提供不适当的或全部个人信息。

5.2.2 维基

维基是一个通过万维网浏览器，使用简化的标记语言，或所见即所得（WYSIWYG）文本编辑器，创建和编辑任何数量的相互链接网页的一个网站。维基通常由 wiki 软件授权，并且经常由多个用户联合使用。应用维基的例子有团体网站、公司内联网、知识管理系统和注记服务，该软件也可用于个人笔记。

维基可用于不同的目的。有些可以控制不同的功能（访问级别），例如，编辑权可对材料进行更改、增加或删除。其他功能则不要求强制访问控制就能进行访问。

维基概念的本质是：

（1）维基吸引所有用户在维基网站内去编辑网页或创建新的网页，它只使用 plain-vanilla 万维网浏览器，而不需要任何额外的附件。

（2）维基鼓励在不同网页之间的有意义的题目联合，方法是使网页链接操作很直观且容易，并显示是否有预想的目标网页存在。

（3）维基对偶尔的访客不是一个精心设计的网站。相反，它试图把访客引入到一种创作和合作的过程中，并经常改变万维网网站的景色。

在 wiki 网站中的单一网页称为"维基页"，而全部网页的集合称为"维基"，这些网页可以通过超链接很好地互联在一起。维基本质上是用于创建、浏览和搜索信息的数据库。

维基技术很明显的一个特性是使用它制作和更新网页很容易。一般来讲，在修改被承认之前无须检查，很多维基是公开允许公众去交流的，不要求他们去注册用户账户。有时为了会话则建议去注册，以便建立一个"维基—签名"网络跟踪器，用于签署自动编辑。但是，很多编辑可以实时构建并能即时在线出现，这可能助长该系统的违规使用。私人维基服务器要求对编辑网页进行用户验证，有时甚至读网页也要用户确认。

维基也是一种应用程序，是一个典型的万维网应用软件。它允许合作式修改、扩充或删除它的内容和结构。在典型的维基中，书写文本可以使用简单的标记语言（称为"维基标记"）或功能强的文本编辑程序。虽然维基是一种内容管理系统，但它还是与博客或其他类似的系统不同，区别在于所书写的内容无须规定拥有者或领导者，而且维基有一点隐含结构，允许在用户需要时才显现出来。

Wikipedia 这一百科项目，是能在公共万维网上浏览网页的最普通维基，但是有很多网站是运行多种不同的维基软件的。维基可以用于多种网站，既可是公众的，也可是私人的，包括知识管理、做笔记、团体网站和因特内联网。有些维基还可以对不同的功能（访问级别）进行控制，例如，某些编辑权限允许对材料进行变更、增加或删除，有一些则允许在没有强制访问控制下进行访问。而另外一些规则会迫使你对内容进行组织。

5.2.3 博客与微博

1. 博客

博客（Web 与 log 的混成词）是一种万维网站或网站的一部分。博客一般由个人维护，并有评论、事件描述的词条或像图形或视频那样的其他材料。词条一般以反时间顺序显示。博客也可以作动词使用，意为维护博客或增加内容到博客上。

大多数博客是交互式的，可以通过博客上的小窗口相互留言，甚至报文。这种交互与其他静态网站的交互是有区别的。

很多博客是按具体主题提供评论或新闻的，而其他博客则负责更个人化的在线日记工作。典型的博客可以把文本和图像组合在一起并链接到其他博客、网页和其他与本题目相关的媒体上。读博客的人可以把评论以交互的格式留下来，这是很多博客的重要功能。虽然某些博客集中在艺术（艺博）、摄影（影博）、视频（视博）、音乐（MP3 博）和语音上（podcasting），但大多数博客主要是文本形式的。微博则是博客的另一种类型，是一种很短的邮件。

截至 2011 年 2 月 16 日，公共博客已超过 1.56 亿。到 2014 年 2 月 20 日，世界上大约有 1.72 亿 Tumblr 博客和 7580 万 WordPress 博客。按照某些评论家和博主的说法，博主是当今最普通的博客服务使用者，但是博主不提供公共统计数字。到 2014 年 2 月 22 日，Technorati 有 130 万博客。

有很多不同类型的博客，他们之间的区别不仅在内容的类型上，分发和书写内容的方法上也有所不同。

2. 微博

微博是一种博客形式的广播媒体。微博与传统博客的不同之处是它的文件规模比较小。微博"允许用户交换像短句、单个图像或视频链接那样内容的小元素"。

与传统博客相比，微博版主发出的题目范围，从简单的，如"我现在正在做什么"到"赛车"这样的主题。也有商业微博，用于促进网站服务或促销产品，并促进一个单位内的合作。

有些微博服务提供隐私设定这样的功能，它允许用户去控制谁可以读他们的微博；另外，除了基于 Web 的界面，还可用另外一些方法公布一些事项。这些可能包括文本消息、即时消息、电子邮件或数字声音。

微博服务已经改变了信息消费的方法。它已经使市民本身强化为数据的感知者或数据源，而数据也可能变成重要的信息片段。人们现在共享在他们周围观察到的东西，共享有关事件的信息，共享某些方面的见解，例如，共享政府的健康管理政策。

还有，这些服务存储了从这些邮件来的各种各样的元数据，如位置和时间。这种数据的汇聚分析包括各个方面，像空间、时间、题目、情趣、网络结构等，并给研究人员一个机会，以便从某些感兴趣的事件中了解人对社会的认知。微博也刺激了著述业，在 Tumblr 微博平台上，再博客功能将邮件链接到原创那里。

微博有可能成为一种新的非正式的信息通信媒体，特别适合于单位内合作工作的情况。过去几年，通信方式已从面对面转移到网络电子邮件、即时消息、文本消息和其他一些方式上。但是，有人认为，现在的电子邮件是一种慢速、低效率的通信方法。例如，耗时的"电子邮件链"可以发展下去，但是，两个或多个人可能会为简单的事情，如安排会议，而消耗在冗长的通信中。微博提供的一对多广播方式，可以认为由于回避了这种不利情况而提高了效率。

用户和单位可以建立他们自己的微博服务；为此可以使用免费和开源软件。托管微博平台也可用于商业和机关。

5.3 云计算

云计算是一种交付服务而不是产品，它提供了一种计算方式，在这种方式中，将共享的资源、软件和信息作为一种实用程序通过网络，特别是因特网提供给计算机或其他设备（这类似于电网）。

1. 概述

云计算在技术上是一种商业术语，这种技术提供计算服务、软件、数据访问和存储服务，而不要求最终用户知道提供这些服务的系统所在的实际位置和配置。与这一概念相似的是电网，电网内的最终用户用电，也不必知道提供服务的供电设备或基础设施的情况。

　　云计算是一种基于因特网协议的 IT 业务的新的补充、消费和交付的新模型，通常包含动态可伸缩的，并且是虚拟化的资源。它是一种由因特网提供的，远程计算站容易访问的副产品和后续产品。这可以采取基于 Web 工具的形式，或者是通过 Web 浏览器，用户能访问和使用的应用程序，如果该应用程序已经安装在用户本地的计算机上了。

　　云计算提供商通过因特网交付应用，这些应用程序可以从 Web 浏览器、台式计算机和移动应用上访问，同时，商业软件和数据存储在远端的服务器上。某些情况下，继承应用（是一种商业应用系列，直到现在，在瘦客户 Windows 计算中，还是很普遍的）是通过共享屏幕技术提供的，而计算资源是统一放在远程数据中心上的；其他情况，整个商业应用是用像 AJAX 那样的基于 Web 的技术编码的。

　　基础会聚（会聚的基础）和共享服务的更广义的概念是云计算的基础。这种类型的数据中心环境能使企业去建立他们自己的应用，并更快地运行起来，而管理比较容易，不需要太多的维护，也能使 IT 业更快地调整 IT 资源（如服务器、存储器和联网），以满足变动的、不可预期的商业需求。

　　大多数云计算的基础设施是由服务组成的，这些服务是通过共享数据中心交付的，并对消费者的计算需求是以单一访问点呈现的。

　　云计算对商业的巨大影响已经促使美国联邦政府把云（计算）视为重构他们的 IT 基础和降低预算的措施。随着顶级政府正式采用云计算，很多代理商已经至少有一个或多个云系统在线了。

　　图 5-3 为云计算逻辑图。

2. 公共云、私有云和混合云

　　公共云是构建在因特网上的，凡是付费给这种服务的用户都可以访问它。公共云归服务提供商所有，通过预约就可以使用。图 5-4 顶部的标注框是典型的公共云架构。可以使用的公共云有很多，包括谷歌应用引擎（GAE）、亚马逊万维网服务（AWS）、微软的 Azure、IBM 的蓝云和 Salesforce.com 的 Force.com。上述云的提供者是商业性的，他们为在他们的专利架构内创建和管理虚拟机，提供一个公共的远程访问接口。公共云提供一套可选择的商业流程。这种应用和架构服务是基于柔性价格的。

　　私有云是构建在一个单位所拥有的内联网范围内的。这样，它就是客户拥有和管理的，对其访问也就受限于它所拥有的客户和伙伴。这种安排并不意味着可以通过公共访问接口在因特网上出售容量。私有云为本地用户提供一个柔性的、灵活的私有架构，在它管辖的范围内承载工作负荷。私有云可以视为一种更有效、更舒适的服务。它可能影响着云的标准化，但它又保持着更多的用户和单位的管理权。基于内联网的私有云连入公共云是为了获取更多的资源。

　　混合云是公共云和私有云二者共同构建的云，如图 5-4 左下角所示。私有云也能支持混合云模型，其支持方式是使外部公共云的计算能力为本地服务。例如，研究计算云（RC2）是由 IBM 公司构建的一种私有云，它连接散布在美国、欧洲和亚洲的八个 IBM 研究中心的计算和信息技术资源。混合云可对客户、合伙人网和第三方进行访问。

总之，公共云激励着标准化，维护资本投资，提供应用的灵活性。私有云则力图做到用户化，为用户提供高效、弹性、安全和保护私密的环境。混合云处于这二者之间，借助于资源共享而采取了很多折中。

3．特性

云计算展示出以下重要特性。

（1）计算资源授权给最终用户，方法是将这些资源放在用户自己的控制之下，而不是由中央式的 IT 服务去控制。

（2）灵活性改善了用户反复使用基础技术资源的能力。

（3）应用程序接口对软件的访问能力，能使该机器与云软件交互，其方法与用户接口能使人与机器交互的方法一样。云计算系统一般使用基于 REST 的 API。

（4）设备和位置独立性能使用户用 Web 浏览器访问系统，而不管系统的位置或使用的是何种设备（PC 机还是手机）。当基础设施不在网站上，(典型的是由第三方提供的设备)，并且要通过因特网访问时，用户可以从任何地方进行连接。

（5）多租用性，允许跨越大的用户群共享资源和分摊费用。

（6）可靠性增加了，如果使用多个冗余网站的话，这种良好设计的云计算很适合于企业运作的连续性和灾害恢复。

（7）性能是可监视的，使用 Web 服务这种系统接口，可构建连续而松散的耦合体系结构。

（8）安全得到改善，这是由于采取了数据集中管理，强化资源的安全等措施。

（9）云计算应用的维护比较容易，因为这些应用软件不需要安装在每个用户的计算机上。

5.4 大数据

大数据是一种表示数据集的、意义广泛的术语。这种数据之大、之复杂，是传统数据处理应用所无能为力的。这种数据的挑战性包括分析、截取、处置、搜索、共享、存储、传送、可视化和信息隐匿。简言之，这一术语是指使用预测分析或其他一些先进的方法，从数据中提取其值，但又很少关注数据集的具体规模。

通过对数据集的分析可以发现"现货企业的趋势，灾害的预防，暴力犯罪等"问题中的新的相关性。在很多领域中，科学家、媒体和广告从业者以及政府都遭遇到大数据集的限制。这些限制影响了因特网的搜索，财务和企业的信息工作。

例如，科学家在电子科学领域，包括气象学、基因组学、神经网络体学、复杂物理模拟和生物及环境科学研究中就受到限制。

数据集在规模上的增长，部分原因是数据持续从各种来源被广泛收集，这些来源包括廉价而大量的移动设备、高空传感技术（遥感）、软件记录、相机、麦克风、射频标识阅读器和无线传感器网。世界技术的发展，人均存储的信息量，自 20 世纪 80 年代以来，大概每 40 个月翻一番。到 2012 年，每天产生 2.5 艾字节个数据。对大型企业的挑战是确定谁将拥有跨越整个机构的大数据的主导权。

图 5-5 为整体信息存储量的增长情况。

1．定义

大数据是用来描述信息（结构化和非结构化的）的指数增长，有效性和使用的通用术语。按国际数据集团的说法，是单位和信息行业的领导们特别关注的，不断增加的信息量、信息种类和传播速度，促成了大数据的形成。

（1）量

很多因素造成了数据量的增加——长年存储的业务数据，经常性的、从社交媒体来的文本数据流，收集的传感器数据等。以前，过多的数据量造成存储问题。虽然现今的存储价格降低了，而其他问题又出来了。例如，如何确定大量数据之间的关联性问题，如何从有关的数据中产生新值的问题。

（2）种类

当今，数据以各种形式出现——从传统数据库到由最终用户和联机分析处理软件产生的层次型数据存储，到文本文档、电子邮件、计量器收集的数据、视频、音频、股票行情指示器和财经业务数据。据估计，约 80%的单位，数据不是数值化的！但这些数据必须包含在分析和决策支撑之中。

（3）速度

按照甘特的说法，速度是指"产生数据的速度有多快，满足要求的数据处理速度有多快。"射频识别标记和智能仪表是使数据接近于实时迸发的推手。

2．大数据的使用

实际问题不是你收集了大量数据（因为很清楚，我们已经处于大数据时代了），而是用那些关系重大的大数据去做什么。大数据的有用之处是，单位有能力去利用相关联的数据，并用这些数据去做出最好的决策。

当今的技术不仅支持大量数据的收集和存储，也提供了解和利用其全部值的功能，这有助于单位的工作更加有效并获益。例如，利用大数据和大数据解析学，可以：

- 分析数据量庞大的库存，以确定最大赢利下的最优价格并清仓。
- 几分钟内重新计算有价证券的风险，了解今后降低风险的可能性。
- 快速确认关系重大的那些客户。
- 在销售点，根据客户当前和以前的购物记录，产生零售优待券，确保较高的偿还率。
- 在恰当的时间，向移动设备发出合适的建议，而客户能在合适的地点利用这些建议。
- 分析从社交媒体来的数据，以便发现新的市场需求趋势。
- 使用单击流分析和数据挖掘技术，检测欺骗行为。
- 通过研究用户行为、网络运行记录和机器传感器，确定产生事故、问题和缺陷的根源。

大数据的一些实例：

- 射频识别系统产生的数据量是传统条形码系统的 1 千倍。
- 全世界每秒钟进行 1 万笔支付卡交易。

- 沃尔玛每小时处理 1 百多万笔客户交易。
- 每天传送的推文多达 3.4 亿个，相当于每秒 4 千个推文。
- 脸书有超过 9.01 亿个产生社交数据的实际用户。
- 超过 50 亿人正在手机上呼叫，传送电文，发推文和浏览网站。

3. 技术

大量新技术的使用，使得很多单位产生了大量大数据，并且需要对这些大数据进行分析，这是因为：

- 廉价的海量存储器和有强大处理能力的服务器。
- 更快的处理器。
- 能买得起的大容量存储器，如 Hadoop。
- 专门为大数据量（包括非结构化数据）设计的新存储和处理技术。
- 并行处理，簇技术，大规模并行处理，虚拟化，大栅格环境，高连通性和高吞吐率。
- 云计算和其他灵活的资源分配和配置。

大数据技术不仅具有收集大量数据的能力，也具有了解大数据和利用它的值的能力。所有单位利用大数据集的目标，应该是利用最相关的数据，进行最优化的决策。

非常重要的一点是，并非所有的数据都是相关或有用的。但如何发现关系最密切的数据点？这是一个大家都认可的问题。"大多数企业已经经历了从大数据中提取有效值的漫长过程。有些公司企图在大数据上采用传统的数据管理方法，只学会了不再使用陈旧的规则"，这是 Dan Briody 2011 年在经济学家智库刊物上说的，"大数据利用游戏变更财富"。

5.5 慕课与翻转课堂

1. 慕课

大规模开放在线课程是一种通过万维网，使无数学生访问开放的在线课程。除了传统的课程材料，如电影课程材料、读物和习题集以外，很多慕课都提供交互式用户论坛，以支持学生、教授和助教之间的团体交互。慕课是最近才发展起来的一种远程教育，2008 年首次推出，2012 年成为一种大众化学习方式。

早期的慕课常强调开放访问的功能，如内容、结构和学习目标的开放许可，以激励资源的再利用和再融合。后来的慕课则对他们的课程材料使用封闭式许可，同时让学生随意访问。

图 5-6 给出 MOOC 每个字母的含义。

（1）技术

不像传统的课堂教学，慕课要求由视频图形制作人员、教育设计者、信息技术专家和平台专家提供的特殊技巧。佐治亚工学院教授 Karen Head 指出，19 个人在为他们的慕课工作，但需要更多的人手。由于加入者众多，像媒体/内容共享的 Web 网站一样，这种平台应能满足各种需求。通常慕课使用云计算并通过创作系统进行制作。制作慕课的创作工具是教育专用软件包，如 Elicitus、IMC Content Studio 和 Lectora，这些都是容易使用的，并支持像 SCORM 和 AICC 那样的电子学习标准的。

授课方式包括对视频和其他学习材料、考试和考评方法的异步访问，以及在线论坛。在 2013 年以前，每个慕课都力图开发他们自己的授课平台。Edx 在 2013 年 4 月与斯坦福大学合作（该大学以前有一个自己的平台，称为 Class2Go），在 XBlock SDk 上工作，这是一个联合开源平台。它是在 Affero GPL 开源许可证下的公共资源，要求对该平台的所有改进都应让公众知晓，并用在同一个许可证下。斯坦福教务副主管 John Mitchell 说，他的目标是提供"Linux 在线学习环境"，这与一些公司，如 Coursera 公司开发他们自己的平台不同。

（2）可能的好处

慕课指南列出了 12 项好处：

- 适合任何连入因特网的场合（Web 或 Wi-Fi）；
- 任何一种语言或多种语言；
- 任何在线工具；
- 避开时间区和具体边界；
- 短时间的制作和授课；
- 所有人都可共享上下文内容；
- 设置简单；
- 对等接触可能引起无意中的学习；
- 更容易跨越学科和制度壁垒；
- 学生加入的门槛低；
- 通过参与使人们的学习环境和网络得到提升；
- 改善了终身学习的技巧。

2. 翻转课堂

翻转课堂是对传统教学的一种颠倒。在这种教学中，学生首先面临的是课堂以外的新材料，通常是阅读或观看视频材料，而课堂上则通过诸如解答问题、讨论或辩论等方法去探讨一些较难理解的知识。

在翻转课堂中，学生和教师的角色和期待，有如下变化：

- 学生在他们自身的学习和钻研核心内容方面有更多的责任，在课前可以单独或分组进行预习，然后采用更高层次的思考，在更大的活动空间上展现他们的知识和技能。
- 通过简单的主动学习，吸引学生，指导学习，纠正理解上的错误以及采取各种教学方法，还有及时进行反馈，就形成了很重要的学习条件。
- 更重要的是在面对面的环境下，集中在对概念的探讨，含义的解释和知识的展示和应用上，如图 5-7 所示。

教育技术（见图 5-8）是翻转课堂的重要基础，因为它可用于：

- 使学生在他们合适的时间和合适的地点去获取重要的内容，（例如学习材料、读物和交互媒体）；
- 以多种形式提供学习材料，以适应不同学习层次和多种方式的学习（如文本、视频、音频、多媒体等）；
- 提供课内外演讲交互的机会（如投票工具、讨论工具、内容创作工具等）；
- 及时向学生传达信息，更新和提示信息（例如，微博、通知工具等）；
- 向教师和学生提供即时和匿名反馈信息（例如测验结果），以标明修正点；
- 获得有关学生的数据，用于分析他们的进步，标明"处于危险境地"的学生（用分析软件）。

第三部分　程序设计语言和数据库

第 6 章　程序设计语言

6.1　程序设计语言概述

　　语言是一种交流的体系。程序设计语言由使人能与机器通信交流的所有符号、字符及使用规则组成。有些程序设计语言的产生是为了服务于特殊的目的(例如控制机器人)，而其他程序设计语言则是比较灵活的通用工具，适用于多种类型的应用。然而，每一种程序设计语言必须按一些确定类型的指令格式去编写，以使计算机系统能够完成大量常见的操作。换言之，每一种语言必须具有以下为人们所熟悉的指令类：

　　（1）输入输出指令。用于 I/O 设备与中央处理器之间的通信。这些指令提供了要完成这种类型的输入或输出操作的细节及操作期间将用到的存储单元。

　　（2）计算指令。在处理过程中实现加、减、乘、除的指令，显然，所有程序设计语言均有此类指令。

　　（3）逻辑比较指令。这些指令用于转移程序控制，在编写程序中用于选择和循环结构。

　　（4）存储检索和传送指令。这些指令用于处理期间数据的存储、检索和传送。数据可以从一个存储单元复制到另一个存储单元或在需要时进行检索。

　　即使所有的程序设计语言都具有执行上述这些操作的指令集，但在机器语言、汇编语言与高级语言中所使用的符号、字符及语法方面仍有明显的区别。

1．机器语言

　　计算机的机器语言由二进制数串组成，并且是唯一能被 CPU 直接"识别"的语言。任何机器语言指令至少由两部分组成。第一部分是命令或操作，它告诉计算机要完成什么功能。每一台计算机为每一操作命令都安排了一个操作码（op code）。指令的第二部分是操作数，它告诉计算机在哪里找到或存储数据或者其他将要完成的指令。一条指令中操作数的数目因计算机不同而异。在单操作数机器中，指令 ADD 0184 的等效二进制指令会让地址 0184 中的值，与算术逻辑部件中存于某一寄存器的值相加。程序员为早期的 IBM 机器书写指令"ADD 0184"的形式为：

00010000000000000000000000010111000

除了记住机器指令集中数十个命令的数字代码外，程序员还必须知道数据和指令的存储单元。最早的编码常常花费数月时间，因此非常昂贵，且常常出错。用检查指令找出错误，与最初编程时一样令人乏味。而且如果程序以后必须修改的话，则要耗费数周的时间。

2. 汇编语言

为了减轻程序员的负担，20 世纪 50 年代初期开发了助记符操作码和符号地址。为了改进程序准备过程，首先要做的一项工作是用字母符号（即助记符）去替代数字的机器语言操作码。现在每一台计算机都有一套助记符代码，当然实际的符号因机器型号而异。计算机仍使用机器语言处理数据，不过汇编语言软件首先把指定的操作码符号翻译成等效的机器语言。

这一改进为日后的发展奠定了基础。如果计算机能方便地将符号翻译成基本操作，那为什么不能再完成其他事务性编码功能，诸如将存储器地址分配给数据呢？符号寻址就是将地址表达为程序员方便使用的符号，而不按照它的绝对数字地址表示。

程序员不再像以前那样将实际地址赋给符号数据项。现在他们只需要规定程序所需的第一个地址即可，汇编语言程序将从这里开始，为指令和数据分配地址空间。

这一汇编程序（或汇编器），还能使计算机将程序员的汇编语言指令翻译成它自己的机器代码。在汇编语言中，由程序员编写的指令程序叫做源程序。源程序由汇编程序转换为机器码后，则称为目标程序。

汇编语言的一个最大的缺陷在于它是面向机器的，即它们是为所用的特定型号的处理器设计的，程序在不同机器上要重新编码才能运行。

3. 高级语言

早期的汇编程序中，一条源程序指令只产生一条机器指令。为了加快编码速度，人们开发出了一种汇编程序，它能为每一源程序指令生成数量不等的机器语言代码。换句话说，一条宏指令可以产生若干行机器语言代码。例如，程序员写了一条"READ FILE"（读文件）程序，翻译软件会自动地提供一系列详尽的、预先准备好的机器语言指令，把从输入设备读入的数据文件中的一个记录，复制到主存储器中。这样，就减轻了程序员的任务，而不必为要执行的每一个机器操作书写一条指令。

助记符技术和宏指令的研发又导致了高级语言的研发，它们通常面向某类特定的处理问题。例如，若干高级语言已经用于处理科学计算性质的问题，而其他一些高级语言则着力于文件处理的应用。

与汇编语言不同，高级语言程序几乎可以不加修改地用于不同型号的计算机。这样，当换用新设备时，重新编程的费用可大大减少。高级语言的其他优点在于：

（1）比汇编语言更易于学习；
（2）书写程序所需的时间较少；
（3）提供较好的文档；
（4）易于维护；

（5）能够熟练使用这种语言编写程序的程序员，将不受某一机器类型的限制。

6.2　C、C++ 和 C#

　　C 语言把结构化高级语言的一些最好的特性与汇编语言组合在一起了——也就是它编程相对容易（至少与汇编语言相比较），能有效地使用计算机资源。虽然 C 最初是作为系统程序设计语言设计的（实际上 UNIX 操作系统的主要程序首先是用 C 语言写的），但它已被证明是可用于各种应用软件的强大的和灵活的语言。C 语言是绝大多数计算机专业人员用来开发软件产品的语言。

　　比较新的、面向对象的 C 版本被称为 C++（见图 6-1）。C++包括 C 的基本特性，所有 C 程序都能被 C++编译器识别。但 C++具有另外一些特性，如对象、类和面向对象程序的其他元素。C++也有可视化版本（VC++）。所有这些表明，C++是用于图形应用的最通用的程序设计语言之一。

　　C++完全支持面向对象的程序设计，包括面向对象开发的四个特征：封装、数据隐藏、继承性和多态性。

　　单元的自包含特性称为封装。使用封装，我们就可以实现数据隐藏。数据隐藏是一个很有价值的特征，这样，用户可以使用一个对象而无须知道或关心其内部的工作机制。这就像你使用冰箱，而不需要知道压缩机是如何工作的一样，你可以使用一个设计得很好的对象，而不用知道它的内部数据成员的情况。

　　通过创建用户自定义的类，C++支持封装和数据隐藏特性。一旦一个良好定义的类，作为完全封装的实体被创建，就可以充当一个整体来使用。这个类的内部真实工作被隐藏起来了。使用良好定义的类的用户，不需要知道类是如何工作的，他们只需要知道如何使用它就行。当 Acme 汽车公司的工程师们想制造新车的时候，他们有两种选择：要么从零开始，要么对现有的车型加以修改。也许他们的星型模型是近乎完美的，但是他们想加一个涡轮充电器和一个 6 速变速器。总工程师则不想从头做起，他说："我们再造一个'星'吧，并且要增加一些功能，我们把它叫做'类星'吧。""类星"车是"星"型车的一种类型，但"类星"具有一些新的特点。

　　C++通过继承支持重用性。可以声明一个新类，该类是一个现有类的扩展。这个新子类可以说是从现有类派生来的，故有时称为派生类。这就好像"类星"车源自"星"车，它继承了"星"车的所有性能，但能按需增加一些新的性能。

　　C++支持这样一种思想，即不同的对象通过所谓的函数多态性（polymorph）和类的多态性都能做"正确的事"。poly 意思是许多，morph 意思是形状。多态性是指同样的名字有多种形式。

　　尽管 C++是 C 语言的超集，而且实际上任何一个合法的 C 语言程序也是一个合法的 C++程序，但是 C++与 C 之间的区别是很明显的。C++凭借与 C 语言的关系而受益多年，是因为 C 语言程序员可以很容易地掌握 C++。然而，为了真正掌握 C++特点，许多程序员发现必须抛弃以前所知道的一些知识，而学习全新的概念和解决编程问题的方法。

　　C++的主要标准版本之一是 C++11（正式称为 C++OX），是 2011 年 8 月 12 日获批并公布的。

2014 年 C++14（也称为 C++1y）作为 C++11 小规模扩展版本面世，其主要特点是错误的修正以及一些小的改进。其目标类似于 C++03 对 C++98 做的一样。2014 年 8 月中旬，C++14 国际标准草案投票过程完成。

在 C++14 之后，主要版本，非正式名称是 C++17，计划 2017 年推出。

C 的最新版本是 C#（读作"C 夏普"）。作为 C 和 C++的混合产品，C#是为直接与 Sun 公司的 Java 语言竞争而开发的最新的程序设计语言。C#是为提高开发 Web 应用软件效率而设计的一种面向对象的程序设计语言。C#5.0 是最新版本，是 2012 年 8 月 15 日发布的。

微软的 Visual C#也遵循 C#规范，它包含在微软 Visual Studio 产品系列中。它是基于 C#语言的 ECMA/ISO 规范的，也是微软公司的产品。虽然该规范有多个品种，但 Visual C#显然是迄今为止应用最广的。

6.3 Java

1．概述

Java 是一种程序设计语言，最初是由 James Gosling 在 Sun Microsystems 公司（现为 Oracle 公司的一部分）开发的，并于 1995 年作为 Sun Microsystems 的 Java 平台的核心软件公布的。该语言很多是来自 C 和 C++语法，但有一个比较简单的对象模型和少量的低层功能。Java 应用程序一般被编译成字节代码（类文件），字节代码可运行在任何 Java 虚拟机（JVM）上，与计算机的体系结构无关。Java 是一种通用的、并发式的、基于类的、面向对象的语言，它是专门设计成具有尽可能少的实现的依赖性的。其目标是使应用程序开发者能"写一次，便可在任何机器上运行"，这表明编译好的 Java 程序可以运行在所有支持 Java 的平台上而无须再次编译。Java 是目前最通用的程序设计语言之一，特别适合于客户-服务器 Web 应用环境，据报告有 900 万开发者。

2．目标

研制 Java 语言有 5 个主要目标：

（1）它应该是"简单、面向对象和通俗的"；

（2）它应该是"健壮和安全的"；

（3）它应该是"体系结构是中立的和可移植的"；

（4）执行时应该是"高性能的"；

（5）它应该是"解释性的，线程的和动态的"。

3．Java 平台

可移植性是 Java 的一种特性，也就是用 Java 语言编写的计算机程序可运行在任何类似的硬件/操作系统平台上。这是通过把 Java 语言代码编译成称为 Java 字节代码的中间表达式实现的，而不是直接编译成由平台规定的机器代码。Java 字节代码指令类似于机器代码，但是能被专为该宿主机硬件写的虚拟机（VM）解释。最终用户通常使用在他们自己机器上的为单独 Java 应用而安装的 Java 运行环境（JRE），或使用用于 Java applets

的 Web 浏览器。

标准化库提供访问宿主机的专用功能（如图形、线程和网络技术）的通用方法。

采用字节代码的主要好处是移植性。但是，解释所需要的开销表明，在本地可执行环境上，运行解释程序总比编译程序要慢。早期就开发出来的即时编译程序，可以把字节代码编译成运行时的机器代码。

4. Java applet

Java applet 是以 Java 字节代码格式提供给用户的一种小应用程序。Java applets 可以用在使用 Java 虚拟机（JVM）的 Web 浏览器上；而在 Sun 的 Applet Viewer 中，它是一个独立的测试小应用程序的工具。

Java applets 的运行速度与其他编译语言，如 C++，有可比性，但速度比 C++等语言慢，但到 2011 年前后，它仍比 JavaScript 快很多倍。此外，Java applets 可以使用 Java 中的 3D 硬件加速功能。这使得 applets 很适合于最重要的，敏感的可视化计算任务。当浏览器获得了以 Canvas 和 Web GL，以及 Just in Time compiled JavaScript 格式表示的，本地硬件加速图形的支持时，其速度的差别便不明显了。

由于 Java 的字节代码是交叉平台或与平台无关的，因此 Java applets 可以用浏览器在很多平台上运行，包括微软 Windows、UNIX、Mac OS 和 Linux。通常 Java applet 可以与其他特别小的程序配合在一起应用。这种方式的优点是可以在离线方式下运行 Java applets，而无须因特网浏览器软件，并且可以直接使用集成开发环境（IDE）。

6.4 标记和脚本语言

有一些不是程序设计语言，而是与应用开发相结合的语言。这些语言大多数与万维网有关，如下面几部分所讨论的。

1. HTML 和其他标记语言

（1）HTML

现今大多数 Web 网页是用标记语言写的。采用标记语言，能用最少的线路容量实现在网上传输文件。标记语言通过使用各种标记定义了 Web 网页的结构和轮廓，而不必传送有关 Web 网页外形的精确描述。最通用的制作 Web 网页的标记语言是 HTML（超文本标记语言）。HTML 使用 HTML 标记。

在制作 Web 网页时——使用字处理软件、文本编辑器或专用的 Web 站点开发程序——HTML 标记是被插入到该 Web 网页文本的相应位置上的。某些标记单独使用，而其他一些则成对使用。例如，<b>将标记后的文本转换为粗体，直到</b>出现为止。因此下面的 HTML 语句：

```
<b> This text is bolded </b>
```

在用大多数 Web 浏览器观看时，可看到如下的粗体效果：

This text is bolded

一种 Web 网页及其相应的 HTML 代码如图 6-2 所示，其中带有一些通用的 HTML 标记。

HTML5 是用于构建和展示万维网内容的，因特网标记语言的一种核心技术。到 2014 年 10 月，公布了 HTML5 万维网联盟（W3C）的 HTML 标准的最终版和第五版。而前一版 HTML4 是 1997 年发布的。

这一核心的目标是改进该语言，以支持最新多媒体，同时保持它的易读性，并能够使计算机和一些设备（万维网浏览器，语法分析程序，等等）更一致地理解。HTML5 力图不仅包括有 HTML4，也包括 XHTML1 和 DOM Level2 HTML。

具体而言，HTML5 增加很多新的语法功能，包括新 <video>、<audio>和<canvas>元件，和可以集成可缩放矢量图形的内容（取代一般的<object>标签），以及使用 MathML 数学公式。这些功能可以很容易地在万维网上处理多媒体和图形内容，而无须求助于相应的插件和应用程序接口。其他的构建页面的新元件，如<main>、<section>、<article>、<header>、<footer>、<aside>、<nav> 和<figure>是为了更丰富文档的语义内容的。为了同一目标引入了新功能，但同时有些元件和功能被去掉。另一些元件，如<a>、<cite>和 <menu>被变更，被重新定义或标准化。APIs 和 DOM 不再是追加部分，而是 HTML5 规范的基础部分。 HTML5 也为无效文档的处理在某些细节上做了规定，因此，语法错误将由所有符合规范的浏览器和其他用户代理统一处理。

（2）XML

XML 是一种标记语言，它定义了一组对文档编码的规则，其格式既可供人读，也可供机器读。它是由 W3C 的 XML1.0 规范和其他几个相关规范定义的，全部为免费开放标准。

XML 的设计目标强调跨因特网时的简洁，通用和可用性。它具有文本数据格式，通过单一编码，对不同的语言有很强的支持能力。虽然 XML 的设计集中在文档上，但可广泛用于任意数据结构，如用于为万维网服务的那些数据结构中。

在定义基于 XML 的语言时，有几个图表系统很有用，同时有很多应用编程接口已被开发出来，用于辅助处理 XML 数据。

（3）.NET

与 XML 非常紧密相关的是微软公司的.NET 技术，是用来增加个人计算与 Web 的融合度的。简单地说，.NET 是微软公司实现基于 XML Web 业务的平台。这些业务允许应用程序通过因特网进行通信和共享数据，而与所用的操作系统或编程语言无关。

2. 脚本语言

（1）JavaScript

HTML 原则上是为活动元素的页面布局设计的，很像桌面出版程序是为打印页面布局而设计的。因此 HTML 只有少量的制作 Web 网页的工具，仅当用户观看这些页面时可以做修改，或者允许用户在屏幕上与 Web 网页交互，而不具备 DHTML 和新的 HTML 增强版的某些功能。如果你希望开发的网页有大量的动态内容，采用像 JavaScript 这样的脚本语言比较合适。这些语言可使你将程序命令或脚本直接写入 Web 网页的代码中，以增加动态内容。例如，JavaScript 脚本语言，经常用于实现在一个菜单项被点击后，能够显示其子菜单或新的图形，如图 6-3 所示。

JavaScript 最初是由网景公司开发的，能使 Web 网页的作者建立交互式 Web 站点。虽然 JavaScript 具有很多丰富的 Java 语言的特性和结构，但它还是独立开发的。当使用 JavaScript 时，必须认识到并非所有脚本命令对所有浏览器都适合。因此，要确保你的网站使用的 JavaScript 的重要性能不是专门针对某个浏览器的。

（2）PHP

PHP 是用于万维网开发的服务器端的脚本语言，但也可作为通用编程语言使用。到 2013 年 1 月，PHP 已经安装在超过 2.4 亿个 Web 网站上（占取样数据的 39%），和 210 万台万维网服务器上了。

PHP 代码可以和 HTML 代码简单地混合使用，也可以与各种模板引擎和万维网框架组合使用。PHP 代码通常由 PHP 解释程序处理，它是作为 Web 服务器本地模块，或作为可执行的通用网关接口（CGI）去实现的。在 PHP 代码被解释和执行以后，Web 服务器就把结果用所产生的网页的部分形式送给它的客户端。例如，PHP 代码会产生一个 Web 网页的 HTML 代码，一幅图像或某些其他数据。PHP 也包含一个命令行接口（CLI），可用于独立的图形应用。

正规的 PHP 解释程序，由 Zend 引擎强化，该程序是在 PHP 授权之下的免费软件。PHP 已被广泛移植，并可在几乎所有 Web 服务器上、几乎每种操作系统和平台上免费使用。

尽管它很通用，但直到 2014 年，还没有编写出 PHP 语言的规范或标准，而将正规的 PHP 解释程序视为事实标准。2014 年以来，PHP 规范的制订工作一直在进行。

也是在 2014 年，新的 PHP，命名为 PHP7 也在开发之中。

第 7 章　数据库

7.1　数据库的概念

人们经常需要快速检索大量数据。一家航空代理在电话上要能快速为客户查到从亚特兰大到多伦多的最低价格机票。一个大学的注册主任可能要快速扫描学生记录，以查出将于 6 月份毕业的其学绩点平均在 3.5 分或以上的那些学生。视频商店的店员可能需要决定某一种电影是否适合于零售。用于这些任务的软件就是数据库管理系统。计算机化的数据库管理系统正在迅速取代纸面上的文件系统，因为在过去，使用纸面上的文件系统，人们要费很大力气才能找到所需的信息。下面就讨论基于 PC 机的关系型数据库的基本特性和概念，以微软的 Access 作为应用的例子。

1. 什么是数据库程序

数据库是按一定方式存储和组织数据的一种数据的集合，在需要时能够检索信息。数据库管理系统——有时也称为数据库软件——能够在计算机上创建数据库，并能很容易地访问存储在数据库中的数据。

虽然不是所有数据库的组织方式都一样，但大多数基于 PC 机的数据库都是按字段、记录、文件去组织的。数据库中的字段是单一类型存储的数据，如一个人的姓名或电话号码。记录是相关字段的集合——例如，费利斯·哈弗曼的身份证号、姓名、地址和行业（见图 7-1）。文件，在 PC 数据库中常称为表——是相关记录的集合（如所有学生的地址数据、年级数据或课程表数据）。相关文件或表（如全体学生数据）的最终集合构成了数据库。

大多数 PC 机的数据库软件是一种关系型数据库管理系统。

2．建立一个数据库

数据库可以包含各种对象（见图 7-2）。在新数据库中最初建立的对象就是一张表，当需要时，可以建立其他对象，并与这个表连接。

在建立一个数据库时，应该确定要包含在数据库中表的数量。应该能识别存储在每张表中的数据项，以便使用相应字段的特性。对于每个字段，应该确定以下内容：

（1）字段名（表中的唯一识别名）

（2）要包含在字段中的数据类型（文本，数字，日期等）

（3）字段大小（存储数据需要多少个字符）

一旦这些技术要求确定下来，包含这些字段技术要求的每张表的结构就建立起来了（见图 7-3）。

表结构建立好以后，就可以往表里输入数据了。数据输入可以在常规表视图中完成，也可以建立和使用表单，常规表视图有时称为数据单视图，因为这种表看起来很像一张电子表格。表单可以使你以更正规的方法查看或编辑表中的内容——通常一次操作一个记录，而不是一页记录，如同在数据单视图中一样。图 7-3 说明，一旦表结构建立起来了，就可以使用这两种方法去输入数据了。

3．修改数据库

一旦建立了数据库表，该表可能需要修改。当需要时可以对表的结构或放在表中的数据进行修改。

（1）修改表结构

只是当需要更改字段的性质时，才去修改表的结构。例如，可能要加宽一个字段，以容纳比以前设定的更长的名字，而错误的字段类型可能是当初选择的，也可能需要增加一个新字段。

（2）编辑、添加和删除记录

要更改表中的具体数据，应先把表打开（使用表的数据单视图或表单），然后对需要更改的数据做修改。为了移动到一个具体记录上，以便编辑它的内容，可以使用屏幕上的箭头和键盘上的其他定向键，也可以使用视窗底部的记录按钮（再请参见图 7-3）。因为通常是在表的末尾添加记录，在一组记录按钮中通常有一个新记录按钮（New Record），此按钮会自动移到表的末尾，成为一个空记录。

要删除一个记录，可以使用键盘上的删除（Delete）键或者使用菜单栏中的删除记录（Delete Record）选项。

4．查询和报告

为了从数据库中检索信息，要使用查询和报告程序。查询是一种提问，在数据库术语中，查询是请求从数据库中找到指定的信息。查询程序用你希望查什么信息的指令，去查找并显示刚刚查到的信息。报告则是更正规地打印出一张表或查询结果。

7.2 万维网与数据库

在万维网上使用数据库是特别普通的。实际上，通过 Web 网站提供产品、合作信息、在线订货或类似活动的所有公司，都使用数据库。最常用的是客户-服务器数据库事务处理方式，在这种情况下，用户的浏览器为客户软件。然而，对等方式信息交换的使用也在不断增加。

1. Web 数据库使用举例

如何在 Web 上使用数据库有很多例子。数据库促进了信息的检索和处理，并允许更多的交互和动态内容的检索和处理。下面这几部分讨论的是 Web 数据库工作的例子，并扼要介绍一下其他与 Web 数据库相关的问题。

（1）信息检索

在 Web 上，用租用的数据库本身去进行数据检索是极其自然的，实质上，它是用作检索的特大数据仓库。数据存储在数据库中，Web 网站上的访问者可以请求和观看这些数据（见图 7-4）。

（2）电子商务和电子企业

另一种在 Web 上广泛使用的数据库应用，是支持并促进了电子商务的软件。目录信息、价目、客户信息、购物卡余额和其他信息都可存储在数据库中，当需要时可使用相应的与 Web 网站数据库连接的脚本文件或程序进行检索（仍见图 7-4）。

（3）动态 Web 网页

静态 Web 网页，每次在为个人显示时，所显示的是同一个信息，直到该网页文件被修改为止。相反，动态 Web 网页的外形和内容是按用户的输入变化的。这种输入可以由该网页上规定的表单选项来决定，也可以用其他方法控制，如 Java applet、ActiveX 控件，或用户已经完成的对该网站的一些动作，如单击一个显示的内容，特别是单击产品的超链接。

2. Web 数据库如何工作

（1）有关数据库和 Web 共同工作的例子

为进一步说明数据库和 Web 如何共同工作，让我们看一个例子。

向 Web 数据库请求检索信息或向该数据库存储数据通常由用户提出。实现数据库请求的通用方法是填写 Web 网页表单，从显示在 Web 网页上的菜单中选择一个选项，或单击屏幕上的项目。该请求由 Web 服务器接收，然后将此请求转换成数据库查询，并通过称为中间件的中间软件的帮助，将其传送到数据库服务器上。该数据库服务器检索出相应的信息并将其返回给 Web 服务器（要再次通过中间件），此 Web 服务器将所检索的信息，作为 Web 网页显示在用户的屏幕上，这些步骤表示在图 7-5 上。

（2）中间件

连接两个完全不同的应用程序——如图 7-5 所示的 Web 服务器和数据库管理系统——

的软件称为中间件。用于数据库和 Web 网页之间接口的，最常用的中间件是 CGI 和 API 脚本程序。而更通用的、比较新的脚本语言是 PHP 和 ASP。

7.3　MySQL

到 2013 年 7 月，MySQL 是世界上第二个应用最广的关系型数据库管理系统（RDBMS），也是应用最广的开源关系型数据库管理系统。名字中 My 是合作创立者 Michael Widenius 女儿的名字。缩略语 SQL 代表结构化查询语言。

MySQL 开发课题允许其源代码用在 GNU General Public 许可证下，也可以用在各种专利协议下。 MySQL 曾经为瑞典的营利公司 MySQL AB 所拥有并负责经营，现在为 Oracle 公司拥有。

在 Web 数据库应用中，大家普遍选择 MySQL，它也是广泛用于 LAMP 开源 Web 的应用软件栈的核心部件。LAMP 是"Linux，Apache，MySQL，Perl/PHP/Python"的缩略语。要求全功能数据库管理系统的免费软件开源课题通常使用 MySQL。

对于这种专利的使用，还有几个付费版本可用，并且还提供附加功能。使用 MySQL 数据库的应用包括 TYP03、MODx、Joomla、WordPress、phpBB、MyBB、Drupal 和其他软件，MySQL 也用于许多高配置的大型网站，包括 Google（虽然不是为了搜索）、Facebook、Twitter、Flickr 和 YouTube。

1．界面

MySQL 是一个关系型数据库管理系统，出售时并不带有管理 MySQL 数据库或管理包含在该数据库内的数据图形用户接口的工具。用户可以使用包含在内的命令行工具，或使用 MySQL "前端"、桌面软件和 Web 应用软件，这些应用软件可以建立和管理 MySQL 数据库，建立数据库结构，备份数据，检查数据记录的状态和工作。MySQL 官方前端工具套件、MySQL 工作台，实际上是 Oracle 开发的，是免费使用的。

2．图形

官方 MySQL 工作台是由 MySQLAB 开发的免费集成环境，这一环境使得用户用图形方式管理 MySQL 数据库，用可视化方式设计数据库结构。MySQL 工作台取代了以前的 MySQL GUI Tools 软件包。类似于其他第三方软件包，始终考虑权威的 MySQL 前端，MySQL 工作台让用户管理数据库设计和建模，进行 SQL 开发（取代 MySQL 查询搜索软件）和数据库管理（取代 MySQL 管理器）。

MySQL 工作台有两个版本，一个是免费和开源社区版，这可以从 MySQL 网站上下载，另一个是专利标准版，它扩展和改善了社区版。

第三方专利和免费图形管理应用软件（或"前端"）是与 MySQL 集成的，能使用户使用数据库结构和数据可视化应用。

3．部署

MySQL 可以用源代码手动创建和安装，但这很繁琐。因此，更通用的方法是用二进制包安装，除非是专门定制的。对于大多数 Linux 用户，用这种二进制包管理系统去下载和安装 MySQL 比较省事。当然，要做进一步的配置，则需要对安全和优化设置做一些调整。

虽然对更强大的专利数据库来讲 MySQL 是作为低端应用开始的，但它渐渐地支持了较大规模的应用。MySQL 一直用于小到中型单服务器场合，可作为基于 LAMP 的 Web 应用的软件，也可作为单独的数据库服务器使用，见图 7-6。MySQL 有这么大的吸引力是因为其相对简单和容易使用，还有它是一个像 phpMyAdmin 一样的开源工具生态系统。MySQL 可以为中等规模环境做扩展，方法是将其安装在比较强的硬件上，如带吉字节存储器的多处理器服务器上。

但是，在一个单服务器上，功能扩展是有限度的。因此，在规模更大时，则需要多服务器 MySQL 配置，以提供更高的性能和可靠性。典型的高端配置包括一台强大的主数据库，用于处理数据的写操作。另外由多个从数据库处理所有读操作。主服务器一直与从服务器同步，以便万一主服务器出故障，一个从服务器就被提升为新的主服务器，使停机时间减到最小。进一步改善性能的方法可以采用存储器高速缓存技术，它是将存储器中的数据库查询结果缓存起来，或把一个数据库分解成称为碎片的更小的组块，这些组块遍布在大量的分布式服务器簇上。

第四部分 应 用 软 件

第 8 章 办公自动化软件

8.1 办公自动化软件基本知识

当你使用任何一种应用软件，如使用字处理软件打印一封信，或为纳税而使用税务处理程序时，一些基本概念和功能需要熟悉。其中包括通用文档处理任务、软件套件和软件所有权的概念以及当程序工作时如何得到帮助等。下面几部分就讨论这些问题。

1. 办公自动化软件基本概念

（1）文档处理操作

虽然某种应用程序是专为处理某些文档而设计的，但像打开一个文档、存储和打印文档的概念是完全通用的。图 8-1 描述了少量最普通的文档处理操作，给出了在 Windows 应用程序中实现这些操作所使用的图标。

一般来讲，完成这些操作的命令在所有的 GUI 程序中是相同或非常相似的，因为绝大多数人需要快速使用文档处理命令，故这些操作通常放在容易单击的菜单中或工具栏上。

（2）软件套件

大多数办公用的程序，如字处理程序和电子表格程序，是与其他相关的应用软件以软件套件形式，捆绑在一起出售。在办公应用套件的销售中，微软 Microsoft Office 稳坐第一把交椅。这一套件的高端版本捆绑了 Word（字处理）、Excel（电子表格）、PowerPoint（演示文稿）、Access（数据库管理），并且还与其他几个程序，如 FrontPage（Web 网站开发）一起出售。

使用软件套件的最大优点之一是能够把文档或部分文档从一个程序中传送到另一个程序中，或共享文档。例如，你正在用字处理程序写一封信，但是希望插入一张电子表格，你可以启动电子表格程序，在存储的表单中找到你需要的那张表格，把它复制并粘贴到你的信中就可以了——整个过程无须关闭字处理程序（见图 8-2）。

（3）在线帮助

大多数人在使用软件程序时会遇到问题，或在某一方面需要帮助。为了提供帮助而又不使你离开计算机屏幕，大多数应用程序有在线帮助功能。程序有各式各样的工具提供在线帮助。某些可能的配置如图 8-3 所示。

（4）所有权和分配权

涉及软件产品的所有权和用户权时是很敏感的。通常一个软件的制作者或发布者开发了一个程序，就在其上面制作一个版本权，并在以后保持这一程序的全部权限的所有权。以后由发布者决定谁可以使用、复制或分发该程序。下面讨论不同类型的所有权和允许使用权。

① 专有软件。

当今所用的很多系统软件和应用程序都是专有软件。这表明某人拥有该程序的权限，并且拥有者希望用户购买这些程序的复制品。

② 共享软件。

某些软件可作为共享软件使用。虽然你不必为安装和试验共享软件付费，但大多数共享软件规定在很短期限——通常为一个月——期限到了以后，你要继续使用该软件，则需要付费。

③ 免费软件。

免费软件，或称公用域软件，是指你可以使用并和其他人共享的免费程序。

2. 办公自动化软件的一般特性

用户接口是你使用的应用程序的一部分。大多数应用程序使用图形用户接口（GUI）去显示称为图标的图形元素，图标代表熟悉的对象和鼠标。鼠标控制屏幕上的指针，用于选择像图标那样的一些项。另一个特性是显示信息窗口的使用。窗口是可以包含文档、程序或消息的矩形区域（不要把窗口这一术语与微软公司的各种版本的 Windows 操作系统相混淆，那些是程序）。在计算机屏幕上一次可以打开和显示多个窗口。

包括微软 Office 2010 在内的大多数软件程序都有菜单、对话框、工具栏和按钮（见图 8-4）。菜单给出命令，通常显示在屏幕顶部的菜单栏内。当选中菜单中的一项时，会出现菜单选项列表或出现附加信息和请求用户输入的对话框。工具栏一般是在菜单栏下，它们包含称为按钮的小图形元素，提供快速访问常用命令的快捷方式。

微软最新的 Office 2010 版又重新设计了界面，旨在使用户更容易找到和使用应用软件的所有功能。这种新设计推出了条形框、上下文选项卡和图库等。

（1）条形框取代了菜单和工具栏，是把常用的命令组成一套选项卡。这些选项卡显示与用户正在处理的任务最相关的一些命令按钮。

（2）上下文选项卡是自动出现的一些选项卡。这些选项卡只是在需要时才出现，并指出用户要完成的下一次操作。

（3）图形库简化了从图形目录中选择的过程。这是由于某些结果采用图形显示而不是对话框之故。

8.2 微软 Office 2013

微软 Office 2013（以前是 Office 15）是微软办公软件的一个版本，是微软 Windows 高效率软件系列。它是微软 Office 2010 的后继者，包括扩展的文件格式，用户界面更新和这些新特性中的触摸功能。Office 2013 适合于 IA-32 和 x64 系统，要求配备 Windows7、Windows Server 2008 R2 或以后的任何版本软件。Office 2013 版也可用在 Windows RT 设备上。

微软 Office 2013 有 12 种不同的版本，包括 3 个用于零售市场，两个用于批量授权，5 个通过微软 Office 365 程序使用的基于订阅的版本。还有称为 Office Web apps 的万维网应用版和为平板电脑与移动设备制作的 Office RT 版。

2014 年 2 月 25 日，微软 Office 2013 Service Pack1（SP1）面世。

1. 新特性

比起以前的版本，Office 2013 更多地基于云；一个域登录，Office 365 账号或微软账号，现在可用于在设备之间同步 Office 应用软件的设定（包括最新的文档），用户也可以把文档直接存在他们的 SkyDrive 账号中。

新特性包括在微软 Word 中的新的读方式，在微软 PowerPoint 中的展示方式和所有 Office 程序中的触摸和喷墨方式。微软 Word 可以插入从在线源那里获得的视频和音频，也可以在万维网上广播文档。Word 和 PowerPoint 也有类似的书签功能，可在不同计算机之间同步文档的位置。

Office Web Apps 簇也对 Office 2013 做了更新，引入了附加的编辑功能和界面的更改。

Office 2013 的其他功能包括：

- 当敲键或进行选择时,会出现漂亮的带状界面和精巧的动画(Word 和 Excel 中)。
- 在 Outlook 中新的调度任务是可视化的。
- 改型的开始屏幕。
- Word 中的新图形选项。
- 诸如图像这样的对象，可以随意移动。这些图像可以很快地放到段落边缘、文档边缘或列的边缘。
- 从 Office.com、Bing.com 和 Flickr 网站来的内容，可支持在线图形（默认的只有公共域中的图像）。
- 在 Word 和 PowerPoint 中，可以返回到最后观看和编辑的位置。
- 在 PowerPoint 2013 中有新的幻灯设计、动画和变换。

- 在 Outlook 中支持 Outlook.com 和 Hotmail.com。
- 支持与 Skype、Yammer 和 SkyDrive 的整合。
- 支持 IMAP 专用文件夹。

图 8-5 为 Microsoft Office 2013 成员图标。

2．版本

与以前的各种版本一样，Office 2013 做成了几个不同的版本，以适应不同的市场需求。所有微软 Office 2013 的传统版本都包含 Word、Excel、PowerPoint 和 One Note，并都授权在一台计算机上使用。

公布的 Office 2013 的 5 个传统版本是：

- 家庭与学生版：这一零售系列包括 Word、Excel、PowerPoint 和 OneNote 的核心应用。
- 家庭与商业版：这一零售系列包括 Word、Excel、PowerPoint 和 OneNote+Outlook 的核心应用。
- 标准版：这一系列只能通过批量许可通道使用，包括 Word、Excel、PowerPoint 和 OneNote+Outlook 和 Publisher 的核心应用。
- 专业版：这一零售系列包括 Word、Excel、PowerPoint 和 OneNote+Outlook、Publisher 和 Access 的核心应用。
- 专业+版：这种系列只能通过批量许可通道使用，包括 Word、Excel、PowerPoint 和 OneNote+Outlook、Publisher、Access、InfoPath 和 Lync 的核心应用。

8.3 微软 Office Mobile

微软 Office Mobile 是微软为移动设备开发的免费办公套件，它与桌面版办公套件是兼容的。

核心应用软件包括 Word Mobile、Excel Mobile 和 PowerPoint Mobile。其他非核心 Office 应用软件，像 Lync Mobile、OneNote Mobile 和 SharePoint Newsfeed，可以从 mobile app 商店下载使用。

Windows Phone 8 与更新了的 Office Mobile 版本套件一起销售，该移动系列包括 Word、Excel、PowerPoint 和 OneNote 的移动版本。与 Windows Phone 7 各版本相比，新版本增加了一个改进了的 Office Hub 界面，它能与最近打开和修订的文档同步（包括更改通过 Office 365 和 SkyDrive 存储的文档），增加了一个带有附加功能（如语音注释）的 OneNote app 以及改进了的文档编辑与观看功能。

2013 年 6 月，微软公布了 Office Mobile 的 iPhone 版，它类似于 Windows Phone 版本，但事先要求 Office 365 的使用许可。2013 年 7 月，安卓的智能手机版公布，它也要求 Office 365 的使用许可。

iPad 和安卓平板电脑的 Apps 分别是 2014 年 3 月和 2015 年 1 月面世的。这些产品与他们的智能手机等，都是个人免费使用的，这还包括商用 apps 的许可，虽然某些额外功能已被禁止，而且还要求 Office 365。

Windows 10 Mobile 于 2015 年后期公布，它有新的 Office apps 商标，并可使用由 Windows 10 开发的"通用 app"平台。

1. Word

Word Mobile 是可以创建和编辑文档的字处理软件，它支持基本格式，如粗体，改变字体大小和改变颜色。它可以加入注释，但不能用跟踪的变换去编辑文档，不能打开设置了密码的文档，不能改变字体，文本对齐，或字形；不能产生加重列表，插入图片；不能进行恢复。如果原来的文档有以下功能的话，仍然给予保留：脚注、结束附注、标题、脚页、分页符，目录的一些凹陷，和某些字体，当在 Word Mobile 的文档中工作时，既不显示也不能插入。除了 2013 版的这些功能以外，Windows Mobile 2007 版也能以富文本格式保存文档，并能打开过去的袖珍字。还有，它也包括拼写检查，字的计数工具，以及"查找和替换"命令。

2. Excel

Excel Mobile 是一种能编辑 XLSX 文件的电子表格程序。它可以编辑和格式化手机中的文本，计算公式，搜索电子表格，对行和列进行分类，冻结显示窗口，列过滤，加注释和制作图表。除了在文档边缘，不能加入列和行，不能重新安排列和行，删除列或行，或增加电子表格标签。Excel Mobile 不支持保护设定、区域缩放设定、自动过滤设定、某种图表的格式化、隐藏表格和其他一些功能。

3. PowerPoint

PowerPoint Mobile 被包含在 Windows Mobile 5.0 中，它是一个演示程序，可以读取和编辑 PowerPoint 演示文稿（尽管编著功能仅限于增加说明，编辑文本及重新安排幻灯片。它不能创建新的演示文稿）。使用 Windows Phone 7 的 PowerPoint Mobile 的几个版本也能从因特网上观看广播流式的演示文稿。

4. OneNote

微软的 OneNote Mobile 首先在 Windows Mobile 6 中公布。微软 OneNote Mobile 被包含在 Windows Phone 7 的"Office Hub"中，但在 Windows Phone 8 中已经独立出来了。它一直作为核心应用出现在微软的 Office Mobile 许可证的目录中。其功能是与 Office Lens 集成在一起的。它是一种可以捕获白板、黑板、数字拷贝和多页文档，并允许用户在微软 OneNote Mobile 中对其进行编辑的一种应用。每个注释都自动加盖时间戳，可以填入字、图片和录音。它包括如"待办事项清单"功能，但对格式化和轮廓只有有限的选项，不包括附加字体。

第 9 章 多媒体

9.1 多媒体及其主要特点

1. 定义

多媒体是指用各种形式的内容综合成的内容（它是相对于仅仅用基本的计算机文本显示、传统的打印形式或手写材料的媒体而言的）。多媒体是由文字、音频、静态图像、动画、视频或交互的内容形式的一种综合，如图 9-1 所示。

多媒体可以通过计算机和电子设备这样的信息处理设备进行记录、播放、显示、动态交互或进行访问，当然也可以是现场演出的一个部分。多媒体设备是用于存储和体验多媒体内容的电子媒体设备。多媒体在艺术方面区别于混合媒体就在于它包含了音频，因此，它具有更广泛的领域。"富媒体"一词与交互式多媒体是同义的。超媒体在多媒体应用中更加扩大了媒体内容的数量。

2. 多媒体分类

多媒体可以宽泛地分为线性媒体和非线性媒体。线性媒体，其活动过程通常对观众不进行导航控制，比如，演电影。非线性媒体则有交互控制过程，比如视频游戏或自定进程的基于计算机的培训。超媒体就是一个非线性媒体的例子。

3. 多媒体的主要特性

观众可以在舞台上观看多媒体演示，可以通过投影仪或传播后观看，多媒体也可以通过媒体播放设备在本地播放。多媒体播出可以是直播，也可以是录制的多媒体节目。广播和录制既可以采用模拟电子媒体技术，也可以采用数字电子媒体技术。数字在线多媒体可以下载，也可以按流式播出。流式多媒体既可以是直播，也可以是按需播出的。

通过多用户在线网络或本地的离线计算机、游戏系统或模拟器，多媒体游戏和模拟可用于有特殊效果的具体环境。

各种各样的技术手段或数字多媒体可以增强用户的体验，比如，更方便更快捷地传递信息。而在娱乐或艺术领域方面可以超越日常的体验。

将多种形式的媒体内容综合起来可以增强交互的水平。在线多媒体越来越多地变成面向对象和数据驱动的，这就能经过一段时间，使多种内容形式的终端用户的协作创新和个性化的应用成为可能。这方面的例子比如，从网站上多种形式的内容，如具有图像（相片）和用户更新标题（文字）的图片库到对其事件、插图、动画或视频的模拟有修改的可能，允许修改多媒体"体验"而无须再编程。除了视听之外，触觉技术还可以感知虚拟物体。新出现的包括味觉和嗅觉在内的技术，也可以增强多媒体体验。

9.2 多媒体应用

多媒体应用遍布各个领域，包括但不仅仅局限于：广告、艺术、教育、娱乐、工程、医疗、数学、商业、科研和时空应用领域。图 9-2 是使用 PowerPoint 做的演示画面。图 9-3 是使用多媒体内容的虚拟现实。图 9-4 是德累斯顿（德国）的一个多媒体终端。以下是一些应用例子。

1. 创意产业

创意产业从美艺术，到娱乐，到商业艺术，到新闻，到媒体各个领域都在使用多媒体，并且为下述任何行业提供软件服务。一个多媒体设计师在其职业生涯中可能涉及所有的相关领域，对他们的技能要求可以从技术，到分析，到创意。

2. 商业应用

商业艺术家和图形设计师使用的大量过往和新的电子媒体都是多媒体。动人心弦的展示用于在广告中抓住眼球。商业对商业之间，办公室之间的交流通常使用创新服务公司提供的先进的多媒体演示，而不是用简单的幻灯片去呈现新理念或吸引人的培训项目。商业多媒体开发者也被聘来设计政务服务和公益服务中的应用项目。

3. 娱乐和美艺术

此外，多媒体还大量地应用于娱乐业，特别是在电影和动画片中使用特效（VFX、3D 动画等）。多媒体游戏是流行的娱乐消遣方式，是以 CD-ROM 或在线方式发布的软件程序。一些视频游戏也使用多媒体功能。让用户积极地参与而非被动地只坐在那里接受信息的多媒体应用，被称为交互式多媒体。在艺术界，有一些多媒体艺术家，他们的想法是将不同媒体技巧进行混合来让观众以某种方式介入交互。

4. 教育

在教育方面，多媒体用于创作基于计算机的培训课程（常常称之为计算机辅助训练）以及制作像百科全书和年鉴这样的参考书。计算机辅助训练可以用各种信息格式让用户浏览一系列演示、某个主题的文本以及相关插图。寓教于乐就是教育和娱乐的结合，特别是在多媒体娱乐方面。

在过去的十年间，有关的学习理论由于引入多媒体而被梦幻般地扩展。研究的各个方向均被涉及（例如，认知负荷、多媒体学习及其他）。学与教的发展潜力几乎永无终点。

5. 新闻

所有的报业集团都试图在其工作中投向这一新领域。当一些报纸还缓慢前行时，其他那些像《纽约时报》、《今日美国》和《华盛顿邮报》等主流报纸已尝试确立在报业全球化过程中的领先地位。

新闻报道已经不仅仅局限于传统媒体，自由新闻工作者可以通过不同的新媒体为他们的新闻故事制作多媒体报道。这也使得利用技术手段来实现听讲故事的全球化，也为媒体制作者和消费者提供新的交流技术。

6. 工程

软件工程可以使用多媒体对从娱乐到培训（如军事或工业培训）中的任何事件做计算机模拟。软件界面中的多媒体通常是创意人才和工程师们有机结合的产物。

7. 工业

在工业领域，多媒体是作为对投资者、上级和合伙人呈现信息的一种手段。在全球范围内，通过几乎无限的基于万维网的技术，多媒体也对员工培训、广告宣传和产品销售帮助良多。

8. 数学和科学研究

在数学和科学研究领域，多媒体主要应用于建模和模拟。比如，一个科学家可以观察一个特殊物质的分子模型并操纵它得到一个新的物质。

9. 医疗

在医疗领域，可以通过观看虚拟外科手术来训练医生，也可以模拟一个人体如何受病毒和细菌的传播而感染疾病的,从而开发新技术去防止这种感染。像虚拟手术这样的多媒体应用也可以帮助医生进行实操训练。

9.3 多媒体网络技术

　　许多应用诸如视频邮件、视频会议和其他协作工作系统都需要网络化的多媒体。在这些应用中,多媒体对象被存储在服务器上并在客户端上进行播放。这些应用需要向各种远程站点广播多媒体数据或访问大容量多媒体资源。多媒体网络即使是在数据压缩的情况下,也需要一个非常高的传输率或者带宽。传统的网络用于提供无误差传输,而大多数多媒体应用能在传输中允许由于损坏或包遗失而产生误差,且无须重传或进行修正。在某些情况下,为满足实时传递需求或实现同步,甚至允许丢弃某些包。其结果是我们可以在多媒体网络上使用轻量级传输协议。由于可能会产生不能容忍的延迟,这些协议没有重传功能。

　　多媒体网络必须为交互操作提供低延时保障。因为当多媒体数据到达目标站点时必须同步,网络应该用低抖晃提供同步传输。

　　在多媒体网络中,与传统网络点对点不同的是大多数多媒体通讯是多点式的。比如,有两个以上参与者参加的多媒体会议,需要以不同媒体对每一个参加者发布信息。会议网络使用多目标广播和桥接发布方法。多目标广播技术复制一个单一输入信号并将它传送给多个目标站点。桥接技术将多个输入信号组合成一个或多个输出信号,然后传送给参与者。

　　在过去,由于受技术手段的限制,多媒体不能广泛应用。可如今,高速计算机的使用和因特网的连通,使得多媒体的实施更加灵活,而其应用的数量也呈指数上升。像自动取款机和信息亭那样越来越多的 Web 站点配备了多媒体(如计算机软件,用户界面也配备了多媒体)。多媒体演示在商业中用得也很普遍。

　　在基于 Web 的应用中,多媒体元素实际上可以加到任何 Web 网页上,以提高人们的兴趣,更好地发送内容或增加功能。虽然有一些专用的多媒体应用软件,但都是 Web 站点的集成组件。基于 Web 的多媒体应用的优势包括:

- 多平台——由于 Web 网页可以用很多种计算机和平台观看,因而不必去开发多种应用程序的版本(例如 PC 和苹果版本)。
- 熟悉的接口——由于基于 Web 的许多应用都可用一种 Web 浏览器访问,因而用户应该感觉到总是很像第一次访问的应用程序。
- 浏览器与计算机的不兼容性——对你的网站的各种浏览器、平台、操作系统和屏幕分辨率进行测试,以保证你的网站在尽可能多的不同配置下工作。
- 容易升级——基于 Web 应用的内容可根据需要经常去升级,因而用户在每次访问 Web 网站时总能看到最新版本。

第 10 章　计算机图形图像

10.1　概述

计算机图形一词用来在一个很广泛的领域里描述"在计算机上除了文字和声音的任何东西"。通常计算机图形是指以下几个不同事情：
- 用计算机来呈现和处理图像数据；
- 用于创建和处理图像的各种技术。

目前计算机图形已广泛普及。计算机影像在电视、报纸、天气预报或在所有的医学研究和手术治疗中随处可见。一个构建精良的图形可以在一个表格里呈现出易于理解的较为复杂的统计数据。在媒体中，"这样的图形用于阐明文件、报告、论文"以及其他演示素材。

可以通过多种方法在计算机上得到图像。它们可以是扫描到计算机中的手工绘画或照片，也可以是用计算机绘图软件直接在计算机上绘制的图像，还可以用计算机定义一个放置了各种物体的想象的世界。然后，用计算机对这些物体进行渲染以产生极其靓丽、逼真的图像，就像在电影《玩具总动员》中看到的一样。渲染的图像是从场景中物体的数学描述开始，并对它定义了光线和颜色以产生最终的图像。

许多功能强大的工具被开发出来用以显现数据。计算机生成的图像可被分成几种不同的类型：二维（2D）、三维（3D）和动画图形。随着技术的进步，三维计算机图形已经变得更加普及，但二维计算机图形依然被广泛使用。计算机图形已经成为计算机科学研究中对可视内容进行数字综合和处理方法的一个分支领域。过去的十年，其他专用领域也像信息可视化一样发展壮大，科学可视化更多关注"三维现象的可视化（建筑、气象、医疗、生物等）"，这其中强调的是物体、表面、光源等伴随着或许是动态（时间）成分的真实渲染。

计算机图形可用于以下领域：
- 计算生物学
- 计算物理
- 计算机辅助设计
- 计算机模拟
- 数字艺术
- 教育
- 图形设计
- 信息图表
- 信息可视化
- 理性药物设计
- 科学可视化
- 电影特效
- 视频游戏
- 虚拟现实
- 网页设计

10.2　各种各样的计算机图形

1. 二维计算机图形

二维计算机图形是基于计算机产生的数字图像的——主要来自于模型，比如数字图像，以及通过特定的技术产生的。

最初，二维计算机图形主要用在从传统打印和绘图技术，例如排版，发展而来的应用领域中。在这些应用中,二维图像不仅是一个现实世界物体的表达,而且可以是添加了语义值的独立人造物体;二维模型因此更有优势,因为它们对图像的直接控制比三维计算机图形更多,其方式更像是摄影而不是排版。

(1) 像素画

大量数字艺术品都是用光栅图像软件创建的像素图,其中的图像都是可以在像素级上进行编辑的。在多数老式(或相对有局限的)计算机和视频游戏、图形计算器游戏和许多手机游戏中的图形基本上都是像素画。

(2) 精灵图

精灵图是集成在一个大场景中的二维图像或动画。最初包含的只是一些图形对象,这些对象可单独在视频显示的内存位图中做处理,现在则包括各种各样的图形覆盖方法。

精灵最初是一个集成不关联位图,使之成为屏幕上普通位图的一部分的一个方法,例如创建一个动画角色,使其可以在屏幕上移动而不改变整个屏幕的数据。这种精灵既可以由电子电路也可以由软件来建立。用电路的话,硬件精灵是一个硬件结构,它用定制的 DMA 通道通过强制加入两个分离的视频源的方式在主屏幕中集成可视元素。用软件的话,则可以通过特殊的渲染方法来进行模拟。

(3) 矢量图形

矢量图格式是对光栅图的补充。光栅图是以像素阵列来表示图像的,并且通常是用来表示图片的。矢量图是由构成图像的相关图形和色彩的编码信息组成的,它在渲染时有更多的灵活性。有很多例证可以说明使用矢量工具和格式更好,当然,也有很多例证可以说明使用光栅工具和格式更好。我们也常将这两种格式结合起来使用。对于每种技术的优点和局限性以及他们之间的关系就像是使用工具产生的效率和效果一样。

图 10-1 为矢量图与光栅图的对比效果。

2. 三维计算机图形

与二维图形相比,三维图形是用三维几何数据来表达图形的。它是为了展现而存于计算机中,它包含将来用于显示和实时观看的图像。

尽管有一些不同,3D 计算机图形依靠与 2D 计算机图形同样的算法在最终的渲染显示中做成帧和光栅图形(与在 2D 中一样)。在计算机图形软件中,2D 和 3D 的界限有时也模糊不清;2D 应用也使用 3D 技术来实现比如照明的效果,3D 也基本上使用 2D 渲染技术。

3D 计算机图形与 3D 模型一样。除了渲染之外,模型是包含在图形数据文件中。然而,不同之处在于包含的 3D 模型是任意 3D 对象的表达。直到可视化显示之前,一个模型并非一个图形。对于打印而言,3D 模型则不仅仅局限于虚拟的表面。3D 渲染就是如何对一个模型进行显示,当然也可用于非图形计算机模拟和计算。

3. 计算机动画

计算机动画是通过计算机创建移动图像的艺术。它是计算机图形和动画的分支领域。它越来越多地通过 3D 计算机图形进行创建,尽管 2D 计算机图形也依然广泛地满足低带宽的和快速的实时渲染需求。有时动画的目标就是计算机本身,但有时目标则是其他媒体,比如电影。尤其是用于电影时,它也被称为 CGI(计算机生成的图像或计算机成像)。

图 10-2 是使用 Motion 捕捉软件制作的计算机动画实例。

虚拟实体可以包含并被各种各样的属性所控制,比如存储于转换矩阵中的转换值(位置、方向和比例)。动画其实就是一个属性随时间而变化的结果。现在有很多创建动画的方法,最基本的形式就是基于关键帧的创建和编辑,每帧在给定的时间上存储一个值,每个属性都可以赋予活力。2D/3D 图形软件将随每一个关键帧进行改变,创建一个可编辑的随时间映射其值的曲线,因而产生动画。其他动画方法包括基于过程的和基于表达的技术:前者将动画实体的相关元素整合到一组属性中,主要用于创建颗粒效果和群体模拟;后者则允许评估的结果从用户定义的逻辑表达式返回,再加上数学运算,以可预见的方式实现动画的自动化(超越骨骼系统中建立的层次结构,更方便于控制骨骼行为)。

为创建运动的假象,可在计算机屏幕上显示一个图像,然后快速地用一个类似于先前图像的新图像来代替,但稍微有所移动。这种技术与在电视和电影中的运动的错觉是相同的。

10.3 图形软件(1)

1. 桌面出版

虽然桌面出版(DTP)作为一项独立业务已有很长一段时间了,但它却是从文字处理发展起来的。近年来,这二者又显示出重新融为一体的趋势。基本的文字处理(WP)和 DTP 的不同之处可通过对传统印刷品作者所做的工作以及印刷工人所做的工作之间的比较来进行了解。在桌面计算机出现之前,作者必须产生一个打字稿——以正确顺序将文字排成文本。然后由印刷工人将这些文字(也许在设计人员或排字人员的辅助下)以特定的方式排好,其间也许有相应的插图,然后进行印刷。现在的作者与其前辈的做法一样,不同的是使用了文字处理软件,因而文字无须由印刷工人重排。DTP 所做的工作就是用桌面电脑来自动完成印刷工人所要完成的大部分工作。这一切来自于 4 项技术的发展:具有图形用户界面(GUI)的桌面电脑、DTP 软件、激光打印机和页面描述语言 PDL。

激光打印机的重要性在于可以获得高质量的最终文件,而无须传统的排字过程(特别是使用活字)。早期的 300~400dpi 的激光打印机还不能与传统印刷相匹敌,但已能以低廉的成本满足日常需要。尽管 DTP 在某些人的心目中仍然是低质量、业余水平的,但高分辨率激光打印机和数字排字(称为激光照排)的发展已可以用来制作出几乎任意质量的产品。

GUI 的重要性在于它能使 DTP 软件去展示预处理的文本和图形,并且 GUI 使用户能立刻看到准确的最终输出结果的样式。这个"所见即所得"(WYSIWYG)对 DTP 来说是极其重要的,尽管这句话似乎有些保留,因为所见的 72dpi 屏显不可能准确地反映你在 300~1200dpi(或更高)的打印机上所得到的效果。这句话也许该改成 WYSIANATTCMTWYG——"所见近乎技术上能做到之所得"。因此,DTP 软件包的一个必备特征就是有缩放工具,它可以用比正常情况更大的尺寸显示文档的某一部分,从而以更接近最终印刷结果的分辨率来显示这些内容。但遗憾的是在这种放大模式下，任一时刻只能查看文档的一小部分。

用软件让使用者以电子方式对各种各样的文本和图形进行排版的概念是第一个DTP软件 Aldus PageMaker（现在是 Adobe PageMaker）的重要特点。屏幕被用来作为一个"粘贴板"，文本和图形元素可拖至页面上，并重新安置直到满意为止。每一个文本和图形元素都在其"框架"——一个矩形框内——就如同在它本身页面上一样，然后移动并粘贴以构成最终设计。

2. 电子出版（CD-ROMs 和因特网）

近年来，越来越多的资料是用电子的方式而不是在纸张上出版的。两个最重要的新媒体就是 CD-ROM 和 Internet。其结果就是许多 DTP 软件包和文字处理软件的新版本，都提供了专门为这两种媒体开发的格式进行文件输出的功能——特别是 HTML（超文本标记语言）和 PDF（可移植文档格式）。

超文本原理已被扩展至链接图片、图形、声音、视频、动画、图表、地图等。给用户带来的好处是：用户可得到大量信息，并且用户可以决定访问的量以及顺序。而印刷出版仍是一个线性过程，读者是以作者规定的顺序从头到尾进行阅读的，超链接信息的头和尾则都不是唯一的。

所有这一切表明，尽管还有许多与纸张出版物明显的相同之处，但电子出版是需要不同技术和不同软件的一个全然不同的媒体。

人们已设计出新型的不同种类的面向图形的软件，以生产各种电子出版物，如网上出版软件、网上图形设计软件、多媒体"制作"软件、PDF 出版软件。在研制此类新软件产品时所面临的一个潜在问题就是标准太多。已提到的两个标准之一就是 PDF 格式——是由 Adobe 公司通过其"Acrobat"软件套件建立的。另一个是 HTML 格式，目前正被扩展成 XML、DHTML 和 VRML。

PDF 文件格式主要用于产生诸如计算机手册和杂志等现存印刷品的 CD-ROM 版。为帮助建立 PDF 格式，Adobe 设计了一个浏览软件（称为 Acrobat "viewer"），它对各主要操作系统均有免费版。实际上 PDF 是 PostScript——Adobe 的页面描述语言（PDL）的扩展。对于简单的文档，产生一个 PDF 文件所要做的就是有一个专用"打印驱动器"，它不打印，或至少不像人们想象的那样打印到纸张上，而是打印到一个文件中。从原理上讲，这种方法的精彩之处，就是任何产生文本输出的软件——像文字处理软件，都可用于产生 PDF 文档。一旦专业打印机驱动软件（Acrobat "PDF Writer"）被安装成操作系统的一个部分，用户就可以在输出时选择该驱动软件，并把以此软件创建的任何文档"打印"成 PDF 文件。

HTML 格式是用来描述网页外观的一种方法。像 Netscape 和微软的 IE 这样的网上浏览器随处可见，可以免费获取或者非常便宜地买到，同样 PDF 浏览器也可免费得到(这些软件厂商都想用自己的软件来左右行业标准）。也可以认为 HTML 是一个 PDL（页面描述语言），用户用它来确定网页的外观以及如何链接到其他网页上。

10.4 图形软件（2）

1. 计算机辅助设计 CAD

CAD 是图形应用的经典范例，它是在功能强大的大型计算机上发展起来的，并且人们曾经认为用桌面电脑完成此类重要任务是很不切合实际的。不错，现在许多重要的 CAD 工作仍然由图形工作站来完成，但越来越多的功能强大的应用系统也在普通桌面电脑上运行。最简单的计算机辅助设计之一就是二维平面制图——这种工作传统的方式是绘图员用丁字尺和圆规在绘图板上进行的。这类工作现在可以很快地由提供诸多工具的软件来完成。这些工具包括：电子丁字尺、橡皮擦、标准零部件和几何成形、示例、栅格、图层和标准图形库。

功能更强大的 CAD 软件包可用来设计和表现三维物体，这些软件包有时被称为"实体模型制造者"，因为它们在计算机上创建了三维物体完整清晰的模型描述，并可以用隐藏线消除和外形阴影来显示图像。由于三维建模涉及面很广，已远远超出了 CAD 的范畴。

2. 绘图（画图）软件

虽然 CAD 软件与工程设计和其他形式的技术绘图相关，但这些软件包所提供的功能在许多其他应用中也被证明是非常有用的。可以认为绘图（或简称画图）软件是一个没有专门目的的 CAD 软件。像 CAD 软件一样，所绘制的图形可以以矢量方式制作和存储，因此，这类软件应和"绘图"软件区分开来，绘图软件通常用来创建和修改位图图像。CAD 和绘图软件的区别在于侧重点和方向的不同而非原理的不同。在这两种软件中，用户都可以对基于线条的几何图形进行组合和编辑；对象可以被单独选择、移动、变形及缩放；有边界的图形可用颜色和图案来进行填充；多个对象可以组合并看作是单个物体，都提供像栅格和快照之类的多种定位辅助手段。典型的用户可以用这种软件来产生较专业的绘图，同时也可以用它制作非专业的广告设计、宣传册和杂志，因此，其区别在于需要很多强大的功能，而不需要太多专业性。通常这些应用需要很多字体、颜色和彩色效果、现成的、全然不同的图形库和更复杂的曲线。这种图形软件包生成的图像会少一些实用性而多一些美感——更像是"艺术品"。

3. 商业演示软件

这是另一种具有明确应用范围的矢量软件——快速生成用于尽可能紧凑地总结一些概念或数据的"演示"（报告或短文）。主要的特点是：简单的绘图功能（像图例说明软件那样）、剪辑功能、具有吸引力的演示、少量文字和图表。与绘图软件或 CAD 的用户不同，此处的用户对象并非设计人员，因此能自动生成具有美感效果的作品是十分重要的，这一点是通过各式各样的"模板"来实现的。通常，模板是用来帮助人们建立某种对象的模式或导引——就像用刻有字母的塑料模板快速地在纸上描字一样。在图形软件中，模板是一种设计好的东西，用户可以用它快速地完成自己的任务。比如，它可以是一屏"虚化"的文本——用户只须用自己的文本进行替换即可。关键在于专业的图形设计师已设计好模板——选择好了字体和颜色，或许还有图案背景——用户无需任何设计技巧就可生成专业水平的效果，其速度也比从头做起快得多。

10.5 数字图像文件格式

为什么会需要如此众多不同类型的图像文件格式呢？简单的回答就是因为有太多不同类型的图像以及众多不同需求的应用。完整一些的回答（在此先不做深入的分析）还要考虑到市场份额、所有权以及在图像产业中缺乏必要的协调。但是，已经有一些标准的文件格式开发出来了，这里所列举的都是目前广泛使用的文件格式。许多其他图像类型可以通过图像转换软件很容易地转换成这些文件格式。

与计算机成像相关的一个领域就是计算机图形学。计算机图形学是计算机科学的一个特殊领域，它指的是通过计算机对可视数据进行再现，这包括为显示或打印而生成的计算机图像，以及为输出到监视器、打印机、照相机或其他能提供图像的设备上生成和处理的任何图像（真实的或人造的）。可以认为计算机图形学是计算机成像技术的一部分，图形设计师所使用的许多工具也正在被计算机成像专业人士所采用。

在计算机图形学中，图像数据主要分为两种类型：位图和矢量。位图图像（也称为光栅图像）可由图像模型 $I(r, c)$ 来表示，其中包含以某些文件格式存储的像素数据以及相应的亮度值。矢量图像是指仅通过存储关键点来表达线条、曲线和形体的方法。这些关键点足以对形状进行定义，而将它变成图像的过程称之为渲染。当图像被渲染之后，它可以被认为是位图格式，其中每一个像素都具有与之相关的特定值。

尽管有些文件被压缩，但这里所讨论的大多数文件格式的类型均属于位图图像的范畴，因此，$I(r, c)$ 值直到文件解压后方可获得。通常，这些类型的图像包含头信息和原始像素数据。头文件所必须包含的信息为：①行数（高）；②列数（宽）；③频带宽度；④每像素位数；⑤文件类型。此外，对于某些更复杂的文件格式，头文件还必须包含有关所用压缩类型的信息，以及创建图像所需的其他必要的参数。

最简单的文件格式就是 BIN 和 PPM 文件格式。BIN 格式就是原始图像数据 $I(r, c)$，这个文件不包含头文件，因此，使用者必须知道必要的参数——大小、带宽和每像素位数——以便将该文件作为图像使用。PPM 格式是广泛应用的格式，并且有许多转换工具供免费使用（Pbmplus）。它们主要包含具有最简单头文件的原始图像数据。PPM 格式包括 PBM（二进制）、PGM（灰度）、PPM（彩色）和 PNM（处理任何上述类型的文件）。这些图像文件格式的头文件都包含一个确定文件类型的"幻数"、图像的宽度和高度、带宽和最大亮度值（它决定了每一带宽的每像素所需的位数）。

两种常用于许多不同计算机平台和万维网的图像文件格式就是 TIFF（标记图像文件格式）和 GIF（图形交换格式）文件格式。GIF 文件受限于每像素最多 8 位，并允许进行一种称为 LZW 的压缩。每像素限制为 8 位并不意味着它不支持彩色图像，它只是表示在一个图像中不允许超过 256（2^8）色。

这通常由查表（LUT）的方法来完成，其中 256 色被存于一个表格中，对每一像素要用一个字节（8 位）作为索引（地址）。GIF 图像头文件有 13 字节长并包含所需的基本信息。

TIFF 文件格式比 GIF 更复杂，并具有许多选项和功能。TIFF 允许最多 24 位/像素，并支持 5 种压缩格式，其中包括 RLE（行程编码）、LZW 和 JPEG（联合静态图像专家组）。TIFF 头文件长度可变并以层次化方式组成。TIFF 是一个综合性最强的格式，并设计成允许用户对特殊应用进行定制。

第 11 章 现代工业自动化

11.1 概述

除非能以较低的价格提供更好的质量及更短的交货时间提供新产品，否则任何企业都无法在当今的国际竞争中立足。因此，他们都试图用计算机的巨大存储能力、快速处理速度及用户友好的交互图形能力，来自动完成并紧密联系其他繁重的和单独的工程或生产任务，从而减少产品开发和生产的时间和成本。计算机辅助设计（CAD）、计算机辅助制造（CAM）和计算机辅助工程（CAE）正是生产周期中为此目的而采用的技术。因此，为了了解 CAD、CAM 和 CAE 所扮演的角色，需要了解在一个产品设计和生产过程中必须完成的各项活动和功能。这些活动和功能被称之为产品周期。对 Zeid（1991）所描述的产品周期稍加改动后示于图 11-1。

正如图 11-1 的实线框所示，产品周期由两个主要过程组成：设计过程和制造过程。设计过程始于由市场人员认定的客户需求，止于对产品的完整描述，通常用图形来表现。制造过程则始于设计的规格说明，止于产品发运。

设计过程所涉及的活动大体上分为两类：综合和分析。如图 11-1 所示，原始设计活动（如设计需求的认定，设计规格说明的规划，根据收集的相关设计信息进行可行性研究以及设计概念化等）都属综合子过程部分。即综合子过程的结果是各个产品部件间的关系以草图或布局图的形式来表示的所期望产品的概念设计。用于实现该产品理念的主要财政承诺和产品功能都在产品周期的这一阶段确定。在综合子过程中产生和处理的大多数信息是定性的，因此在计算机系统中很难得到。

一旦概念设计形成，分析子过程就要开始对设计进行分析和优化了。首先要导出一个分析模型，因为分析子过程适用于该模型而非设计本身。尽管在工程上，计算机的能力和可用性迅速提高，但对分析模型的抽象依然会在可预见的未来仍有待人去完成。分析模型可以通过以下几个环节获得，如从设计中去除不必要的细节，缩小尺寸和识别并使用对称性等。例如：尺寸缩小表示一张薄的材料由具有厚度属性的等效曲面来代表，或一个细长区域由具有截面特性的线段来表示。在几何和负荷上具有对称性的物体通常用模型的某一部分进行分析。实际上，在机械基础课程上对一个结构进行分析时，早就实践过这种自然的抽象过程了。

一旦设计完成，在经过优化或一些权衡之后，就开始了设计评价阶段。为此目的可进行原型制作。一种用于构建原型并称之为原型速成的新技术现已非常普及。这种技术是通过从底部到顶部层层地构建起原型的。由于原型基本上只需要产品的截面数据，因此可以从设计直接构建原型。如果对原型的设计评价表明这种设计不尽如人意，那么，就以新的设计重复进行上述过程。

当设计评价结果令人满意时,就要准备设计文档了。设计文档包括图纸、报告和材料清单的准备。按常规,应将设计图纸制成蓝图交送制造部门。

如图 11-1 所示,制造过程根据设计过程的图纸始于工艺计划,止于实际的产品。工艺计划的结果就是生产计划、材料订单和加工程序。其他诸如设计钻模和夹具这样的特殊要求都在这一阶段进行处理。工艺计划与制造过程的关系就如同综合与设计过程的关系;它涉及许多经验和定性的判断,这表示很难对工艺计划进行计算机化。一旦工艺计划结束,便开始生产实际产品并根据质量要求进行检查。对通过质量检查的零件进行组装、功能测试、包装、贴标签并发运给客户。

11.2　CAD、CAM、CAE 的应用

上面已经描述了一个典型的产品周期,下面来看一下在此周期内如何使用计算机或 CAD、CAM 和 CAE 技术。正如前面所指出的,在设计过程的综合阶段,计算机用得并不广泛。这是因为计算机不能很好地处理定性信息。但是,在综合子过程中,比如,设计人员可以通过商业数据库很好地为可行性研究搜集相关的设计信息,并以同样的方法搜集目录信息。

同样很难想象计算机如何应用于设计概念化阶段,因为目前计算机还谈不上是智能创造过程的强大工具。在这一阶段,计算机有助于高效率地完成各式各样的概念设计。参数建模、计算机辅助绘图的宏编程能力或几何建模,在完成此类任务时会很有用。这些软件包是 CAD 软件的典型例子。可以想象出来,一个几何模型系统相当于一个三维绘图系统,也就是说,通过这种软件包就可以用三维形体来取代二维图片。

设计过程中的分析子过程最能体现计算机的价值。实际上,对于应力分析、干扰检验和运动分析有许多现成的软件包。这些软件包归入 CAE 一类。使用这些软件包的一个问题就在于如何提供分析模型。如果能从概念设计中自动导出分析模型的话,那就根本不是问题。但是,正如前面说过的,分析模型与概念设计不同,它是通过从设计中忽略了不必要的细节或缩小尺寸而得来的。根据分析模型的类型和所期望求解精度的不同,就有不同的抽象水平。因此,很难对抽象过程进行自动处理;而分析模型通常是分别建立的。常用的做法是,用计算机辅助绘图系统或几何建模系统或有时使用分析软件包的内置功能,为设计过程多建几个抽象模型。

分析子过程可以嵌入到优化循环中以获取最优设计。获取最优解决方案的各种算法都已开发出来,并且许多优化程序亦可在市场上购买到。优化过程可以认为是 CAD 软件的一个部分,但最好还是单独对待。

设计评估阶段使用计算机也会受益匪浅。如果为了设计评价而需要一个设计原型的话,则可以用自动生成驱动原型速成机的软件包对一个给定的设计构建一个原型。这些软件包都归类为 CAM 软件。当然,要做的原型形状应事先以某种数据类型存在。对应于该形状的数据应该由几何建模系统来创建。尽管原型可以方便地用原型速成方法来构建,但如果使用虚拟原型就会更加理想。虚拟原型通常称为数字模型,它提供同样有价值的信息。

随着用于评价数字模型的分析工具的功能变得足够强大，其给出的分析结果与在真实原型上相同实验所获得的结果同样精确时，数字模型就有了取代真实原型的趋势。随着虚拟现实技术能够使数字模型获得与真实原型相同的效果，这种趋势将不断增强。建立数字模型的工作称为虚拟原型设计。虚拟原型也可由专为此设计的某类几何建模系统来产生。

设计过程的最后阶段是设计文档。在这一阶段，计算机辅助绘图是一个强大的工具。计算机绘图系统的文件处理能力也允许对文档进行系统地存储和检索。

计算机技术也同样可用于制造过程。制造过程包括：生产计划、新工具的设计和采购、材料订购、数控编程、质量控制和包装等活动，因此应用于这些活动的计算机技术都属 CAM。例如，用于辅助工艺计划的计算机辅助工艺计划（CAPP）软件就是 CAM 软件的一种。前面曾提到过，工艺计划很难自动完成，因此 100%自动化的 CAPP 软件目前还拿不到。但是，有很多优秀的软件包能够产生驱动数控设备的数控程序。当某种形状以数据形式存在于计算机中时，此类设备就可以加工出给定的形状。这类似于驱动原型速成机。此外，同属 CAM 的还有：对机器人运动进行编程，以便按零部件进行组装或将其送至各个制造活动中的软件包，或对坐标测量机（CMM）编程，以便对产品进行检查的软件包。

11.3 产品生命周期管理

1. 历史

在工业领域，产品生命周期管理（PLM）是一个管理产品从最初计划，经过工程设计和制造，到服务以及所制造出产品报废的整个产品生命周期的全过程。PLM 结合了人、数据、过程和业务系统并且是公司和他们延伸企业的一个产品信息供给中心。

2. 形式

PLM 系统帮助企业在全球竞争的市场里，应对在开发新产品的过程中不断增加的复杂性和工程挑战。

产品生命周期管理（PLM）应该区别于"产品生命周期管理（营销）"（PLCM）。PLM 描述的是产品从它的开发一直到使用寿命这个过程中的管理描述和产品性能的工程部分；而 PLCM 则是产品在商业市场上有关成本和销售措施的营销生命管理。

可以认为产品生命周期管理是生产制造企业信息技术架构的四个基石之一。所有的公司都需要管理他们与用户的通信和信息（CRM 客户关系管理），管理他们的供应商及其实施（SCM 供应链管理），管理他们的企业资源（ERP 企业资源计划）和管理他们的产品计划和开发（PLM）。

有一种被称为以人为中心的 PLM 形式。传统的 PLM 工具只应用于产品发布或发布阶段，以人为中心的 PLM 目标则在设计阶段。

截至 2009 年，ICT 发展（欧盟资助 PROMISE 项目 2004—2008）已经允许 PLM 超越传统 PLM，并将传感器数据和实时"生命周期事件数据"集成到 PLM 中，而且允许在一个产品的整个生命周期中对不同参与者呈现该信息（闭合整个信息环）。这样就将

PLM 扩展至闭环生命周期管理（CL2M）中。

3．效益

可查阅的产品生命周期效益包括：
- 减少进入市场的时间；
- 加大全价销售量；
- 提高产品质量和可靠性；
- 减少原型设计和制作成本；
- 更精准和及时的报价；
- 快速确定潜在销售机会和营收的能力；
- 通过重复使用原始数据带来的节约；
- 产品优化的框架；
- 减少消耗；
- 完整的工程流程集成带来的节约；
- 能将中央产品记录提供给合约制造商；
- 季节性波动管理；
- 改善预测方法以减少材料费；
- 供应链合作的最大化。

4．PLM 应用领域

在 PLM 中有五个主要领域：
- 系统工程（SE）；
- 产品组合管理（PPM）；
- 产品设计（CAX）；
- 制造过程管理（MPM）；
- 产品数据管理（PDM）。

5．开发过程介绍

PLM 的核心是所有产品数据的创建和集中管理以及用于访问这些信息和知识的技术。PLM 虽然是一个自 CAD、CAM 和 PDM 这些工具出现之后的一项技术，但也可以看成是这些工具在产品生命的所有阶段，与方法、人和过程的集成。它不仅是一个软件技术，而且也是一种商业战略。

图 11-2 为产品生命周期管理。

为方便起见，可以按传统的顺序工程工作流的方式来描述每个阶段。事件和任务的精确顺序会随产品和行业的差异有所不同，但主要的过程如下：
- 构想
  - 技术参数　-概念设计
- 设计
  - 详细设计　-确认和分析（模拟）　-工具设计
- 实现
  - 计划制造　　-制造

- 建造和组装   - 测试（质量检测）
- 服务
   - 销售和配送   - 使用
   - 维护和支持   - 报废

主要关键节点事件是：
- 订单；
- 概念；
- 启动；
- 设计冻结；
- 发布。

然而现实更加复杂，人和部门不能孤立地完成他们的任务，不能简单地完成一个任务再启动另外一个。设计是一个迭代过程，设计常常需要根据制造中的一些限制或需求冲突进行修改。在这种情况下，应根据工业类型以及例如，产品是否按订单制造、按订单设计、按订单装配这些条件，按时完成客户订单提出的任务。

6. 生产系统金字塔结构

在生产系统中有五个长期目标需要考虑，如图 11-3 所示：
- 可以用货币单位考量的成本，通常考虑固定成本和可变成本。
- 在一个时间周期内可以用产品数量进行考量的生产率。
- 可以用比如用户满意度考量的产品质量。
- 可以用比如系统生产不同产品的能力来考量的灵活性。
- 可以用生产系统对生态和环境的影响来考量的生态健康情况。

这五个目标之间的关系可以由一个金字塔来表示，它的顶端就是最高的生产率、最好的质量、最经济有效、最灵活和可持续性。其中的各个点可以是这五个准则的不同组合。金字塔的最顶端是理想点，但无法完全实现，最下端当然就是最差点。

11.4 工业 4.0（第 4 次工业革命）

1. 名称

该术语 2011 年在汉诺威工业博览会上首次使用。2012 年 10 月由 Siegfried Dais（罗伯特博世股份有限公司）和 Kagermann（德国科学与工程院）主持的工业 4.0 工作组向德国联邦政府提交了一套第四次工业革命实施建议书。2013 年 4 月 8 日，工业 4.0 工作组在汉诺威工业博览会上提交了其最终的报告。

2. 工业 4.0

工业 4.0 是基于信息物理系统、物联网和互联网服务技术概念的一种价值链组织的技术和概念的一个集合术语，它促进了智慧工厂的愿景。在工业 4.0 模块化结构的智慧工厂里，信息物理系统监控着物理过程，创建一个物理世界的虚拟拷贝并实现分布式决策。信息物理系统之间和人之间，跨越物联网，相互实时沟通与合作。通过互联网服务，价值链的参与者为组织内部和组织之间提供服务和应用。

与此同时，美国也在致力于未来制造业的研究工作，并发起建立智能制造领导者联盟。该联盟是一个由制造业从业者、供应商以及技术公司、制造业联盟、大学、政府机构以及研究实验室组成的非营利组织。该联盟旨在为制造业中利益相关者建立负责合作研发、项目实施的团队，用以研究方法、标准、平台和基础设施共享，从而促进制造智能得到更为广泛的应用。

同样地，美国通用电气公司一直都在致力于创建"工业互联网"。工业互联网的目标是汇聚之前两次变革性革命中取得的进步：工业革命创造出的无数的机器、设施设备、飞机轮船以及运输网络，和近期的互联网革命中在计算、信息和通信系统上所取得的巨大进步。通用电气公司称，将这些成果集聚在一起，就构成了工业互联网的三个基本要素，即智能机器、高级分析法和工作的人，这就是工业互联网的本质。

3．内涵

在第四次工业革命中，工业生产的特点就是在高度灵活化（大规模）的生产条件下，实现了高品质的产品定制。在日益复杂的工作中，由于机器的自我优化、自我配置、自我诊断，再配以工人们的认知能力和聪明才智，使得所需的自动化技术得以完善。目前工业4.0中最大的项目是德国联邦教育与研究部的尖端"智能科技系统Ost WestfalenLippe（简称OWL）"集群项目。此外，还有联邦教育与研究部的RES-COM以及"致力于打造高薪国家的综合生产技术"精英集群项目。

4．差异

今天的工厂和第四次工业革命时期的工厂之间是有一些区别的。在当前工业环境下，成功的关键在于提供高端品质的服务或将产品的成本降至最低。工厂都在尽可能多地运行生产以增加盈利、提高声誉。这样，各种各样的数据资源就可以为工厂的各个方面提供有价值的信息。在这一阶段，利用数据了解现状并检测差错、故障是一个重要的研究课题。例如，在生产中，系统会存在各种问题并可能出现一些错误，为了突显其产生根源，可以利用多种多样的商业工具来获取工厂管理的设备综合效率信息。相比较而言，在第四次工业革命中，工厂除了状态监控和故障诊断外，各个部件和系统都有自我意识和自我预知功能，这就能在工厂管理中更为敏锐地洞察工厂的运行状况。此外，通过对等比较和汇总融合来自各个零部件的健康信息，就可以对零部件以及整个系统的健康状况做出精确的推测，并加强工厂管理，以在最恰当的时机开启所需的维修程序，从而实现实时维护和保持机器的死机时间几乎为零。

5．面临哪些挑战

- 缺少足够的技术技能来加快第四次工业革命的前进步伐。
- 企业IT部门冗余带来的威胁。
- 利益相关者普遍不愿意做出改变。

6．大数据和分析学所扮演的角色

信息物理系统、大数据或云计算等现代信息和通信技术将会有助于预测制造业中是否存在提高生产率、提升产品质量、增加生产灵活性的空间，从而明确自己在竞争中的优势所在。

在第四次工业革命和信息物理系统的环境中，大数据分析由6个C组成，即：联系

（传感器和网络）、云（计算和数据按需即时提取）、计算机网络（模型&记忆）、内容（意义和关联）、社区（共享&合作）以及客户定制化（个性化和价值观）。在这一情境下，为给工厂管理提供有用的参考意见、获取正确的信息，必须利用先进的工具（分析法和演算法）处理数据以获取有意义的信息。鉴于工厂中出现的问题既包括可见的，也包括潜在的，这就要求信息生成算法必须能够探测到并解决诸如工厂车间内机器老化、部件磨损等潜在的问题。

7. 第四次工业革命的影响

预计随着第四次工业革命的到来，很多领域都将会受到影响。其中受影响最大的四个领域为：

- 机械安全；
- 工业价值链；
- 工人；
- 社会经济。

11.5　3D 打印

3D 打印（或称增材制造）是用来制作三维物体的各种加工技术。在 3D 打印中，使用增材加工技术，在这种技术中，在计算机控制下，对材料的连续层进行铺设。这些物体可以是各种形状或几何结构的，并能从 3D 模型中产生，或从其他电子数据源产生。3D 打印机是一种工业机器人。

图 11-4 为一种 3D 打印机。

按原始术语的意思，3D 打印是指用喷墨打印头，连续往粉末床上沉积材料。该术语最新的意思已扩展到包含更为广泛的技术，诸如基于工艺的挤压和黏结等。从更广的意义上讲，技术标准应采用增材制造术语。

1. 基本原理

（1）建模

可以用 CAD 软件包，或 3D 扫描仪，或普通的数码相机和摄影测量软件去创建 3D 打印模型。

为 3D 计算机图形准备几何数据的人工建模过程，类似于造型美术，像雕刻一样。3D 扫描是对一个实体在形状和外观上采集和分析数字数据的过程。在这种数据的基础上，被扫描物体的三维模型就产生出来了。

与所使用的 3D 建模软件无关，3D 模型（通常用.skp、3ds 或其他格式）然后要转换成 .STL 或 .OBJ 格式，以便打印（又称"CAM"）软件去读它。

图 11-5 为三维模型切片。

（2）打印

在从 STL 文件打印 3D 模型前，首先要检查"复印误差"，这一步可以称为"修复"。特别是通过 3D 扫描获取的模型所产生的 STL，通常有很多复印误差需要修正。例如，复印误差是表面的，则不需要在模型中去连接或补缺。可用于修正这些误差的软件有

netfabb 和 Meshmixer，甚至可使用 Cura 或 Slic3r。

一旦做完上述工作，.STL 文件需要一个称为"切片机"的软件去处理，该软件会把上述模型转换成一系列薄层，并产生一个 G-code 文件，此文件含有专为某种 3D 打印机（FDM 打印机）定制的一些指令。这种 G-code 文件可用 3D printing 客户软件打印（该软件装入 G-code，并在 3D 打印过程中指导 3D 打印机）。应该注意的是，实际上这种客户软件和切片机通常组成一个软件程序。现在有几种开源切片机程序，包括 Skeinforge、Slic3r 和 Cura-engine 以及闭源程序，包括 Simplify3D 和 KISSlicer。3D 打印机客户软件包括 Repetier-Host、ReplicatorG、Printrun 和 Cura。

（3）结束

虽然打印机打出的结果，分辨率能满足很多应用的要求，但在标准分辨率内打印时，其尺寸应稍微超过所需物体尺寸，这样，在其后采用更高分辨率的削减材料过程中，会获得更高的精度。

某些可以打印的聚合物，允许对表面做平滑处理，用化学蒸发工艺做进一步改进。

某些增材制造技术能够在构造零件过程中使用多种材料。这些技术能同时打印多种颜色和组合颜色，且不要求涂色。

所有商业化的金属 3D 打印机都包含，在沉积之后，把金属部件从金属衬底上切掉。一种新的用于 GMAW 3D 打印的工艺，可以对衬底表面做修正，用人工小锤去掉铝制部件。

2. 应用

20 世纪 80 年代，增材制造在产品开发、数据可视化、快速成型和专用制造中开始应用。此后几十年，在开发中扩展到了生产（如制品生产、批量生产和分布式制造）。在 2010 年代初，工业生产首先在金属加工行业获得了长足进步。从 21 世纪开始，增材制造机器销售量大幅增长，而价格大体上是下降的。按照某个咨询公司 Wohlers 助理的说法，2012 年世界范围内，3D 打印机及服务的市场价值为 22 亿美元，比 2011 年增长 29%。增材制造技术有很多应用，包括建筑、施工、工业设计、自动化、航天、军事、工程、牙科和医药行业、生物技术（人组织置换）、服装、鞋业、珠宝饰品、眼睛饰品、教育、地理信息系统、食品和其他很多领域。

图 11-6 为表明工业中 3D 打印技术优越性的涡轮模型。

2005 年，随着开源 RepRap 课题的启动，3D 技术很快就有了广泛的嗜好者和家庭市场。实际上，迄今为止，所公布的所有家用 3D 打印机都在 RepRap 课题和相应开源软件项目中有他们的技术根基。在分布式制造中，一项研究表明，3D 打印会变成一个海量市场产品，能让客户在购买一般家用物品时，节省大量钱财。例如，可以不去商店购买工厂注入成型的产品（如量杯或漏斗），人们可以在家中，从下载的 3D 模型中将其打印出来。